OXFORD SERIES ON NEUTRON SCATTERING IN CONDENSED MATTER

OXFORD SERIES ON NEUTRON SCATTERING IN CONDENSED MATTER

1. G. Williams: *Polarized neutrons*
2. E. Balcar and S. W. Lovesey: *Theory of magnetic neutron and photon scattering*

POLARIZED NEUTRONS

W. Gavin Williams

Rutherford Appleton Laboratory

CLARENDON PRESS · OXFORD

This book has been printed digitally and produced in a standard specification in order to ensure its continuing availability

OXFORD
UNIVERSITY PRESS

Great Clarendon Street, Oxford OX2 6DP
United Kingdom

Oxford University Press is a department of the University of Oxford. It furthers the University's objective of excellence in research, scholarship, and education by publishing worldwide. Oxford is a registered trade mark of Oxford University Press in the UK and in certain other countries

Reprinted 2012

British Library Cataloguing in Publication Data
Data available

Library of Congress Cataloging in Publication Data
Data available

ISBN 978-0-19-851005-5

Preface

The scattering of polarized thermal neutron beams has made many important contributions to condensed matter physics, particularly in helping to elucidate the nature of atomic magnetism. All the major neutron research establishments are now equipped with polarized-beam instruments, and these are used in a wide variety of ways. Some of these applications include the study of magnetism, magnetic excitations, disordered magnetic systems, and surface or interface magnetism. The scope for experiments is enormous, and the reader who progresses will find that this list is by no means exhaustive.

This book has been written primarily to inform researchers in condensed matter physics and chemistry of the diversity of scientific problems that can now be investigated using polarized neutron beams. Polarized neutrons are also an important tool in nuclear physics, but this was considered to be too large a subject for inclusion in this volume. It is a book for the experimentalist. However, a basic amount of theory is an essential ingredient to a proper understanding of scattering methods, and this is all presented in Chapter 2. As in all the experimental sciences, the techniques and results are most easily explained by using illustrations, and the author has borrowed liberally from the existing literature. Polarized neutron scattering has benefited greatly from the improvements that have taken place in instrumentation, particularly over the last decade, and these developments are described in Chapter 3. The presentation and discussion of experimental results is treated mainly in Chapters 4 and 5. Chapter 6 is devoted to showing how the precession property of the neutron polarization in a magnetic field can be utilized, either to provide information on the internal magnetic fields in materials (through neutron depolarization), or to measure inelastic neutron scattering processes at very high resolutions (neutron spin echo spectroscopy).

I owe a great debt to past and present co-workers: F. F. Freeman, J. Penfold, J. B. Hayter, T. J. L. Jones, J. Mayers and R. Cywinski who, in chronological order, have been my principal collaborators in developing polarized neutron beam techniques. Many of the methods described in Chapter 3 are a result of their expertise. I have several reasons for acknowledging the advice given by my colleague Professor Stephen W.

Lovesey: his illumination of many of the theoretical aspects of the subject, his reading and help with the whole manuscript, and not least his encouragement during the writing stage. Individual sections of the book have been much improved by the perceptive comments of Professor J. B. Forsyth and Drs J. Mayers, J. Penfold and M. Th. Rekveldt. It is to be hoped that any remaining errors are minor, but in any event they are the sole responsibility of the author.

I was most fortunate that Miss Caroline Monypenny agreed to type the manuscript; her skill and accuracy has greatly facilitated my task. Finally, I wish to express my gratitude to a very special group of people who, to the best of my knowledge, are unaware of the marvels of polarized neutrons. I refer to my family: Anne, Derwyn, Bryan, and Juliet, who have always been most generous with their support and friendship. I dedicate this book to them.

Letcombe Regis, Oxfordshire W.G.W.
February 1987

Contents

Acknowledgements

I am most grateful to the following publishers and learned societies for permission to use their illustrations as figures in this book: Academic Press, American Physical Society, American Institute of Physics, J. C. Baltzer AG, Institut Max Von Laue–Paul Langevin, The Institute of Physics, International Atomic Energy Agency, International Union of Crystallography, Journal de Physique, North-Holland Publishing Company, Oxford University Press, Pergamon Press, Physical Society of Japan, The Royal Society, The Royal Society of Chemistry, and Springer-Verlag.

The following authors have also kindly allowed me free use of their original material: A. Abragam, J. P. Boucher, G. Badurek, P. J. Brown, J. W. Cable, R. Cywinski, J. W. T. Dabbs, J. C. Dore, Y. Endoh, G. P. Felcher, J. B. Forsyth, D. Givord, H. Glättli, J. B. Hayter, T. M. Holden, G. H. Lander, R. D. Lowde, F. Mezei, H. A. Mook, R. M. Moon, G. Parette, J. Penfold, H. Postma, R. Pynn, B. D. Rainford, M. Th. Rekveldt, D. Richter, T. Riste, O. Schärpf, J. Schweizer, G. Shirane, C. G. Shull, C. Stassis, U. Steigenberger, O. Steinsvoll, A. D. Taylor, F. A. Wedgwood, Y. Yamada, C. M. E. Zeyen, and K. R. A. Ziebeck. It is a pleasure to record my sincere thanks to them all.

Glossary of symbols

$\text{Å} = 10^{-10}$ m	(angstrom)
b	Bound nuclear coherent scattering amplitude
$b_{\pm}$	Bound nuclear coherent scattering amplitudes for compound nuclear states with spins $(i \pm \frac{1}{2})$
b_{NS}	Bound nuclear spin incoherent scattering amplitude
$\boldsymbol{B}$	Magnetic induction vector, or the applied magnetic field vector in a vacuum
B_s	Magnetic induction at saturation
$\mathcal{B}$	Nuclear spin scattering parameter $= (b_+ - b_-)/(i + \frac{1}{2})$
c	Velocity of light, or the atomic fractions of different components in a mixture or alloy
d_j	Layer thickness of the jth layer in a thin-film neutron optical device
d_{hkl}	Spacing between the (hkl) planes in a crystalline solid
D	Depolarization factor which changes the neutron beam polarization from P to $P(1 - D)$
D_m	Inverse of the mean free path for a neutron spin reversal in a depolarizing material
e	Electronic charge
E	Neutron energy
E_R	Neutron energy at the peak of a nuclear resonance
f	The efficiency of a neutron spin-flipper which reverses the beam polarization from P to $P(1 - 2f)$
$F(\boldsymbol{\kappa})$	Magnetic form factor
$F_M(\boldsymbol{\kappa})$	Unit cell magnetic structure factor
$F_N(\boldsymbol{\kappa})$	Unit cell nuclear structure factor
g	Landé splitting factor
g	Statistical weighting factor in a resonance capture cross-section $= (2J + 1)/(4i + 2)$
$g_M(r)$	Magnetic pair correlation function
$g_{AB}(r)$	Space correlation function between atoms A and B
h	Planck's constant

$\hbar$	Planck's constant divided by 2π
$\hat{\boldsymbol{i}}$	Nuclear spin angular momentum operator
J	Resultant angular momentum quantum number in a magnetic atom, or of a compound nucleus following neutron capture
$J_{\pm}$	Total angular momentum quantum number of a compound nucleus following $(i \pm \frac{1}{2})$ neutron capture
$\boldsymbol{k}$	Neutron wave vector, usually of a beam incident at a scattering sample
$\boldsymbol{k}'$	Neutron wave vector of a scattered beam
k_B	Boltzmann's constant
L	Orbital angular momentum quantum number in a magnetic atom
m	Neutron mass
m_e	Electron mass
m_p	Proton mass
M	Nuclear mass
$\boldsymbol{M}$	Magnetization vector for an atom
$\tilde{\boldsymbol{M}}$	Unit vector in the direction of the magnetization of a magnetic atom
$n_{1,2}$	Neutron refractive index for a neutron wave passing from medium 1 to medium 2
N	Number of atoms in a scattering sample, or the atomic density of a specimen or filter in units of atoms cm^{-3}
N_c	Number of unit cells per unit volume in a crystal
NSF	Non-spin-flip scattering (used as subscript)
p_{λ}	Population factor for a target in state λ
P	Neutron beam linear polarization, or the polarizing efficiency of a neutron polarizer
P'	Polarization of a scattered neutron beam
P_A	Polarizing efficiency of a spin analyser
$\boldsymbol{P}$	Neutron polarization vector
$\boldsymbol{P}_N$	Nuclear polarization vector
$\boldsymbol{q}$	Magnetic interaction geometric factor $= \tilde{\boldsymbol{\kappa}}(\tilde{\boldsymbol{\kappa}} \cdot \tilde{\boldsymbol{\eta}}) - \tilde{\boldsymbol{\eta}}$, or the reduced wave vector associated with a phonon or a magnon $(\boldsymbol{\kappa} - \boldsymbol{\tau})$
$\boldsymbol{Q}_{\perp} = \tilde{\boldsymbol{\kappa}} \times (\hat{\boldsymbol{Q}} \times \tilde{\boldsymbol{\kappa}})$	Magnetic interaction vector
r_0	$= \gamma_n \times$ (classical electron radius) $= \gamma_n e^2/m_e c^2 = -0.5391 \times 10^{-12}$ cm

R	Flipping ratio in a polarized neutron beam experiment
$\boldsymbol{R}$	Position vector of a scattering centre
R_S	Depolarizing shim ratio in a polarized neutron beam experiment
$\mathscr{R}_{\pm}$	Bragg peak or optical reflectivities for (+) and (−) spin neutrons
$\hat{\boldsymbol{s}}$	Electron spin operator
$\hat{\boldsymbol{s}}'$	Neutron spin operator
$\hat{\boldsymbol{S}}$	Spin angular momentum operator in a magnetic atom
S	Efficiency of a neutron depolarizing shim
$\boldsymbol{S}_{\perp}$	Atomic spin vector perpendicular to the scattering vector $\boldsymbol{\kappa}$
$S(\boldsymbol{\kappa})$	Static structure factor
$S_{AB}(\boldsymbol{\kappa})$	Partial structure factor between atoms A and B
$S(\boldsymbol{\kappa}, \omega)$	Scattering law
$S_{inc}(\boldsymbol{\kappa}, \omega)$	Incoherent or self-scattering law
$S_{coh}(\boldsymbol{\kappa}, \omega)$	Coherent scattering law
SF	Spin-flip scattering (used as subscript)
t	Sample or filter thickness, or time.
T	Temperature in absolute units, or the transmittance of a neutron polarizing filter
T_c	Critical temperature
$T_{\pm}$	Transmittances of a neutron polarizing filter for (+) and (−) spin neutrons
U^{++}, U^{+-}, etc.	Spin-dependent scattering amplitudes connecting $(+)\rightarrow(+)$, $(+)\rightarrow(-)$, etc., neutron spin changes during scattering
v	Neutron velocity
v_0	Volume of a unit cell
$W(\boldsymbol{\kappa})$	Exponent in the Debye–Waller factor
α	Beam divergences or collimation angles (FWHM)
β	$e\hbar/2m_e c$ (Bohr magneton)
β_N	$e\hbar/2m_p c$ (Nuclear magneton)
γ	Neutron gyromagnetic ratio $= -1.833 \times 10^8$ rad $s^{-1}\,T^{-1}$
γ	Ratio of magnetic to nuclear unit cell structure factors $= F_M/F_N$
γ_n	Neutron magnetic moment in units of $\beta_N = -1.913$

γ_N	Nuclear gyromagnetic ratio
Γ	Total energy width of a nuclear resonance, or the half-width at full maximum of a Lorentzian peak profile in angular frequency units.
Γ_n	Neutron width of a nuclear resonance
Γ_γ	Radiation width of a nuclear resonance
$\delta(x)$	Dirac delta function
$\delta_{l,m}$	Kronecker delta function
ε	Efficiency of a neutron detector
η	Standard deviation of the mosaic block distribution in a crystal monochromator
$\tilde{\boldsymbol{\eta}}$	Unit vector in the direction of the spin of a magnetic atom
ξ	Isotope label
$\boldsymbol{\kappa} = \boldsymbol{k} - \boldsymbol{k}'$	Neutron scattering vector
λ	Neutron wavelength
λ^*	Characteristic cut-off wavelength of a neutron guide or a reflecting Soller
$\boldsymbol{\mu}_n$	Neutron magnetic moment vector
μ_N	Nuclear magnetic moment of nucleus N
μ_N^*	Pseudomagnetic nuclear magnetic moment
$\mu_{A(B)}$	Magnetic moment of atom A(B)
θ	Reflection angle in neutron optics
θ_B	Bragg reflection angle in crystal diffraction
θ_c	Limiting or critical angle for the total reflection of neutrons
ρ	Statistical spin weighting factor for the absorption of a neutron by a nucleus
$\rho(x, y, z)$	Unpaired electron spin density at position (x, y, z)
$\hat{\boldsymbol{\sigma}}$	Pauli matrix
$\frac{1}{2}\hat{\boldsymbol{\sigma}}$	Neutron spin operator
σ_A	Absorption cross-section per nucleus
σ_{coh}	Coherent scattering cross-section per nucleus
σ_p	Polarization cross-section per nucleus or atom
σ_s	Total scattering cross-section per nucleus
σ_T	Total cross-section per nucleus (scattering plus absorption)
σ_{NS}	Nuclear spin incoherent scattering cross-section per nucleus

τ	Lifetime of a compound nucleus resonant state
$\boldsymbol{\tau}$	Reciprocal lattice vector, containing a factor 2π, i.e. $\{2\pi[(h/a_0), (k/b_0), (l/c_0)]\}$
τ_D	Double transmission effect
τ_S	Single transmission effect
ϕ	Neutron flux (n cm^{-2} s^{-1})
$\phi_\pm$	Fraction of (+) and (−) spin neutrons in a polarized neutron beam
ϕ_L	Neutron spin (or polarization) Larmor precession angle
ω	Angular frequency (rad s^{-1})
$\hbar\omega$	Neutron energy transfer $\hbar^2(k^2 - k'^2)/2m$
$\omega_{\boldsymbol{q}}$	Phonon or magnon angular frequency
Ω	Solid angle

1

Introduction

The object of this introductory chapter is to familiarize the reader with the neutron beam scattering experiment. It is mainly concerned with unpolarized neutron beams, and we first discuss the special properties which make thermal neutrons such a valuable probe for investigating condensed matter on an atomic scale. A section is then devoted to describing the experimental techniques available for using neutron beams. Following this the concept of spin-polarized beams is introduced, and a brief historical survey is made of the development of polarized neutron methods. Polarized neutron scattering has contributed more than any other single experimental technique to our understanding of the nature of magnetism at the atomic level, but a discussion of this, as well as other areas in which it has had a major impact, must await later chapters.

1.1 PROPERTIES OF THE NEUTRON

Neutron beams are now established as a unique probe for studying the nature of condensed matter. Their value stems from the fundamental properties of the neutron, which are its mass $m = 1.67495 \times 10^{-27}$ kg, its spin angular momentum with quantum number $\frac{1}{2}$, its magnetic moment $\mu_n = -1.913$ nuclear magnetons (β_N), and its zero net charge.

The mass of the neutron has several important consequences. At a practical level it means that the highly energetic neutrons produced by either the fission or spallation of heavy nuclei can be slowed down, or moderated, to a point at which they have the same energy or temperature as the atoms in the moderating medium. Their energies are then suitable for studying a wide range of dynamic phenomena in solids and liquids. Moderation by inelastic neutron–proton collisions are especially effective because of their nearly identical masses, and the energy spectrum of neutrons that emerge from a moderator can be tailored by changing their temperature. Neutron energies are most frequently expressed in electron volt units ($1\ \text{eV} = 1.60219 \times 10^{-19}$ J), and the energies which are useful in condensed matter physics are often assigned to one of the following, somewhat flexible, spectral bands.

Neutron energy band	*Temperature description*
0.1–10 meV	Cold neutrons
10–100 meV	Thermal neutrons
100–500 meV	Hot neutrons
>500 meV	Epithermal neutrons

Table 1.1 shows other neutron energy units which are also in common usage, together with their numerical relationships to the energy E given in meV units.

Once the neutron (kinetic) energy is defined, then the neutron mass determines both the neutron velocity v,

$$v = (2E/m)^{1/2}, \tag{1.1.1}$$

and the de Broglie wavelength λ,

$$\lambda = h/mv = h(2mE)^{-1/2}, \tag{1.1.2}$$

where Planck's constant h is 6.62618×10^{-34} J s.

Another useful neutron parameter, particularly when treating its scattering in a reciprocal crystal lattice, is the wave vector $\boldsymbol{k}$, which is defined to have a magnitude

$$k = 2\pi/\lambda \tag{1.1.3}$$

in the direction of $\boldsymbol{v}$.

The velocities of all neutrons up to kilovolt energies are such that their energies can be defined by the well-known 'time-of-flight' method, in which the neutron transit time over a known flight path is measured. Of profound significance, moreover, is the fact that the neutron mass gives thermal neutrons a de Broglie wavelength of the order of the interatomic distances in solids and liquids. Thus, interference effects are observable which provide information on the structure of the scattering system. It is the fact that neutrons enable both the structure and dynamics of condensed matter to be studied simultaneously which gives them their uniqueness.

TABLE 1.1 Relationships between energy units commonly employed in neutron scattering physics[a]

Energy unit	Symbol	Relationship with E in meV
Frequency (Hz)	ν	$\nu = E/h = 2.418 \times 10^{11} E$ Hz
Angular frequency (rad s^{-1})	ω	$\omega = E/\hbar = 1.519 \times 10^{12} E$ rad s^{-1}
Wave Number (cm^{-1})	$\bar{\nu}$	$\bar{\nu} = E/hc = 8.066E$ cm^{-1}
Temperature (K)	T	$T = E/k_B = 11.606E$ K

[a] The fundamental constants used in these equations are Planck's constant $h = 6.62618 \times 10^{-34}$ J s; $\hbar = h/2\pi$; the velocity of light $c = 2.998 \times 10^8$ m s^{-1}; Boltzmann's constant $k_B = 1.38066 \times 10^{-23}$ J K^{-1}.

TABLE 1.2 Typical values of ν, v, λ and k in the cold, thermal and hot neutron ranges

Neutron Energy E (meV)	5	50	300
Frequency ν (THz)	1.209	12.091	72.54
Velocity v ($\mathrm{m\,s^{-1}}$)	978	3093	7576
Wavelength λ (Å)	4.0444	1.2790	0.5221
Wave vector k (Å^{-1})	1.5535	4.9127	12.034

The numerical relationships between these different parameters are

$$E = 2.0717k^2 = 81.787\lambda^{-2} = 4.1354\nu = 5.2276 \times 10^{-6}\, v^2, \quad (1.1.4)$$

where E is expressed in meV or as frequency (ν) units in Hz, k is in Å^{-1}, λ is in Å, and v is in $\mathrm{m\,s^{-1}}$. Table 1.2 shows typical values of these quantities at energies close to the centre of the cold, thermal and hot neutron ranges.

Although the neutron itself carries a zero net charge, it does contain charged constituent particles, and it is widely accepted that it is the spinning of its charged internal structure which is responsible for the neutron's magnetic moment. The neutron's magnetic moment interacts strongly with the unpaired electron spins in magnetic atoms, and neutron scattering has contributed enormously to our present-day understanding of magnetism. The neutron spin angular momentum has a magnitude $\pm\frac{1}{2}\hbar$, and it is fortunate that neutron beams can be prepared to contain predominantly either $+\frac{1}{2}\hbar$ or $-\frac{1}{2}\hbar$ spin neutrons. These spin-polarized neutron beams are the subject of this book.

The zero net charge of the neutron means that it penetrates deeply into the scattering sample, and it can often be transmitted through bulky instrument components without suffering too large an attenuation penalty. More importantly, since there is no Coulomb barrier to overcome, neutrons can readily interact with nuclei, and this is in fact the principal mechanism for neutron scattering. This scattering, in contrast to that which occurs for electromagnetic radiation, does not depend on the number of electrons in the atom, and varies randomly between the elements, and even isotopes of the same element. Isotopic substitution can sometimes have a profound effect on neutron scattering profiles, and can result in a 'contrast' which is often exploited in structural studies.

1.2 NEUTRON SOURCES AND INSTRUMENTATION

All neutron scattering instruments have a minimum of three 'components': the source, a monochromator to define the neutron energy or wavelength, and the detector. This section is intended to provide a brief description of some of the types of sources and instrumentation now in use. Polarization devices of course add a further degree of complication to the instruments, but these will be discussed in some detail in later chapters.

1.2.1 Sources

There are two main classes of neutron source that are useful for condensed matter research. The most common is the research reactor, which is a steady-state source in which the neutrons are produced by fission in an enriched uranium core. More recently a number of pulsed sources based on particle accelerators have come into operation. Examples include spallation sources in which high-energy proton beams cause a heavy nuclear target such as depleted uranium to break apart to produce the neutrons. The natural method of using pulsed sources is by the time-of-flight technique.

The most powerful steady-state source in the world at present is the 58 Mw high-flux reactor at the Institut Laue–Langevin, Grenoble. The fission neutrons which escape from the core are moderated in a cold, thermal or hot moderator. These provide the spectral distributions shown in Fig. 1.1 at positions about 45 cm from the core, i.e. at the inner ends of the beam tubes (see Maier-Leibnitz (1972)). The D_2O thermal source

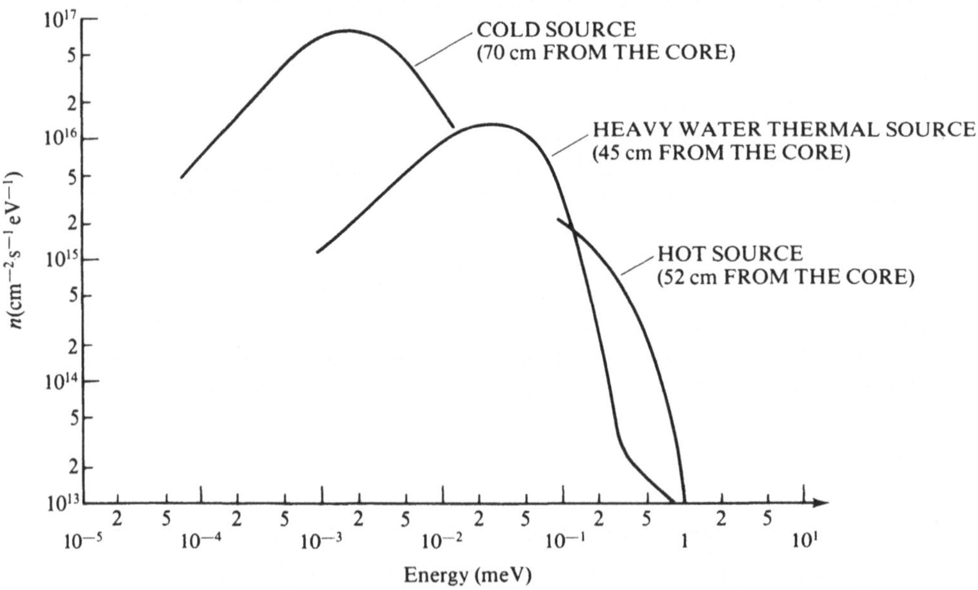

Fig. 1.1 Flux spectra in the thermal moderator 45 cm from the core, near the cold source, and near the hot source of the high-flux reactor in Grenoble (from Maier–Liebnitz 1972).

gives essentially a Maxwellian spectrum at $T = 330$ K, i.e.

$$\Phi(E)\, dE = 2\Phi_0[E/(k_B T)^2] \exp\{-E/(k_B T)\}\, dE, \qquad (1.2.1)$$

where Φ_0 is the total flux. The cold source consists of 25 litres of liquid deuterium and is placed slightly outside the maximum thermal flux region. The hot source is a graphite cylinder with diameter 0.2 m and height 0.3 m which is maintained at 2000°C by nuclear heating. The spectral tailoring referred to in Section 1.1 is self-evident. An important feature of the Grenoble high-flux reactor is its extensive use of cold and thermal neutron guide tubes which use total reflection to transport neutrons away from the immediate vicinity of the reactor with only small losses in flux; these are discussed by Jacrot (1970).

Next we consider pulsed sources. The first type is the pulsed reactor in which the reactivity of the core is varied mechanically. These sources have been developed exclusively in the USSR, and are described by Shabalin (1979). Other pulsed neutron sources use accelerated particle beams, and in one of these the neutrons are produced by (γ, n) reactions with the nuclei in a heavy-metal target. The main facility is a pulsed electron linac which accelerates electrons up to an energy ~100 MeV; these then produce bremsstrahlung as they are rapidly decelerated in the electromagnetic fields of the target nuclei. An example of this type of source is the Harwell linac HELIOS described by Lynn (1980). The second type of accelerator-based facility is the spallation neutron source in which protons accelerated up to energies ~1 GeV interact directly with a heavy-metal nucleus, causing it to disintegrate and produce neutrons. A detailed comparison of all three methods has been given by Windsor (1981). We now present a short description of the main features of the Rutherford Appleton Laboratory spallation neutron source ISIS.

ISIS produces fast neutrons through the interactions of 800 MeV protons with a depleted uranium target (see Fender *et al.* (1980)). The protons are accelerated to this energy in a high-intensity proton synchrotron which has been designed to deliver individual 400 ns bursts containing 2.5×10^{13} protons at a repetition frequency of 50 Hz; the fast neutron yield is about 4×10^{16} n s^{-1}. There are four moderators close to the target, and embedded in a beryllium reflector, which slow these neutrons to epithermal, thermal, and cold neutron energies whilst maintaining pulse lengths in the range 1–100 μs. These are a 4.5 cm thick liquid CH_4 moderator at 95 K poisoned with gadolinium, a 8 cm thick supercritical H_2 moderator at 25 K, and two 4.5 cm thick ambient water moderators, one of which is poisoned symmetrically and the other asymmetrically. The design optimization for a moderator on a pulsed source essentially involves a trade-off between maximizing the flux while keeping a pulse width which is consistent with the resolution demands of the time-of-flight spectrometers which it serves. The 4π-equivalent fluxes

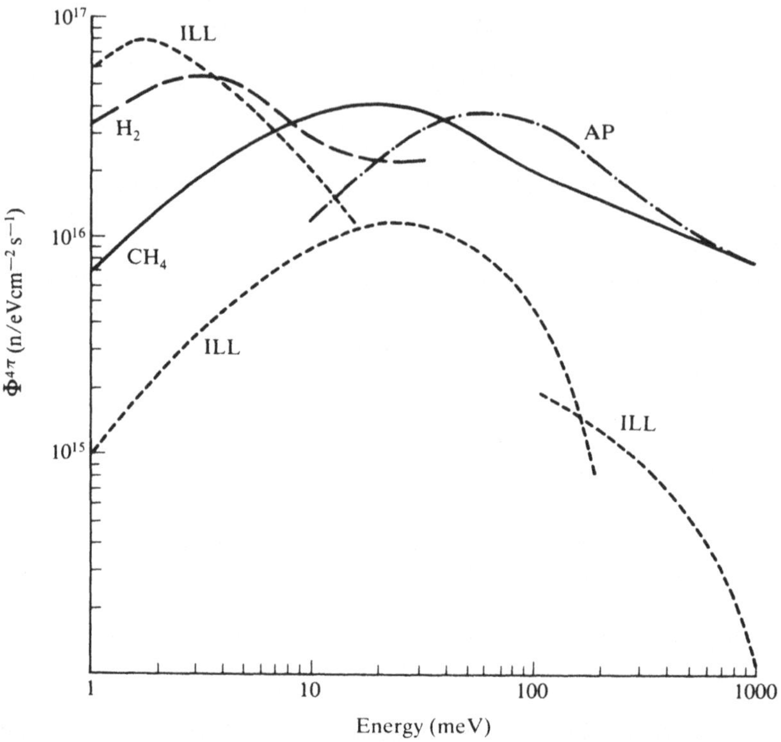

Fig. 1.2 The predicted 4π equivalent fluxes for three ISIS moderators compared with the fluxes of the three Grenoble high-flux reactor moderators (from Taylor 1984).

calculated by Taylor (1984) for three of the ISIS moderators are shown in Fig. 1.2; the designation AP refers to the characteristics when viewing the high-resolution or short-depth-face asymmetrically poisoned ambient water moderator. The equivalent parameters for the moderators at the Grenoble high-flux reactor have also been included for comparison. A full assessment of spectrometer performances, however, requires several other quantities to be introduced, particularly the pulse widths in the pulsed-source case, and this (see also Windsor (1981)) is very case-specific.

In pulsed-source moderators the effect of cooling the moderator is two-fold. Cooling shifts the spectral distribution, just as in the case of the steady-state-source moderators, but in addition there is a reduction in the neutron pulse widths. The main difference between pulsed and steady-state sources is that the former contain copious quantities of epithermal and fast neutrons which can be used in experiments. This affects the design of the scattering instrument components as well as the emphasis of the scientific programmes followed at these sources.

1.2.2 Energy or wavelength selection

It has been shown that all the neutron sources used for condensed matter research yield a continuous range of energies, and we now discuss some of the methods for defining the neutron energy in a scattering experiment.

The most obvious method is to determine the neutron's velocity by measuring its time-of-flight as it traverses a known distance. This is the natural technique to use on a pulsed source, from which all neutrons are emitted from the moderator at the same time (within limits). In elastic scattering experiments, when the scattering leaves the neutron velocity unchanged, all that is required is to determine the arrival time at the detector. One method of performing inelastic scattering experiments is to select a monochromatic beam from the incident white beam and then measure the different arrival times at the detector. This can be done by using a mechanical chopper whose opening is phased with respect to the start pulse (see for example Carlile *et al.* (1985)). Time-of-flight selection is also frequently used at steady-state sources. Here a first chopper replaces the function of the pulsed source by providing a pulsed white beam, and a second chopper which is opened at a well-defined later time transmits only a pulsed monochromatic beam; the principle is illustrated in Fig. 1.3. The so-called mechanical velocity selector (see Bampton and Meardon (1977)), is an extension of this method. It consists of a cylinder rotating about an axis which is parallel to the neutron beam and which contains helical slots cut along its curved surface where it intersects the beam. Only neutrons with the correct velocity can travel along the slots without being absorbed; it is an efficient monochromating device for cold neutrons.

By far the most common technique for selecting monochromatic neutrons is that using Bragg diffraction from a set of lattice planes of a single crystal placed in a white neutron beam; it is used at steady-state

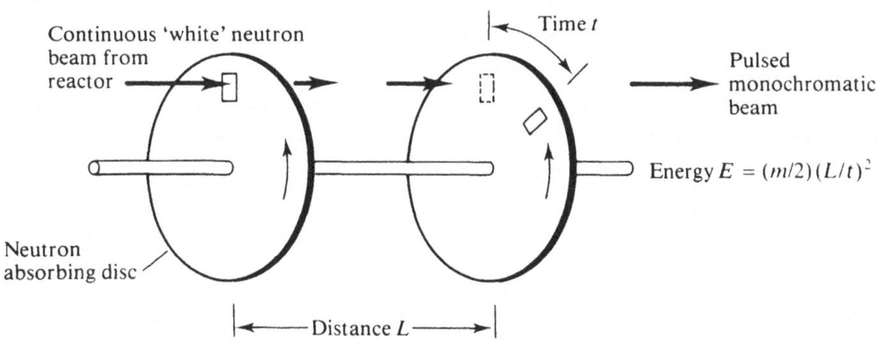

Fig. 1.3 The principle of neutron energy selection on a continuous neutron source using mechanical choppers and the neutron time-of-flight.

and pulsed sources. A monochromator crystal set with its (*hkl*) planes at a mean reflection angle $\bar{\theta}_B$ with respect to the mean direction of the incident beam selects mean wavelengths from the beam according to Bragg's law:

$$\bar{\lambda} = (1/n)2d_{hkl} \sin \bar{\theta}_B, \quad (1.2.2)$$

where d_{hkl} is the interplanar spacing and n is the order of the reflection. Great care is taken to use crystals and reflections for which either higher-order reflections are absent or their intensities are tolerably low. The range of wavelengths reflected, or more precisely the wavelength resolution $\Delta\bar{\lambda}/\bar{\lambda}$, is of great importance in scattering experiments, and differentiation of Bragg's law gives

$$\Delta\bar{\lambda}/\bar{\lambda} = \cot(\bar{\theta}_B) \cdot \Delta\bar{\theta}_B, \quad (1.2.3)$$

where in practice $\Delta\bar{\theta}_B$ depends on the full width at half-maximum (FWHM) incident and reflected beam angular divergences in the scattering plane (α_1 and α_2), and the mosaic spread of the crystal β. This latter parameter is a measure of the imperfection of the crystal lattice, and if this is expressed as the FWHM spread in a crystal rocking curve for a parallel incident beam, Carlile and Ross (1975) showed that

$$\Delta\bar{\theta}_B = \{(\alpha_1^2\alpha_2^2 + \alpha_1^2\beta^2 + \alpha_2^2\beta^2)/(\alpha_1^2 + \alpha_2^2 + 4\beta^2)\}^{1/2}. \quad (1.2.4)$$

Good resolution is achieved by minimizing $\Delta\bar{\theta}_B$ and/or maximizing the Bragg reflection angle $\bar{\theta}_B$. We shall see that this is a limiting feature with polarizing crystal monochromators, since the best polarizing reflections have small Bragg angles in the upper thermal and fast neutron energy regions. In assessing the effectiveness of a crystal monochromator, several other parameters, such as peak reflectivities, incoherent scattering, absorption, etc., are important, and these have been reviewed by Freund and Forsyth (1979). As well as their ubiquitousness as the primary beam monochromators at reactors, crystal monochromators are now also being used as energy analysers in the scattered beam of pulsed-source inelastic scattering spectrometers (see for example Windsor and Heenan (1980)).

Another device for selection of neutron energy which uses Bragg scattering, and which is suitable at long neutron wavelengths, is the polycrystalline scattering filter. In polycrystalline materials the measured and calculated scattering cross-sections as a function of wavelength show discrete jumps corresponding to reflections at $\theta_B = 90°$ for their different (*hkl*) scattering planes in equn (1.2.2). The lowest-index reflection provides the largest interplanar spacing d_{max}, and the filter therefore becomes transparent at wavelengths greater than $2d_{max}$. The cross-sections and wavelength cut-offs for a range of polycrystalline substances are shown in Fig. 1.4. In the most efficient filters all loss processes (e.g.,

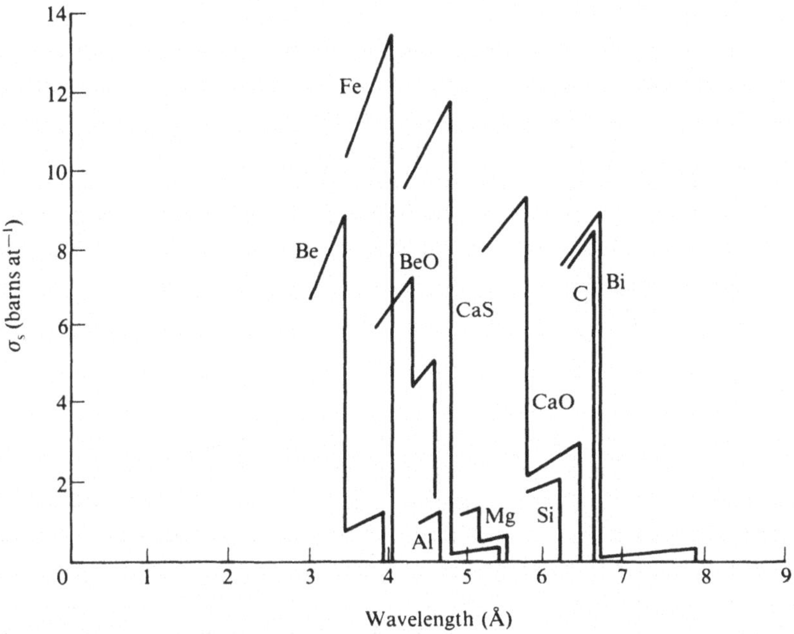

Fig. 1.4 Cross-sections and wavelength cut-offs for various polycrystalline materials (from Hughes 1953). σ_s is defined in Section 1.3.

inelastic and incoherent scattering, absorption) other than loss by coherent elastic scattering should be small, and cooled beryllium is ideal in this respect. The cut-off wavelength is 3.9 Å and a filter of length ~30 cm only transmits neutrons with energies less than 5.2 meV. Cooled beryllium filters can be used to define the scattered neutron energy in both steady-state (see Collins, Haywood and Stirling (1969)), and pulsed-source (see Richardson (1982)) spectrometers. Another application of the beryllium filter is as a scatterer for higher-order reflected neutrons in crystal spectrometers when the fundamental wavelength selected is longer than 3.96 Å; it is commonly used in combination with pyrolytic graphite crystal analysers.

A different type of filter which can also be used to define neutron energies is one containing a nucleus with a 'sharp' and well-isolated nuclear resonance absorption peak; see Section 2.5. The useful resonances are in the 1–10 eV energy range and have transmission widths ~100 meV; they are therefore only really useful for high-energy and relatively coarse-resolution experiments on a pulsed source. There are two basic methods for using nuclear absorption resonances in eV neutron spectroscopy. The first is the resonance detector spectrometer, as for example described by Allen *et al.* (1980), in which a scintillation detector is used to register directly the gamma ray cascade which follows the neutron capture. The second method is a difference technique using neutron detectors, in which the energy-sensitive filter is placed in either the incident or scattered beam of a pulsed-source total scattering spectrometer. The difference time-of-flight spectra collected with and without the absorption filter then either give the sample's response to the resonance-energy incident neutrons or identify those neutrons scattered at the resonance energy. The technique is described by Newport *et al.* (1984), and two of the most suitable resonances are those of ^{149}Sm and ^{181}Ta, which have resonance maxima at energies $E_R = 0.872$ eV and $E_R = 4.28$ eV respectively.

1.2.3 Detectors

There has been an enormous activity in developing efficient neutron detectors for a wide range of applications and the principles used are well-documented (see for example Cocking and Webb (1965)). An updated assessment of the different methods now available has been compiled in the proceedings of a workshop on the position-sensitive detection of thermal neutrons held at the Institut Laue–Langevin, Grenoble, in October 1982, and edited by Convert and Forsyth (1983).

Primary ionization by neutrons is negligible and their detection can only be made indirectly by using induced nuclear reactions. The usual products are charged particles or γ-rays which are then detected in some form of ionization chamber or by scintillators. The main reactions used are ^{10}B(n, α)^{7}Li, ^{3}He(n, p)^{3}H, ^{6}Li(n, α)^{3}H, the fission of ^{235}U or ^{239}Pu, and the production of internal-conversion electrons in a strongly absorbing foil such as Gd. The total interaction cross-sections for some of these nuclei, which may be termed neutron converters, are shown in Fig. 1.5. Note the striking feature that all the cross-sections decrease rapidly towards higher neutron energies, so detection efficiencies degrade with increasing energies.

The most commonly used neutron detectors are gas ionization detectors based on either BF_3 (enriched in ^{10}B) or ^{3}He. Operating pressures vary between 0.5 and 15 atmospheres for ^{3}He, and up to 3 atmospheres for BF_3. The basic assembly consists of a single-wire central anode with a coaxial cylindrical cathode, and these detectors are generally operated in either the ionization plateau or the proportional region of their curves of ionization current versus applied voltage. Their efficiencies depend on the product (gas pressure $\times$ effective thickness $\times$ neutron wavelength) and for a 70% efficiency this product needs to be 16 atm cm Å for ^{3}He, and 23 atm cm Å for $^{10}BF_3$.

In the foil detection method, the converter, which may be any of the absorbing nuclei given above other than ^{3}He, is placed in close proximity to the charged-particle detector. The foil material with the highest

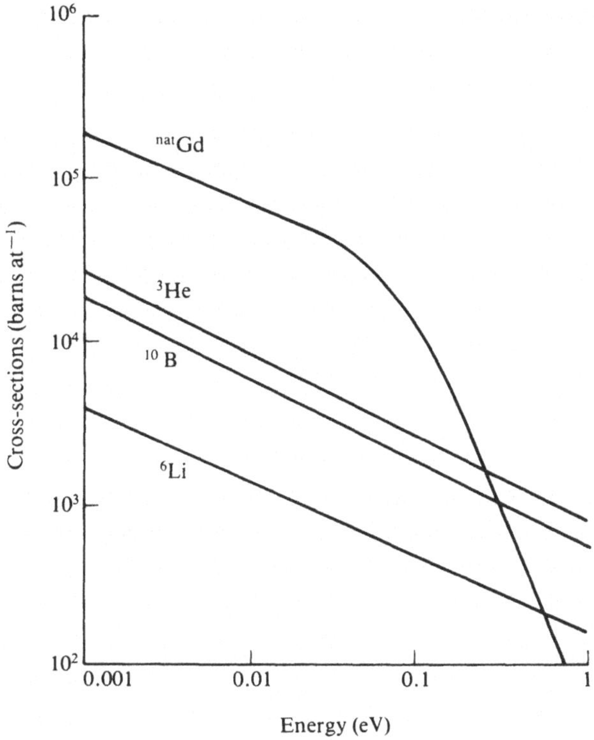

Fig. 1.5 Cross-sections as a function of neutron energy for several converter nuclei commonly used in neutron detectors.

cross-section is gadolinium and its (n, γ) reaction produces conversion electrons with a ratio of captured neutrons to conversion electrons of just less than unity. In a closely coupled geometry, when the electrons can escape easily into their detector, Jeavons *et al.* (1978) have shown that a detection efficiency of about 37% may be achieved at a neutron wavelength of 1 Å with ^{157}Gd foils.

Thirdly, we describe a scintillator detector in which the product particles following the neutron capture cause electronic excitations in the scintillator which then decay with the emission of photons. The example chosen is the position-sensitive detector system developed by Davidson and Wroe (1983) primarily for pulsed-source instruments. There are four basic requirements: (a) the detector thickness must be less than ~20 mm to avoid path-length uncertainties in the time-of-flight measurement; (b) the response time must be fast (~100 ns) in order to accommodate the high instantaneous counts rates which occur at high-intensity pulsed sources; (c) detection efficiencies should be adequately high, say 30% at a neutron energy 1 eV, and (d) the detector should provide positional information to a spatial resolution ~10 mm. The scintillator which most closely satisfies this specification is lithium-loaded glass. Lithium silicate glasses doped with Ce^{3+} (the optically active ion) have fast decay times, are transparent to their light emission which is centred around a 400 nm optical wavelength, produce about 10^4 photons per absorbed neutron, and for a 2 mm thick scintillator give an 80% absorption efficiency at a wavelength $\lambda = 1$ Å. The system developed by Davidson and Wroe uses a fibre-optic encoder to couple a large number of detector elements to a much smaller number of photomultipliers. Each scintillator element is connected by three fibre optic channels to a unique combination of three photomultipliers out of a bank of N. Triple coincidence is then used to label the resolution element where the neutron capture event took place. The number of detector or resolution elements N that can be coded by n photomultipliers is

$$N = n!/[3!\,(n-3)!]. \qquad (1.2.5)$$

This provides great economy in the use of photomultiplier tubes; for example, 40 tubes will code 9880 detector elements. Many different sizes and shapes of detector elements can be used, the detector can operate in air; and it does not have the containment difficulties associated with high-pressure gas detectors. Its principal disadvantage is its sensitivity to gamma radiation, though this can be reduced by interspacing 0.5 mm thick scintillator wafers with plain glass.

1.3 THE NEUTRON SCATTERING EXPERIMENT

The cross-sections as measured in a neutron scattering experiment can be defined by reference to Fig. 1.6. A monochromatic beam with wave vector $\boldsymbol{k}$ travelling in the Z direction is scattered by the specimen containing N atoms, whose centre is placed at the origin. The scattered beam has a wave vector $\boldsymbol{k}'$ at an angle $2\theta_s$ with respect to the Z direction (polar angle), and its projection on to the X–Y plane makes an azimuthal angle ϕ with respect to an arbitrarily chosen X direction.

In the most general experiment we measure the number of neutrons scattered per second (n_1) into a small solid angle $\mathrm{d}\Omega$ about $2\theta_s$, and which have an energy between E' and $E' + \mathrm{d}E'$. The double differential cross-section is defined by

$$(\mathrm{d}^2\sigma/\mathrm{d}\Omega\,\mathrm{d}E') = n_1/(\Phi\,\mathrm{d}\Omega\,\mathrm{d}E'), \qquad (1.3.1)$$

where Φ is the incident beam flux with a dimension area^{-1} time^{-1}. The double differential cross-section is commonly expressed in units of barns sterad^{-1} eV^{-1}, where 1 barn $= 10^{-24}$ cm^2.

Of the many types of neutron scattering instruments that can be used to measure $(\mathrm{d}^2\sigma/\mathrm{d}\Omega\,\mathrm{d}E')$ we now describe the one that is universally regarded to be the most versatile, the triple-axis spectrometer. This

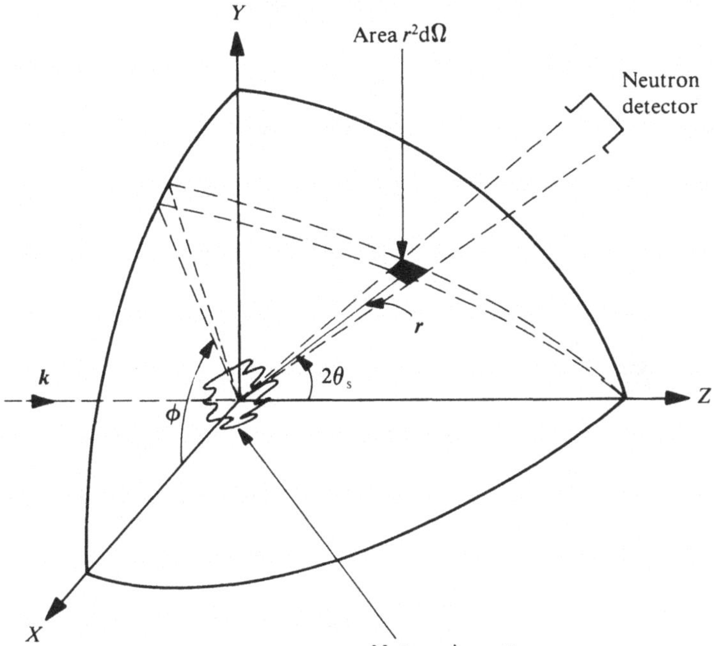

Fig. 1.6 Definition of the geometry in a neutron scattering experiment.

instrument was pioneered by Brockhouse (1961) and is shown schematically in Fig. 1.7. The initial and scattered neutron wave vectors $\boldsymbol{k}$ and $\boldsymbol{k}'$ are defined by Bragg reflection from single-crystal monochromators, and a series of collimators C serve to define the resolution. At a selected scattering angle $2\theta_s$ the scattering triangle defines the scattering vector

$$\boldsymbol{\kappa} = \boldsymbol{k} - \boldsymbol{k}'. \tag{1.3.2}$$

The scattering vector $\boldsymbol{\kappa}$ is of fundamental importance in defining a scattering event; note that it is independent of the individual values of the wave vectors $\boldsymbol{k}$ and $\boldsymbol{k}'$, and depends only on their difference.

The energy transfer $\hbar\omega$ is given by

$$\hbar\omega = (\hbar^2/2m)(k^2 - k'^2), \tag{1.3.3}$$

where m is the neutron mass, $k = |\boldsymbol{k}|$, and $k' = |\boldsymbol{k}'|$. By changing the sample orientation direction ϕ_s, the scattering vector $\boldsymbol{\kappa}$ can be positioned anywhere in the scatterer's reciprocal lattice. The most frequently used scan gives a particular energy transfer a constant $\boldsymbol{\kappa}$, and involves keeping $\boldsymbol{k}'$ constant by simultaneously changing θ_m, θ_A, θ_s and ϕ_s. The triple-axis spectrometer has been used extensively to measure the energy dispersion of coherent excitations in single-crystal specimens; a polarized beam

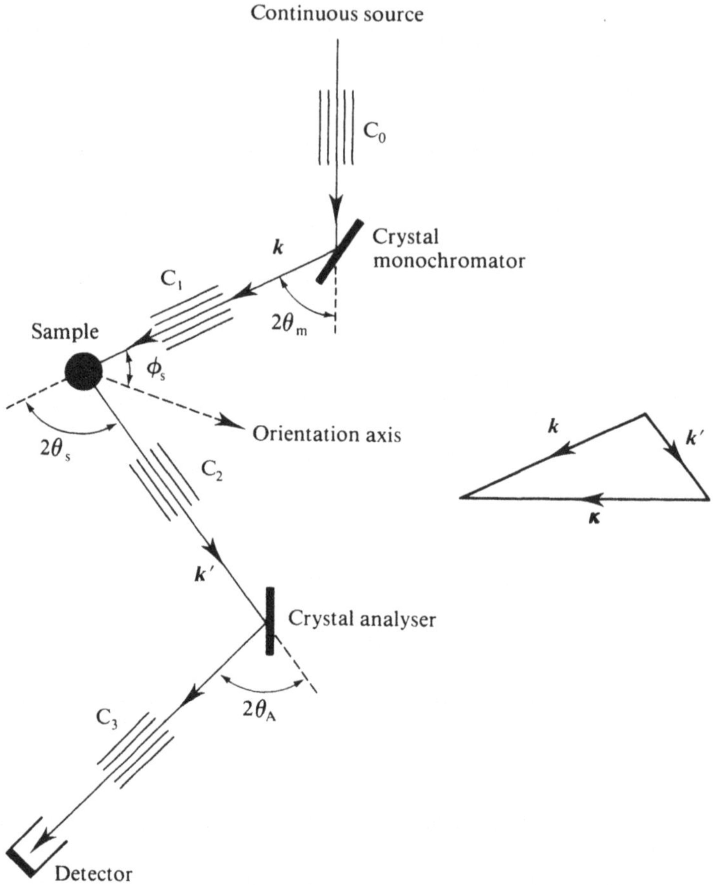

Fig. 1.7 Schematic diagram of a triple-axis neutron spectrometer.

version can be constructed by using polarizing crystal monochromators at the first and third axial positions.

In some experiments it is not necessary to analyse the scattered neutron energy. In this instance one simply measures the total number of neutrons (n_2) scattered per second into $d\Omega$. The cross-section determined in these measurements is the differential cross-section

$$(d\sigma/d\Omega) = n_2/(\Phi\, d\Omega). \tag{1.3.4}$$

The usual unit for the differential cross-section is barns sterad^{-1}.

The most frequently used instrument for measuring $(d\sigma/d\Omega)$ is the 2-axis diffractometer on a steady-state source, and this can be visualized by replacing the analyser in Fig. 1.7 by a neutron detector. This type of instrument is used to measure elastic cross-sections, as for example in a single-crystal diffractometer in which the Bragg peaks identify the elastic scattering, or sometimes total cross-sections when the sample scatters both elastically and inelastically. A liquids diffractometer is one example of the latter. The magnitude of the scattering vector for elastic scattering is

$$\kappa = 2\,|\boldsymbol{k}|\,\sin\theta_s. \tag{1.3.5}$$

The third type of cross-section is the total scattering cross-section σ_s

$$\sigma_s = n_3/\Phi, \tag{1.3.6}$$

where n_3 is the total number of neutrons scattered per second at all energies and in all directions; σ_s is usually measured in barns (10^{-24} cm^2).

In the absence of absorption, the easiest method of measuring σ_s is to measure the transmittance T of a scatterer with constant thickness t. If the atomic density is N (e.g. atoms cm^{-3}) then

$$T = \exp(-\sigma_s N t). \tag{1.3.7}$$

Such measurements are often complicated by an absorption cross-section σ_A, although this can sometimes be separated out by using its neutron energy dependence,

$$T = \exp[-(\sigma_s + \sigma_A)Nt] = \exp(-\sigma_T N t). \tag{1.3.8}$$

The three scattering cross-sections are inter-related as follows:

$$(\mathrm{d}\sigma/\mathrm{d}\Omega) = \int_0^\infty (\mathrm{d}^2\sigma/\mathrm{d}\Omega\,\mathrm{d}E')\,\mathrm{d}E' \tag{1.3.9}$$

and

$$\sigma_s = \int_\Omega (\mathrm{d}\sigma/\mathrm{d}\Omega)\,\mathrm{d}\Omega \tag{1.3.10}$$

and are usually expressed per atom of scatterer. This allows comparisons of the scattered neutron intensities to be made between different experiments. For magnetic scatterers, the scattering amplitudes which determine the cross-sections can be calculated from first principles.

In formulating the above cross-sections we have deliberately ignored the fact that the neutron has a spin. There are subsets of all three cross-sections which are neutron-spin-dependent, and their measurement and interpretation are the major subject of this book. Spin-dependent cross-sections can only be studied by using spin-polarized beams, and we now define the characteristics of these beams.

1.4 DEFINITIONS AND PROPERTIES OF SPIN-POLARIZED NEUTRON BEAMS

We represent the spin of the neutron by the vector $\hat{\boldsymbol{s}}'$, which is an operator for which only one component along an arbitrary axis can be determined at a given instant. If we imagine making a large number of measurements of this spin component and then take the average, then this quantity will be finite only for a polarized beam. It follows that each single neutron, when considered in isolation, has its own polarization $|\boldsymbol{P}| = 1$ along its spin direction. The polarization vector $\boldsymbol{P}$ may be described as the expectation value of the spin operator $\langle \hat{\boldsymbol{s}}' \rangle$ divided by its maximum value, and we may write as a general formula

$$\boldsymbol{P} = \langle \hat{\boldsymbol{s}}' \rangle / (1/2) = 2(\langle \hat{s}'_X \rangle \tilde{\boldsymbol{i}}_X + \langle \hat{s}'_Y \rangle \tilde{\boldsymbol{i}}_Y + \langle \hat{s}'_Z \rangle \tilde{\boldsymbol{i}}_Z), \tag{1.4.1}$$

where $\tilde{\boldsymbol{i}}_X$, $\tilde{\boldsymbol{i}}_Y$, $\tilde{\boldsymbol{i}}_Z$, are unit vectors in the directions of an arbitrary (X, Y, Z) coordinate system along which the three spin components can be measured. In a neutron beam which contains N spins, each with a polarization $\boldsymbol{P}_i$, the beam polarization may be defined as the vector average over the polarization of each neutron, i.e.

$$\boldsymbol{P} = \left(\sum_i \boldsymbol{P}_i \right) \Big/ N. \tag{1.4.2}$$

In equn (1.4.1) we have used the fact that the neutron is a spin-$\frac{1}{2}$ particle with two possible spin angular momentum values $\pm\frac{1}{2}\hbar$. In the majority of the polarized beam experiments so far performed, it is usually sufficient just to consider the linear polarization vector in the direction of an applied magnetic field. The N neutrons can then be considered to be subdivided into two subsets: N_+ with $(+)$ or 'up' neutron spins, and N_- with $(-)$ or 'down' neutron spins. The beam polarization may then be expressed by

$$\boldsymbol{P} = [(N_+ - N_-)/(N_+ + N_-)]\tilde{\boldsymbol{h}}, \tag{1.4.3}$$

where $\tilde{\boldsymbol{h}}$ is a unit vector in the applied field direction, which is normally assigned as the Z direction. If a neutron beam has a fraction ϕ_+ of its neutrons in the spin 'up' state, and a fraction ϕ_- of its neutrons in the 'down' state, then the magnitude of its polarization in the Z direction is

$$|\boldsymbol{P}| = \phi_+ - \phi_- = 2\phi_+ - 1 = 1 - 2\phi_-. \tag{1.4.4}$$

This description of a polarized beam in terms of just two spin components is perfectly adequate when discussing one-dimensional polarized beam and polarization analysis experiments. It will be used extensively in this book when, for example, describing the action of instrument components like polarizers and spin-flippers, and in the scattering experiments themselves we shall need to distinguish between spin-flip (e.g. $(+) \rightarrow (-)$) and non-spin-flip (e.g. $(+) \rightarrow (+)$) events.

It is important to understand how linear polarization differs from two other quantities with which it is often confused, orientation and alignment in spin systems. The interaction of a generalized spin S with a magnetic field in a direction Z generates $2S + 1$ energy levels each

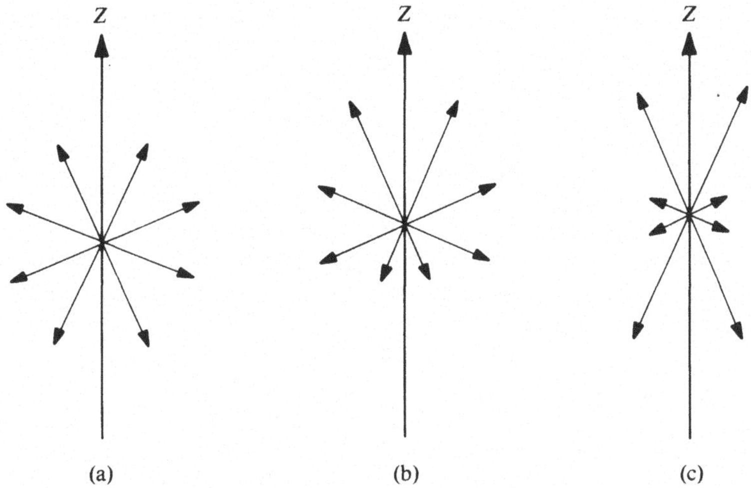

Fig. 1.8 Spin systems quantized in the Z direction; the arrows give the orientation directions of the spins and their lengths are drawn in proportion to the populations of the spin levels. (a) is an unoriented assembly, (b) is an oriented and polarized assembly, and (c) is an oriented and aligned assembly.

characterized by a value $m_s = \langle S_Z \rangle$, the expection value of S in the Z direction. The term 'orientation' simply means that the population of each m_s state depends on the value of m_s. Fig. 1.8(a) shows a system of unorientated spins. The spin system is polarized if the number of spins pointing in the $+Z$ direction differs from that in the $-Z$ direction, as shown in Fig. 1.8(b). The linear polarization is just the average value of $\langle S_Z \rangle / S$ for the whole system and is proportional to the magnetization. Alignment is another type of orientation, in which the population of the m_s states depends on m_s^2, and an aligned system is in fact unpolarized if the populations of each pair of $+m_s$ and $-m_s$ states are equal. This is depicted in Fig. 1.8(c). Polarization and alignment are therefore subsets of the general phenomenon of orientation.

Although essentially all polarized neutron experiments performed up to about 1970 were concerned only with one-dimensional polarization, several techniques have been introduced since which use vectorial, or three-dimensional, polarization. These methods involve the production, manipulation, scattering and analysis of polarization vectors which are not parallel to the local magnetic field, and a review of the possibilities offered by the extension has been published by Mezei (1986). One of the methods for describing vectorial polarization is to use a density matrix description, see Lovesey (1986), and we now reproduce this formulation.

The spinning neutron possesses a magnetic moment μ_n with its operator $\hat{\boldsymbol{\mu}}_{\mathrm{n}}$ given by

$$\hat{\boldsymbol{\mu}}_{\mathrm{n}} = \mu_{\mathrm{n}} \hat{\boldsymbol{\sigma}}, \tag{1.4.5}$$

where the components of the 'spin' operator $\hat{\boldsymbol{\sigma}}$ are the Pauli matrices

$$\hat{\sigma}_X = \begin{bmatrix} 0 & 1 \\ 1 & 0 \end{bmatrix}; \quad \hat{\sigma}_Y = \begin{bmatrix} 0 & -\sqrt{(-1)} \\ \sqrt{(-1)} & 0 \end{bmatrix}; \quad \hat{\sigma}_Z = \begin{bmatrix} 1 & 0 \\ 0 & -1 \end{bmatrix}. \tag{1.4.6}$$

The expectation value of $\hat{\boldsymbol{\sigma}}$ is equivalent to the spin polarization vector $\boldsymbol{P}$ given in equn (1.4.1), i.e. $\boldsymbol{P} = \langle \hat{\boldsymbol{\sigma}} \rangle$. The introduction of $\hat{\boldsymbol{\sigma}}$ provides a convenient way of quantum mechanically relating the three polarization components P_X, P_Y, P_Z, and this can be done (see Lovesey (1986)), by introducing a density matrix operator

$$\hat{\boldsymbol{\rho}} = \frac{1}{2}(\mathcal{I} + \boldsymbol{P} \cdot \hat{\boldsymbol{\sigma}}) = \frac{1}{2}\begin{bmatrix} 1 + P_Z & P_X - \sqrt{(-1)}P_Y \\ P_X + \sqrt{(-1)}P_Y & 1 - P_Z \end{bmatrix} \tag{1.4.7}$$

with

$$\langle \hat{\boldsymbol{\rho}} \rangle = \mathrm{Tr}\, \hat{\boldsymbol{\rho}}\hat{\boldsymbol{\rho}} = \frac{1}{2}(1 + |\boldsymbol{P}|^2). \tag{1.4.8}$$

Hayter (1978a) used this concept in his description of the precessional behaviour of the neutron polarization vector in an uniform magnetic field. The Hamiltonian for the interaction of the neutron moment with a magnetic field $\boldsymbol{B}$ is

$$\hat{V}_\mathrm{M} = -\hat{\boldsymbol{\mu}}_\mathrm{n} \cdot \boldsymbol{B} = -\mu_\mathrm{n}\hat{\boldsymbol{\sigma}} \cdot \boldsymbol{B}, \tag{1.4.9}$$

and in a representation in which $\hat{V}_\mathrm{M}$ is diagonal, which corresponds to B being parallel to Z, the time evolution of the matrix elements $\rho_{lm}(t)$ are given by

$$\rho_{lm}(t) = \rho_{lm}(0) \exp[\mathrm{i}(E_m - E_l)t/\hbar], \tag{1.4.10}$$

where E_l, E_m are the energy eigenvalues of the interaction $\hat{V}_\mathrm{M}$, and $i = \sqrt{(-1)}$. The time variation of the individual polarization components in a uniform local field can then be obtained by combining equns (1.4.7) and (1.4.10) to give

$$\begin{aligned} P_X(t) &= \cos(\omega_\mathrm{L}t)P_X(0) - \sin(\omega_\mathrm{L}t)P_Y(0), \\ P_Y(t) &= \sin(\omega_\mathrm{L}t)P_X(0) + \cos(\omega_\mathrm{L}t)P_Y(0), \\ P_Z(t) &= P_Z(0), \end{aligned} \tag{1.4.11}$$

where ω_L is the well-known Larmor frequency

$$\omega_\mathrm{L} = (E_m - E_l)/\hbar = 2\,|\mu_\mathrm{n}B|/\hbar. \tag{1.4.12}$$

This result, where $P_Z(t)$ is time-invariant, and both $P_X(t)$ and $P_Y(t)$ rotate in the (X, Y) plane at the Larmor frequency, is the classical equation of motion of either a magnetization or a spin vector in an

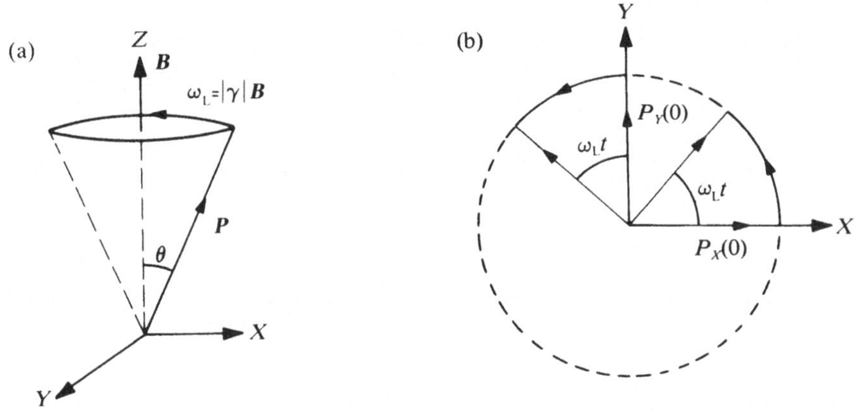

Fig. 1.9 (a) The neutron polarization vector $\boldsymbol{P}$ precessing about $\boldsymbol{B}$ with a frequency $\omega_L = |\gamma|\, B$; $\dfrac{d\boldsymbol{P}}{dt} = -\gamma(\boldsymbol{P} \times \boldsymbol{B})$. (b) The time evolution of the polarization components P_X and P_Y in the XY plane. $P_X(t) = P_X(0)\cos\omega_L t - P_Y(0)\sin\omega_L t$; $P_Y(t) = P_Y(0)\cos\omega_L t + P_X(0)\sin\omega_L t$.

applied magnetic field. This is depicted diagrammatically in Fig. 1.9(a), and in Fig. 1.9(b) we show how equn (1.4.11) is consistent with the two polarization components $P_X(0)$ and $P_Y(0)$ rotating in the (X, Y) plane at the frequency ω_L. It is also evident from equn (1.4.1) that the expectation value of the neutron spin vector must also precess about the local field $\boldsymbol{B}$ in precisely the same way as $\boldsymbol{P}$, therefore we have

$$d\langle \hat{\boldsymbol{s}}' \rangle / dt = -\gamma \langle \hat{\boldsymbol{s}}' \rangle \times \boldsymbol{B} = \langle \hat{\boldsymbol{s}}' \rangle \times \omega_L, \qquad (1.4.13)$$

with $\omega_L = -\gamma \boldsymbol{B}$. The neutron gyromagnetic ratio γ is the ratio of its magnetic moment μ_n to its spin angular momentum $\frac{1}{2}\hbar$, and numerically $\gamma = -2.916 \times 10^7\ \mathrm{Hz\,T^{-1}}$. The average value of the neutron spin vector $\langle \hat{\boldsymbol{s}}' \rangle$ therefore precesses with a Larmor frequency

$$\omega_L = |\gamma|\, B = 2\,|\mu_n B|/\hbar, \qquad (1.4.14)$$

in agreement with equn (1.4.12) for the polarization. By substituting constants we obtain

$$\omega_L = 1.84 \times 10^8 B\ \mathrm{rad\,s^{-1}}, \qquad (1.4.15)$$

with B given in tesla ($=10^4$ oersted). This precessional property of the neutron spin polarization has several important applications, for example neutron spin echo spectroscopy, the study of magnetic domains, and the determination of spin-dependent neutron scattering amplitudes. These methods are to be discussed in Chapter 6.

A note is warranted at this point on the use of $\boldsymbol{B}$ in this book to denote the 'local' magnetic field which is effective in the magnetic neutron interaction. This quantity is used to represent either an applied field in vacuum, or the magnetic induction in a ferromagnetic medium. The quantity $\boldsymbol{B}$ is, of course, identical with the applied field parameter $\boldsymbol{H}$ for the *in vacuo* condition.

1.5 DEVELOPMENT OF POLARIZED NEUTRON BEAM METHODS

We complete this introductory chapter by making a brief historical survey of the development of polarized neutron beam applications. A recurring theme in this section will be the enormous improvement in the sophistication of polarized-beam experiments that has followed advances in the technologies for producing both neutron sources and polarizing devices.

Bloch (1936) was the first to make an extensive investigation of the interaction between magnetic atoms and what were supposed to be magnetic neutrons. Bloch's model for the neutron's magnetic moment was a point magnetic dipole, and he was able to make several remarkable predictions: that the strength of the magnetic scattering was comparable with that for nuclear scattering, and that neutron transmission measurements through magnetized iron would supply the magnitude of the neutron moment, provide details about atomic magnetization, and also produce polarized neutron beams. This was shortly followed by the detailed quantum mechanical treatment of magnetic scattering of Schwinger (1937), in which the origin of the neutron moment was described as a microscopic amperian current. Schwinger, like Bloch, also recognized that coherent interference between the magnetic and nuclear scattering could be applied to produce polarized beams. The question of the model for the neutron's magnetic moment was to be resolved over 10 years later in favour of Schwinger.

Probably the most remarkable set of theoretical papers on the magnetic scattering of neutrons are those of Halpern and Johnson (1937a,b, 1939), Halpern and Holstein (1941), and Halpern *et al.* (1941). These revealed the importance of paramagnetic scattering experiments, in which the results are relatively straightforward to interpret, and progressed to deal with ferromagnetic scattering, the production of polarized beams, and the application of neutron depolarization measurements for investigating ferromagnetic domain structures. The theory of the neutron spin dependence of nuclear scattering was also examined. These publications contain most of the theory needed to understand the principles of magnetic neutron scattering experiments, with both polarized and unpolarized neutron beams.

The first significant polarized neutron beam experiment was published by Alvarez and Bloch (1940). These authors used two magnetized iron plates, one as a neutron polarizer and the second as a polarization analyser, with a magnetic resonance spin-inverter between the two to determine the magnitude of the neutron moment; they found $|\mu_n| = 1.93 \pm 0.02\ \beta_N$, the first accurate determination of this quantity. Their method for flipping the neutron spin, the radiofrequency (r.f.) spin-flipper, is still used today in some polarized-beam diffractometers. During the 1940s, as more intense neutron beams from reactors became available, there were several attempts at obtaining quantitative values for the magnetic scattering cross-sections in iron. The early measurements all used the single transmission effect to estimate the polarization part σ_p of the total cross-section. The polarization cross-section σ_p is defined by

$$\sigma_{0\pm} = \sigma_0 \pm \sigma_p, \tag{1.5.1}$$

where σ_0 is the normal transmission cross-section for demagnetized iron, and the $\pm$ signs denote the cross-sections for the two neutron spin states. The single transmission effect τ_S is the fractional increase in the transmitted intensity observed when the material is magnetized:

$$\begin{aligned}\tau_S &= (I_{\text{Mag}} - I_{\text{Demag}})/I_{\text{Demag}} \\ &= \cosh(Nt\sigma_p) - 1 \simeq \tfrac{1}{2}(Nt\sigma_p)^2,\end{aligned} \tag{1.5.2}$$

where Nt is the atomic thickness (atoms cm^{-2}) of the magnetic specimen, and σ_p is the wavelength-dependent polarization cross-section. The σ_p values at various monochromatic wavelengths were determined in this manner by Hughes *et al.* (1948), and by Fleeman *et al.* (1949). Both results were in good agreement with the values of σ_p calculated by Steinberger and Wick (1949), which are reproduced in Fig. 1.10. The discontinuities in σ_p correspond to the cut-off wavelengths of the different crystallographic planes in iron as described in Section 1.3. Fleeman *et al.* (1949) also showed that by observing the specimen thickness dependence of τ_S it was possible to determine a neutron depolarization term which is related to the size of the magnetic domains in iron. The use of polarized neutrons to study magnetic domain structures has progressed steadily from these first measurements, and is to be discussed in Chapter 6.

As higher-intensity neutron beams became available Burgy *et al.* (1950) were able to use the double transmission effect, τ_D, which allowed the polarization (or depolarization) of a neutron beam to be determined directly, and this greatly facilitated domain size investigations. In this technique a second polarizing ferromagnetic block is used to analyse the polarization transmitted from the first. Application of the cross-section equation (1.5.1) for two blocks of equal thickness t shows that when the fields on the two blocks are parallel

$$I_{\text{Par}} = I_0 \exp(-2Nt\sigma_0)\cosh(2Nt\sigma_p), \tag{1.5.3}$$

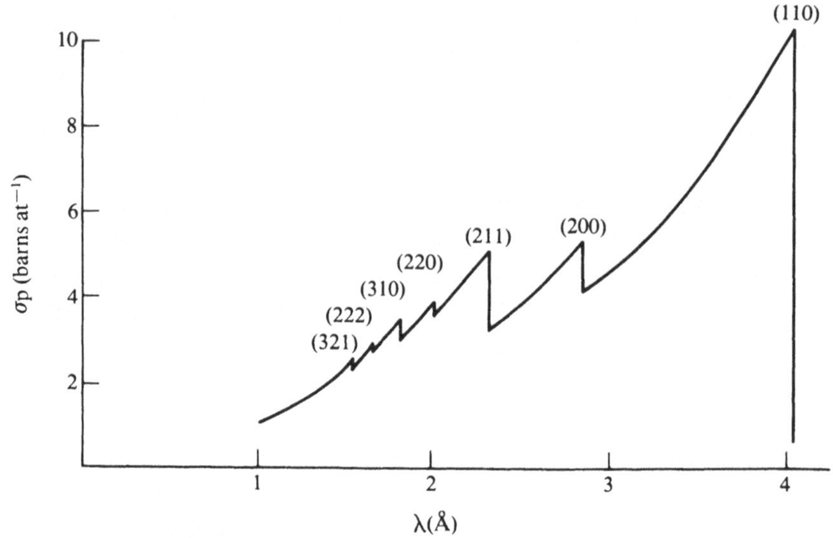

Fig. 1.10 Variation of the polarization cross-section as a function of neutron wavelength for magnetized polycrystalline iron as calculated by Steinberger and Wick (1949).

whereas when they are antiparallel

$$I_{\text{Antipar}} = I_0 \exp(-2Nt\sigma_0). \tag{1.5.4}$$

The double transmission effect is defined by

$$\begin{aligned}\tau_D &= (I_{\text{Par}} - I_{\text{Antipar}})/\bar{I}_{\text{Av}} \\ &= 2(I_{\text{Par}} - I_{\text{Antipar}})/(I_{\text{Par}} + I_{\text{Antipar}}) \\ &= 2\tanh^2(Nt\sigma_p) = 2(Nt\sigma_p)^2.\end{aligned} \tag{1.5.5}$$

Great care is needed in designing the magnetic field profile between the polarization blocks in the antiparallel field arrangement, since an intrinsic assumption in the above analysis is that the neutron spin must cross the magnetic field regions between the two polarizing elements non-adiabatically. The neutron spin effectively undergoes a spin-flip with respect to its local field in the antiparallel geometry, though this was not explicitly stated by the original authors. The technique of 'field-flipping' to reverse the neutron polarization direction has subsequently been used in many polarized-beam instruments. Burgy *et al.* (1950) were able to show that, as expected, the double transmission effect is much more sensitive than the single transmission effect in determining magnetic domain properties.

In parallel with these investigations of magnetic scattering, there was a smaller effort aimed at investigating the neutron-spin-dependence of nuclear scattering cross-sections. The basic theory had again been derived by Halpern and Johnson (1939), and Borowitz and Hamermesh (1948) extended this in a theoretical analysis of the neutron depolarization or spin-flip scattering that occurs in the scattering of a polarized beam from a nuclear scatterer in slab geometry. First experiments were reported by Meyerhof and Nicodemus (1951) using magnetized iron filters as the polarizer and the polarization analyser; the spin-flip scattering was evaluated from a combination of single and double transmission effect determinations. They measured the following spin-flip scattering probabilities: -0.09 ± 0.21 for C, 0.56 ± 0.10 for H, and 0.73 ± 0.15 for P. These were the first polarization analysis experiments on nuclear scatterers.

The first method to produce fully polarized beams was total reflection from magnetized mirrors. Halpern *et al.* (1941) had already shown that the two different neutron spin states have different refractive indices in a ferromagnetic medium, and the proposal to use total reflection to prepare polarized beams was made independently by Halpern (1949) and by Achieser and Pomeranchuck (1948). A vitally important development was the suggestion by Hamermesh (1949) that completely polarized beams were obtainable from magnetized cobalt surfaces since the two spin refractive indices were on opposite sides of unity, and only one of the spin states could be externally reflected. Experiments on magnetized iron and cobalt mirrors were soon to follow (see Hughes and Burgy (1951)), and confirmed the earlier predictions. Figure 1.11 shows the reflectivity data obtained by these authors for iron, which illustrate that there are two critical angles for reflection. This experiment was also of fundamental significance since it enabled the Bloch and Schwinger models of the neutron's magnetic moment to be distinguished. The curve in Fig. 1.11 labelled Bloch shows the expected behaviour for the point dipole model where the interaction potential depends on the $\boldsymbol{H}$ field in the medium. The second curve corresponds to a Dirac amperian current representation as used by Schwinger, a consequence of which is that the magnetic interaction depends on the induction $\boldsymbol{B}$. Experiments on a magnetized thin film of cobalt deposited on to copper gave a reflected beam with 100% neutron polarization. Polarization by total external reflection is especially well-suited to cold neutron beams, since the critical reflection angles, which are proportional to the neutron wavelength, then become sizeable. The method has been widely applied ever since its inception.

During the second half of the 1940s, with the advent of more intense beams from reactors, neutron diffraction was developed as a complementary tool to X-ray and electron diffraction for determining the structures of polycrystalline and single-crystal materials. Much of the early magnetic structures work was carried out at the Oak Ridge National Laboratory,

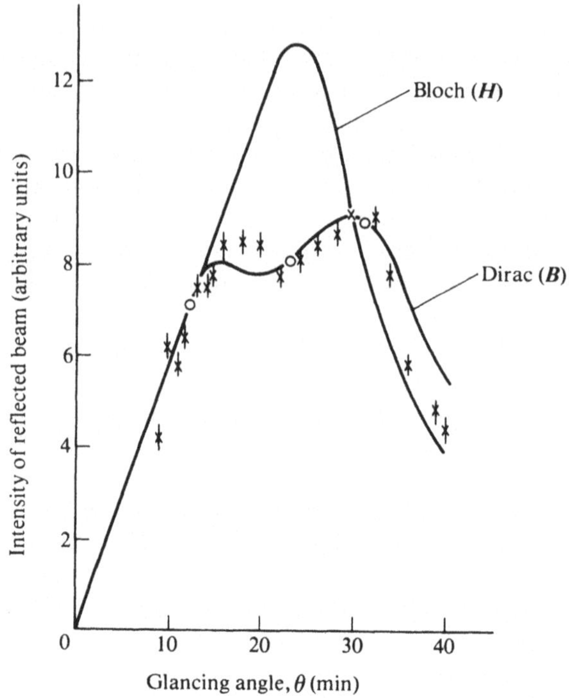

Fig. 1.11 Reflected intensities from a magnetized iron mirror as a function of glancing angle. The 'Bloch' curve is calculated for the dipole description of the neutron moment, and the 'Dirac' curve corresponds to the amperian current model as used by Schwinger (from Hughes and Burgy 1951).

notably by Shull, Wollan and their collaborators. The paper by Shull *et al.* (1951) on neutron scattering and polarization by ferromagnetic materials was, in retrospect, to have a very profound effect on later applications of polarized neutron beams in condensed matter and materials research. We now reproduce some of the observations made by these authors on the (220) reflection from magnetite (Fe_3O_4).

The directionality of the coherent elastic scattering cross-section in a ferromagnet was known, following Halpern and Johnson (1939) to be given by

$$(\mathrm{d}\sigma/\mathrm{d}\Omega) \propto N^2 + 2NM\boldsymbol{q}\cdot\tilde{\boldsymbol{P}} + M^2q^2, \tag{1.5.6}$$

where N and M are nuclear and magnetic scattering amplitudes associated with a set of Bragg reflections in a crystal. $\tilde{\boldsymbol{P}}$ is a unit vector in the direction of the polarization of the neutron being scattered, and the magnetic interaction vector $\boldsymbol{q}$ is

$$\boldsymbol{q} = \tilde{\boldsymbol{\kappa}}(\tilde{\boldsymbol{\kappa}}\cdot\tilde{\boldsymbol{\eta}}) - \tilde{\boldsymbol{\eta}} \tag{1.5.7}$$

with

$$q^2 = 1 - (\tilde{\boldsymbol{\kappa}} \cdot \tilde{\boldsymbol{\eta}})^2, \tag{1.5.8}$$

in which $\tilde{\boldsymbol{\kappa}}$ is the unit scattering vector, and $\tilde{\boldsymbol{\eta}}$ is a unit vector parallel to the spin direction of the ferromagnetic ion.

The first notable experiment on magnetite involved examining the (220) reflection peak intensity in its powder diffraction pattern in two applied field geometries. In a zero field, and with unpolarized neutrons ($\boldsymbol{P} = 0$), powder averaging gives a cross-section which is proportional to the amplitude factor $(N^2 + \frac{2}{3}M^2)$, whereas in strong magnetizing fields this becomes $(N^2 + M^2)$ for $\tilde{\boldsymbol{\kappa}} \cdot \tilde{\boldsymbol{\eta}} = 0$, and N^2 for $\tilde{\boldsymbol{\kappa}} \cdot \tilde{\boldsymbol{\eta}} = -1$. The results of the intensity measurements obtained by Shull *et al.* in these geometries are shown in Fig. 1.12. The important observation is that the nuclear and magnetic scattering amplitudes are approximately equal, and this led these authors to examine the polarizing properties of this reflection in a single-crystal specimen. Equations (1.5.6) and (1.5.7) show that, when the magnetization is perpendicular to the scattering vector ($\boldsymbol{q} = -\tilde{\boldsymbol{\eta}}$ and is parallel to $\tilde{\boldsymbol{P}}$), the cross-sections for the two neutron polarization states in an unpolarized beam, denoted here by $\pm$, become

$$(\mathrm{d}\sigma/\mathrm{d}\Omega)_{\pm} \propto (N \pm M)^2. \tag{1.5.9}$$

Thus an exact balancing of N and M, as appears to occur for the magnetite (220) reflection, should produce a 100% polarized beam. The

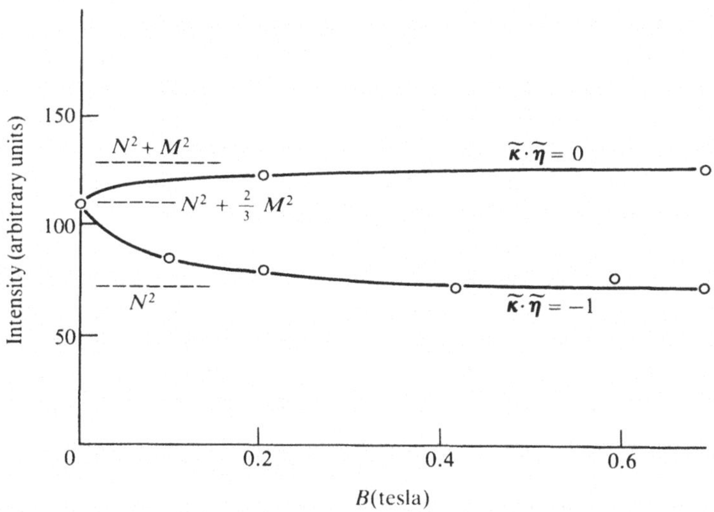

Fig. 1.12 Scattering intensities observed for the (220) reflection in polycrystalline magnetite as a function of a magnetic field applied parallel ($\tilde{\boldsymbol{\kappa}} \cdot \tilde{\boldsymbol{\eta}} = -1$) and perpendicular ($\tilde{\boldsymbol{\kappa}} \cdot \tilde{\boldsymbol{\eta}} = 0$) to the scattering vector (from Shull *et al.* and Koehler 1951b).

Bragg scattered (220) peak from a magnetite single crystal, magnetized in a geometry such that $\tilde{\kappa} \cdot \tilde{\eta} = 0$, was later spin-analysed by an iron polarizing filter which was alternately magnetized and demagnetized. The measured polarization was 104% with an estimated error $\sim \pm 5\%$. Thus a method for producing monochromatic and highly polarized beams had been invented, and it was estimated that $\sim 10^5$ monochromatic neutrons per second could be obtained using this reflection and the radiation from a reactor with a 10^{12} n cm^2 s^{-1} central flux. It is of interest to quote the last sentence of Shull *et al.*'s famous paper: 'With regard to the three methods of polarization, viz. (1) transmission through polycrystalline iron, (2) critical reflection from magnetized mirrors, and (3) Bragg scattering from ferromagnetic crystals, there appear to be unique characteristics for each, and the selection of the optimum method should be decided by the requirements of the use to which the polarized beam is to be put.' This statement is no less true today and we are now able to add other polarizing methods to this list; see Chapter 3.

It is now appropriate to make a brief reference to the contributions made by Sherwood, Stephenson and Bernstein (1954) in advancing polarized beam methods by building a Stern–Gerlach magnet system. This device provides a high field gradient so that neutrons in the two spin states experience oppositely directed forces in the magnet, and can be spatially separated; field gradient devices are discussed in Chapter 3. The technique is most usefully applied as a spin analyser for cold neutron beams, and its particular advantages are that the measurement is direct, and that the deflection direction unequivocally supplies the sign of the spin polarization of the neutron beam. Some of the important results established by the experiments of Sherwood *et al.* (1954) were that (a) the polarization of the beam reflected from an iron mirror is positive, (b) the polarization of the beam transmitted through an iron filter is negative, and (c) the Bragg-reflected beam from the (220) planes of magnetized magnetite is negatively polarized.

The discovery of an efficient polarizing crystal monochromator created an enormous upsurge in the number and range of polarized neutron beam experiments. An early application, which exploited the monochromating properties of this polarizer, was to re-examine the wavelength dependence of the polarization cross-section in the polycrystalline magnetized iron polarizer. Stanford *et al.* (1954) used the magnetite (220) reflection to analyse the wavelength-dependent polarization $P_1(\lambda)$ obtained from a magnetized iron filter using the 'depolarizing shim' method. In this technique, the shim is used to depolarize the beam between two polarizers. The ratio of the count rate C_1 without the shim to that C_2 with a depolarizing iron shim placed after the first polarizer gives the shim ratio

$$R_S = (C_1/C_2) = 1 + P_1(\lambda)P_A, \qquad (1.5.10)$$

where P_A is the known polarizing efficiency of the magnetite analyser. The efficiency of the filter polarizer $P_1(\lambda)$ is related to $\sigma_p(\lambda)$ by (see for example Williams (1975))

$$P_1(\lambda) = -\tanh[Nt\sigma_p(\lambda)], \tag{1.5.11}$$

and the $\sigma_p(\lambda)$ values determined in this way were shown to be in good agreement with those found by the single transmission effect.

Another type of measurement to benefit from the availability of polarizing monochromators, and which was of major importance in nuclear physics, was the determination of the spins of resonant nuclear states formed by neutron capture. Rose (1949) had shown that the absorption cross-sections for polarized neutrons (±) by polarized nuclei at a nuclear resonance energy is given by

$$\sigma_\pm = \sigma_A(1 \pm \rho P_N), \tag{1.5.12}$$

where σ_A is the normal unpolarized beam absorption cross-section, and P_N is the nuclear polarization. The numerical constant ρ is a statistical spin weighting factor which takes the values -1 when the angular momentum J of the compound nucleus is $(i - \frac{1}{2})$, or $i/(i+1)$ when $J = (i + \frac{1}{2})$; i is the nuclear spin quantum number of the absorbing nucleus. It is evident that an examination of the transmittance of a polarized beam through a target of polarized nuclei at a nuclear resonance energy can furnish the angular momentum J of the compound state, provided that the signs of the neutron and nuclear polarizations are known. One of the earliest experiments was that by Roberts *et al.* (1954) who studied the 97 meV thermal resonance level in ^{149}Sm. The sample was the paramagnetic salt samarium ethyl sulphate, which was cooled to ~0.12 K by adiabatic demagnetization; the experimental arrangement used is shown in Fig. 1.13. The magnetic field shims provided a slowly rotating guide field for the neutron spins, so as to ensure zero instrumental depolarization. The ^{149}Sm spins were known from hyperfine field data to be positively polarized, whereas the magnetite (220) reflection was known from a Stern–Gerlach experiment to reflect (−) spin neutrons. The measured shim ratios R_S, which were less than unity, were therefore interpreted as being consistent with a $J = (i + \frac{1}{2})$ thermal neutron resonance in ^{149}Sm. Many more determinations of nuclear resonance spins were to follow in subsequent years. Twenty years later, Williams (1973) proposed using this principle as a basis for producing an efficient broad-band polarizer for thermal neutrons; this work is discussed in Chapter 3.

It is in the application of polarized beam diffraction that polarizing monochromators have had their greatest impact. The polarized-beam technique, as it is now known, is to be discussed at length later in this book; a schematic view of the basic experimental arrangement is shown

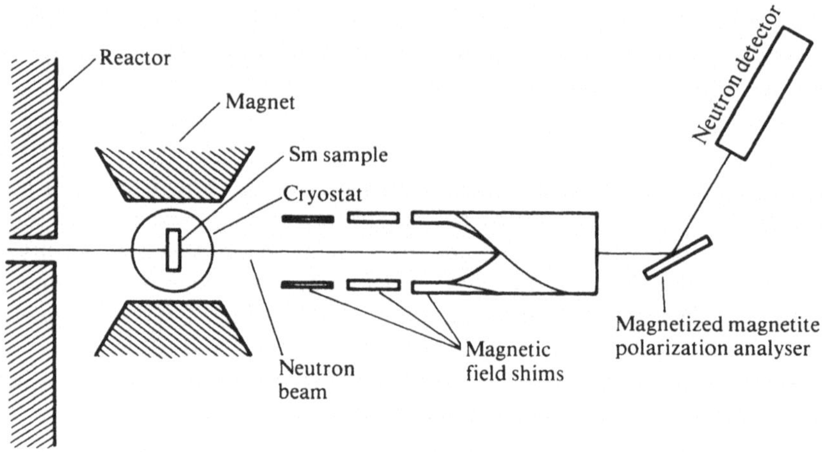

Fig. 1.13 Schematic diagram of the experimental layout used to examine the neutron polarization produced after an unpolarized beam was transmitted through a target containing polarized ^{149}Sm nuclei (from Roberts *et al.* 1954).

in Fig. 2.6. It is used to measure both the size and the sign of the interference term between nuclear and magnetic scattering amplitudes; this provides a much more accurate measurement of the magnetic scattering than is possible by unpolarized-beam methods. The angular distribution of the neutron magnetic scattering amplitude is given by a form factor which is the Fourier transform of the magnetic moment density and therefore precise measurements of the magnetic scattering allows the spatial distribution of the atomic moments to be determined.

The first publication on the polarized-beam technique was that of Nathans *et al.* (1959). Provision was made either to depolarize the beam reflected from the incident beam polarizing monochromator with an iron shim, or to reverse its polarization with a radiofrequency spin-flipper. These authors, in addition to determining the magnetic form factors for iron and nickel, developed a new polarizing crystal monochromator $Co_{0.92}Fe_{0.08}$ and showed that its (111) and (200) reflections give positively polarized neutrons. The applied field dependence of the magnitude of the polarizing efficiency for the (111) reflection of this crystal polarizer is compared with that for the (220) magnetite reflection in Fig. 1.14. The (200) reflection in $Co_{0.92}Fe_{0.08}$ gives a calculated polarizing efficiency close to 100%, compared with 98.1% for the (111) reflection. $Co_{0.92}Fe_{0.08}$ is still one of only two polarizing crystal monochromators in common usage, though one of its disadvantages is the absorption loss due to cobalt. It has been superseded, especially at longer neutron wavelengths, by the Heusler Alloy crystal polarizer Cu_2MnAl introduced by Delapalme *et al.* (1971).

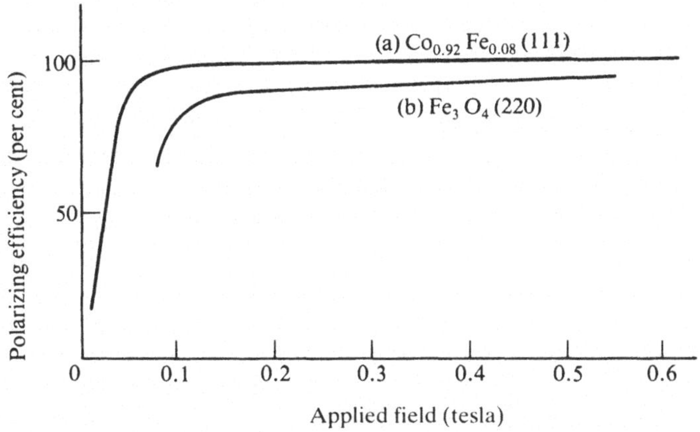

Fig. 1.14 The polarizing efficiencies of the magnetite (220) and $Co_{0.92}Fe_{0.08}$ (111) polarizing reflections as a function of the applied field (from Nathans *et al.* 1959).

During the 1960s a new form of neutron polarizer based on the differential scattering of the two neutron spin states in a filter containing polarized protons was developed by Shapiro and his colleagues in the USSR. This filter is an effective polarizer at all neutron energies up to about 10^5 eV, and has been usefully employed in nuclear physics experiments though, partly owing to its technical complexity, it has not found favour with condensed matter physicists. It is, however, the only true white-beam polarizer and it may yet find applications at pulsed neutron sources.

First tests on the feasibility of the method were made by Draghicescu *et al.* (1964), and a fuller account of the principles involved together with further results were published by Lushchikov *et al.* (1970). The two scattering cross-sections σ_+ (triplet) and σ_- (singlet) for neutrons polarized parallel (+) and antiparallel (−) to the polarization of the protons are substantially different, e.g. for slow neutrons the σ_- cross-section is twenty times greater than the σ_+ cross-section. We can express these cross-sections in terms of a polarization cross-section $\sigma_p = -\sigma_p' P_H$ as follows:

$$\sigma_\pm = \sigma_0 \mp \sigma_p' P_H, \tag{1.5.13}$$

where P_H is the proton polarization. By analogy with our previous discussion on the iron polarizing filter, an unpolarized beam has a filter transmittance

$$T = \exp(-Nt\sigma_0)\cosh(-Nt\sigma_p' P_H) \tag{1.5.14}$$

and emerges from the filter with a neutron polarization

$$P = \tanh(Nt\sigma_p' P_H), \tag{1.5.15}$$

where Nt is the proton atomic thickness (atoms cm^{-2}).

Lushchikov *et al.* (1970) in fact argued that this cross-section formula is only valid at neutron energies $E \gtrsim 1\,\text{eV}$, and also gave expressions appropriate for cold and thermal energy neutrons. The polarization cross-section σ_p is strongly neutron-energy-dependent, and it is this term together with the proton polarization value which dominate the filter's properties. As with all other polarizing filters, an increase in thickness results in improved polarization but a reduction in the total transmittance. The proton material first used was a single crystal of lanthanum magnesium nitrate (LMN) in which the protons in the water of crystallization can be dynamically polarized by using the 'solid-state effect'. This nuclear polarization technique had been developed for high-energy physics experiments and is described in Chapter 3. Figure 3.7 shows typical polarization and transmittance values for a polarized proton filter as a function of the neutron energy.

Towards the end of the 1960s and in the early 1970s a new generation of high-flux beam reactors became available and this brought with it a resurgence of interest in polarized-beam methods. The first new technique to appear was one-dimensional polarization analysis (see Moon *et al.* (1969a)). The original instrument, see Fig. 2.9, was a polarized triple-axis spectrometer with polarizing monochromators on the first and third axes. This enabled the scattered beam to be both energy- and spin-analysed. Such an experiment, which uses monochromating crystals with relatively poor reflectivities, is only viable at a high-flux reactor source. The introduction of the method allowed the four partial cross-sections (++), (+−), (−−), and (−+) which connect the neutron spin states before and after the scattering to be determined. Many exciting new experiments on systems with no preferred axes, like randomly oriented nuclear samples, paramagnets and spin glasses, as well as antiferromagnets, became possible for the first time. This new experimental technique is described in Chapters 2 and 4. The main areas of application are the separation of (a) spin incoherent scattering from other nuclear scattering, (b) paramagnetic scattering from nuclear incoherent scattering, (c) nuclear from magnetic coherent elastic scattering, and (d) magnon from phonon scattering.

A polarization analysis spectrometer for cold neutrons using iron filters as the polarizer and analyser was built by Hicks and coworkers (see Ahmed *et al.* (1974)). Despite the poor efficiencies of magnetized iron filter polarizers ($P \sim 50\%$), this instrument has provided excellent data on magnetic diffuse scattering cross-sections in a range of materials.

The other major advances in the early 1970s resulted from the construction of highly efficient cold neutron sources and neutron guide tubes, especially at the Grenoble high-flux reactor, and involved the production and utilization of cold neutron polarized beams. The principle of polarizing cold beams by total reflection from magnetic surfaces had been known, as we have seen, for over twenty years. However, few practical devices had been built owing to the unfavourably small critical glancing angles below which reflection occurs. As an example, the commonly used high-permeability alloy Permendur, $Co_{49}Fe_{49}V_2$, has a critical angle $\theta_c = 7.3 \times 10^{-3}$ radians at a neutron wavelength $\lambda = 4$ Å, and a single mirror ~4 m long would be required to polarize a 3 cm wide beam. This might form a useful polarizer, but it would be unacceptably bulky as an analyser, for which mobility and compactness are important. A major breakthrough in the realization of practical devices came with the use of thin (~1000 Å) evaporated films of magnetic alloys rather than the use of bulk metals (see for example Penfold *et al.* (1975)). Good polarizing characteristics were observed for cobalt–iron films magnetized in applied fields ~0.1 T or less, and deposited on plastic foils ~0.05 mm thick whose neutron refractive indices closely matched that of the (−) spin neutrons in the magnetic alloy. A series of such foils could then be assembled to form a compact polarizing Soller bender for cold neutrons, as first reported by Hayter *et al.* (1978).

There were further notable developments in the use of evaporated films. Lynn *et al.* (1976a) showed that two-dimensional multilayer structures made from alternate layers of a (magnetized) magnetic material whose refractive index for (−) spin neutrons was matched to that for the second (non-magnetic) layer, acted as a polarizing long-wavelength monochromator. Their polarizing efficiency measurements for a magnetized Fe–Ge multilayer at $\lambda = 4.05$ Å, and obtained by the depolarizing shim method following initial polarization using a Heusler alloy monochromator, are shown in Fig. 1.15.

For many applications, however, a broad-band polarized beam is required, and this became possible following the invention of 'supermirrors' by Mezei (1976a). These consist of evaporated multilayer films which are deposited, in a controlled way, to provide a lattice spacing gradient. Thin-film interference then results in a reflectivity angular profile which, for a particular neutron wavelength, extends to approximately three times the simple mirror critical glancing angle. Polarizing supermirrors contain a magnetic material as one of the alternating thin-film layers. All these thin-film coatings can be prepared on glass, and polarizing guides, Soller benders, etc., for cold neutrons are now used routinely in a wide range of neutron instruments.

The production of high-quality cold neutron polarizers allowed the implementation of a second new polarized-beam technique during the mid 1970s. This was the method of neutron spin echo spectroscopy first proposed by Mezei (1972), and largely developed by him in collaboration with Hayter (1978b). The real value of spin echo spectroscopy is not in the study of polarization-dependent scattering cross-sections themselves, but as a very high-resolution inelastic or quasielastic scattering spectro-

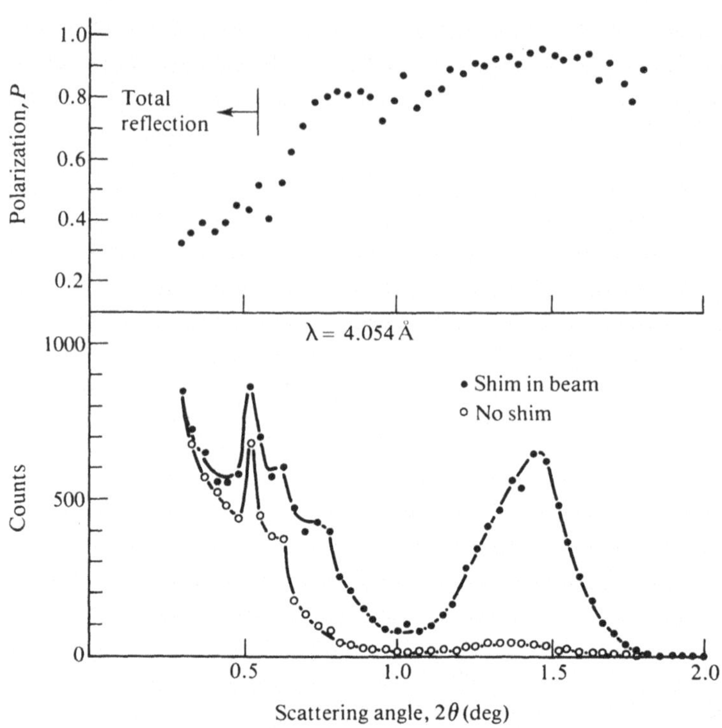

Fig. 1.15 Counts and polarizations as a function of twice the angle of reflection observed in a depolarizing shim experiment on a magnetized Fe–Ge polarizing multilayer (from Lynn *et al.* 1976a).

meter; energy resolutions in the neV region are currently attainable. It is essentially an elegant time-of-flight method which uses the fact that each 'polarized' neutron precessing in a magnetic field B provides its own 'clock' through its Larmor frequency $\omega_L = 2\,|\mu_n B|/\hbar$. By carefully manipulating the precession fields B in the incident and scattered beam flight-paths, small changes in the neutron velocity which may take place on scattering can be inferred from the phase shift $\delta\phi_L$ in the Larmor precession angle. Spin echo spectroscopy is now established as a major technique in experimental neutron scattering, and will be described in Chapter 6.

During the time that neutron spin echo was being implemented at the Grenoble high-flux reactor, two other neutron-precession techniques were being developed at reactors with very modest fluxes; these became successful because they involved transmission rather than scattering measurements. The first was the three-dimensional depolarization technique, discovered independently by Rekveldt (1973) and the Leningrad group (see for example Okorokov *et al.* (1976)). The experimental arrangement allows a 3×3 depolarization matrix to be evaluated which describes the depolarization of polarized beams as they traverse magnetic materials. From this matrix, various quantities, such as the direction and amplitude of the mean magnetization, the mean domain size, and the mean local magnetization direction within the domains, can be determined. The other neutron precession technique which appeared in the early 1970s was the spin precession method in specimens containing polarized nuclei. This was developed by the Saclay group (see for example Abragam and Goldman (1982)), and from the phase shift in the precession angle the method enables nuclear spin scattering amplitudes to be ascertained. Both these precession techniques are also described in Chapter 6.

This brings us to the close of our brief summary of the development of polarized neutron experimental methods. We have purposely avoided discussing the advances which have taken place in spin-flipper technology, as well as the further progress which is now taking place at the new pulsed neutron sources. Both will be given ample coverage in the following chapters. The aim of this section has been to concentrate on the highlights, and to show that the advances have evolved largely as a result of (a) the introduction of new methods for producing and manipulating polarized neutron beams, and (b) the improved intensities available from the neutron sources themselves.

2

Basic theory

The fundamental principles underlying polarized neutron scattering experiments, together with some of the theory needed to understand the action of certain neutron polarizing devices, are collected in this chapter. We begin by discussing the scattering of unpolarized neutrons by nuclei and atomic electrons, and progress to describe the equivalent experiments with polarized neutron beams. This is followed by a description of polarization analysis experiments in which the spin polarization of the scattered beam is measured. Finally, brief accounts are given of the basis of neutron-optical phenomena and the characteristics of compound nuclear resonances in which the slant is towards how these can be applied to provide neutron beam polarizers.

2.1 SCATTERING CROSS-SECTIONS WITH UNPOLARIZED NEUTRONS

Before discussing the more complicated but more interesting situation of polarized beam scattering, it is necessary to establish a basic scattering formalism for unpolarized beams. In this section we present expressions for the unpolarized beam scattering cross-sections for nuclear and magnetic specimens in both crystalline and non-crystalline states. Specific examples will be introduced later, but we first reproduce the derivation of the master cross-section equation as given by Lovesey (1986) in terms of a generalized potential $\hat{V}$.

Let us first consider the differential cross-section $d\sigma/d\Omega$, introduced in Section 1.3, for scattering from a plane wave with wave vector $\boldsymbol{k}$ to a scattered plane wave $\boldsymbol{k}'$, with $|\boldsymbol{k}| = |\boldsymbol{k}'| = k$ and energy $E = \hbar^2k^2/2m$. The probability for this process is given by Fermi's Golden rule,

$$W_{k\to k'} = (2\pi/\hbar)\left|\int \mathrm{d}\boldsymbol{r}\, \psi_{k'}^{*}\hat{V}\psi_{k}\right|^2 \rho_{k'}(E), \tag{2.1.1}$$

where $\hat{V}$ is the interaction potential, and $\rho_{\boldsymbol{k}'}$ is the density of $\boldsymbol{k}'$ states per unit energy change. By normalizing the wave functions in a box of volume L^3, and defining the matrix element

$$\langle \boldsymbol{k}'|\, \hat{V}\, |\boldsymbol{k}\rangle = (L^3m/2\pi\hbar^2)\int \mathrm{d}\boldsymbol{r}\, \psi_{k'}^{*}\hat{V}\psi_{k}, \tag{2.1.2}$$

the general result for the differential scattering cross-section becomes (see Lovesey (1986)):

$$(\mathrm{d}\sigma/\mathrm{d}\Omega) = (k'/k)\,|\langle \boldsymbol{k}'|\,\hat{V}\,|\boldsymbol{k}\rangle|^2. \tag{2.1.3}$$

Since $(\mathrm{d}\sigma/\mathrm{d}\Omega)$ has the dimension of area, the matrix element of equn (2.1.3) has the dimension of length. Note also that this matrix element depends only on the neutron wave vector difference $\boldsymbol{\kappa} = \boldsymbol{k} - \boldsymbol{k}'$, and not on $\boldsymbol{k}$, $\boldsymbol{k}'$ individually. This feature of the scattering process reflects the fact that the interaction is weak, and the neutron wave functions can be represented to a good approximation by plane waves. In view of this the matrix element for elastic scattering, for which $(k'/k) = 1$, may be written as

$$(\mathrm{d}\sigma/\mathrm{d}\Omega) = |\langle \boldsymbol{k}'|\,\hat{V}\,|\boldsymbol{k}\rangle|^2 = |f(\boldsymbol{\kappa})|^2, \tag{2.1.4}$$

where $f(\boldsymbol{\kappa})$ is simply a scattering amplitude with a dimension of length. The determination of the phase of $f(\boldsymbol{\kappa})$ requires more detailed theory and the result is

$$f(\boldsymbol{\kappa}) = -\langle \boldsymbol{k}'|\,\hat{V}\,|\boldsymbol{k}\rangle. \tag{2.1.5}$$

In extending this analysis to inelastic scattering, in which $k \neq k'$, the initial state of the neutron and scattering system are described by a composite state function $|\boldsymbol{k}\lambda\rangle$ with a target system energy E_λ, and the final combined state function by $|\boldsymbol{k}'\lambda'\rangle$ with a system energy $E_{\lambda'}$. For initial and final neutron energies E and E' respectively, energy conservation gives

$$E + E_\lambda = E' + E_{\lambda'}. \tag{2.1.6}$$

If the energy transfer $\hbar\omega$ is defined as positive for neutron energy loss, i.e.

$$\hbar\omega = E - E', \tag{2.1.7}$$

the scattered neutron energy distribution vanishes unless $\hbar\omega = E_{\lambda'} - E_\lambda$, and the double differential cross-section is defined to be

$$\left(\frac{\mathrm{d}^2\sigma}{\mathrm{d}\Omega\,\mathrm{d}E'}\right)_{\lambda\to\lambda'} = \frac{k'}{k}|\langle \boldsymbol{k}'\lambda'|\,\hat{V}\,|\boldsymbol{k}\lambda\rangle|^2\,\delta(\hbar\omega + E_\lambda - E_{\lambda'}). \tag{2.1.8}$$

The delta function expressing the conservation of energy relation has the dimension of $(\text{energy})^{-1}$, so the cross-section expressed in equn (2.1.8) has the dimension of (area/energy).

In a real experiment the measured cross-section is the result of scattering from a number of initial system states $|\lambda\rangle$ with population factors p_λ given by

$$p_\lambda = \exp(-E_\lambda/k_\mathrm{B}T)/Z, \tag{2.1.9}$$

where T is the absolute temperature of the scatterer. The partition function Z is defined by the condition

$$\sum_{\lambda} p_{\lambda} = 1. \tag{2.1.10}$$

Moreover, all final states will contribute to the measured cross-section, so it is necessary to sum equn (2.1.8) over all λ'. The experiment will in general not only probe some average over different nuclear isotopes or nuclear spin states, but will also be strongly dependent on the nuclear positions. The double differential cross-section is therefore expressed more completely as

$$\left(\frac{\mathrm{d}^2\sigma}{\mathrm{d}\Omega\,\mathrm{d}E'}\right) = \frac{k'}{k} \overline{\sum_{\lambda\lambda'} p_{\lambda}\,|\langle \boldsymbol{k}'\lambda'|\,\hat{V}\,|\boldsymbol{k}\lambda\rangle|^2\,\delta(\hbar\omega + E_{\lambda} - E_{\lambda'})}, \tag{2.1.11}$$

with the horizontal bar denoting this averaging.

This relation, derived from Fermi's Golden Rule, is equivalent to the Born approximation to the cross-section, and is described here in terms of a general interaction potential $\hat{V}$. There are in fact two main forms of neutron scattering: that caused by nuclei, with an interaction potential $\hat{V}_{\mathrm{N}}$, and that due to unpaired or magnetic electrons, with an interaction potential $\hat{V}_{\mathrm{M}}$. The scattering due to these potentials are universally referred to as nuclear and magnetic scattering. The two interaction potentials are fundamentally different; we now discuss their main features and then proceed to derive specific cross-section formulae for each case.

2.1.1 Bound nuclei

The range of the neutron–nucleus interaction is approximately one-tenth of a typical nuclear radius, and orders of magnitude less than the wavelengths of even epithermal neutrons. The scattering therefore only contains s-wave components, and is spatially isotropic. Its strength can be described by a constant b, the bound scattering length. This parameter can be positive or negative and, particularly at neutron energies in the region of a resonance absorption (see Section 2.5), can sometimes contain an imaginary component. An important feature of b is that, unlike in the case of X-rays, it does not vary systematically with atomic number, and it frequently happens that isotopes of the same element have widely different scattering lengths.

In the following discussion of cross-sections we shall restrict ourselves to the bound nuclear scattering length b, i.e. the case in which the nucleus is so tightly bound to, say, a crystal lattice, that it cannot undergo a recoil. The appropriate quantity to use in a gas is the free scattering length a, which equals $bM/(m+M)$, where m and M are the neutron and nuclear masses respectively. The bound scattering length is observed directly in a single-crystal Bragg scattering experiment, where the appropriate mass is that of the crystal sample.

Monoisotopic specimens

It can be shown that the only interaction potential V_N which can result in isotropic scattering within the Born approximation is the Fermi pseudopotential.

$$\hat{V}_N(\boldsymbol{r}) = (2\pi\hbar^2/m)b\,\delta(\boldsymbol{r}-\boldsymbol{R}), \tag{2.1.12}$$

where $\boldsymbol{r}$ and $\boldsymbol{R}$ are the neutron and nuclear position vector coordinates respectively. Let us first consider the simplest situation in which all the nuclei in the scatterer have the same scattering length b. Substituting the potential (2.1.12) into equn (2.1.2) gives

$$\langle \boldsymbol{k}'|\,\hat{V}_N\,|\boldsymbol{k}\rangle = b\int \mathrm{d}\boldsymbol{r}\,\exp(-\mathrm{i}\boldsymbol{k}'\cdot\boldsymbol{r})\,\delta(\boldsymbol{r})\exp(\mathrm{i}\boldsymbol{k}\cdot\boldsymbol{r}) = b. \tag{2.1.13}$$

Hence the differential cross-section per atom is

$$(\mathrm{d}\sigma/\mathrm{d}\Omega) = b^2 \tag{2.1.14}$$

and the net coherent scattering cross-section over all angles is

$$\sigma_{\mathrm{coh}} = 4\pi b^2. \tag{2.1.15}$$

Multi-isotopic scatterers

Next we consider an assembly of nuclei with different scattering lengths b_j, located at positions defined by the vectors $\boldsymbol{R}_j$. The equations derived are also valid for a single-isotope scatterer which has a nuclear spin i, since the nucleus–neutron interaction can then produce two compound nuclear states with spins $(i \pm \frac{1}{2})$ and different scattering lengths b_+ and b_-. Substitution of the Fermi pseudo-potential into the elastic scattering form of equn (2.1.11) gives the differential cross-section per atom,

$$(\mathrm{d}\sigma/\mathrm{d}\Omega) = \sum_{jj'} \exp[\mathrm{i}\boldsymbol{\kappa}\cdot(\boldsymbol{R}_j - \boldsymbol{R}_{j'})]\overline{b_j b_{j'}}, \tag{2.1.16}$$

where $\overline{b_j b_{j'}}$ is in general the value of $b_j b_{j'}$ averaged over all the random isotope distributions and nuclear spin orientations in the scattering sample.

There is no correlation between the values of b_j and $b_{j'}$, so that if j and j' correspond to different nuclear sites

$$(\overline{b_{j'} b_j})_{j' \neq j} = \bar{b}_{j'}\bar{b}_j = |\bar{b}|^2 \tag{2.1.17}$$

and this gives the coherent component of the differential cross-section

$(\mathrm{d}\sigma/\mathrm{d}\Omega)_{\mathrm{coh}}$

$$(\mathrm{d}\sigma/\mathrm{d}\Omega)_{\mathrm{coh}} = |\bar{b}|^2 \left|\sum_j \exp(\mathrm{i}\boldsymbol{\kappa} \cdot \boldsymbol{R}_j)\right|^2. \tag{2.1.18}$$

This is an example of the more general expression for the coherent component of the scattering, which is the square of the average amplitude, the average being taken with respect to all relevant variables, e.g. isotope, mass, and nuclear spin distributions.

For $j = j'$ we obtain

$$(\overline{b_{j'} b_j})_{j=j'} = \overline{|b_j|^2} = \overline{|b|^2} \tag{2.1.19}$$

and combining this with equn (2.1.17) gives the general relation

$$\overline{b_{j'} b_j} = |\bar{b}|^2 + \delta_{j,j'}(\overline{|b|^2} - |\bar{b}|^2). \tag{2.1.20}$$

The second term in this expression generates an incoherent component to the scattering with a cross-section given by

$$(\mathrm{d}\sigma/\mathrm{d}\Omega)_{\mathrm{incoh}} = N(\overline{|b|^2} - |\bar{b}|^2) = N\,\overline{|b - \bar{b}|^2}, \tag{2.1.21}$$

where N is the number of scattering nuclei.

Here we should sound a note of caution regarding terminology, since the difference between the total and coherent cross-sections is sometimes called the diffuse cross-section. In this book we adopt the more common practice of assigning the term 'diffuse' to that part of the scattering which is due to some disorder in a multicomponent crystal, whereas incoherent scattering is due simply to a random distribution of scattering lengths. The former is frequently described as the Laue diffuse scattering; together with Bragg scattering it can be considered to be a component of the coherent cross-section, and is treated at the end of this section.

The total differential cross-section therefore becomes

$$(\mathrm{d}\sigma/\mathrm{d}\Omega) = N(\overline{|b|^2} - |\bar{b}|^2) + |\bar{b}|^2 \left|\sum_j \exp(i\boldsymbol{\kappa} \cdot \boldsymbol{R}_j)\right|^2. \tag{2.1.22}$$

The physical interpretation of this mathematics is that coherent interference effects, which contain all the information on the scattering system's cooperative properties, arise only from the average scattering potential experienced by the neutron, i.e. the result is the same as if all the nuclei were assigned the same scattering length $\bar{b}$. Deviations from this average potential are randomly distributed and produce a cross-section which, for elastic scattering, is isotropic. The incoherent scattering cross-section can only be zero for single isotopes with zero nuclear spin. Some examples with their natural abundancies given in parenthesis are ^{12}C (98.89%), ^{16}O (99.76%), ^{28}Si (92.21%), ^{40}Ca (96.97%) and ^{56}Fe (91.66%).

The most trivial form of incoherent scattering occurs in scattering samples which contain more than one isotope where all the nuclear spins are zero, but which have different scattering lengths; this scattering is called isotopic incoherent scattering. Of greater relevance to polarized neutron scattering is the incoherence which arises due to the nuclear spin. Let us consider a single isotope specimen which possesses a nuclear spin with a spin quantum number i. The interaction of the spin $\frac{1}{2}$ neutron with the nuclear spin produces two compound nuclear states with total spins $(i+\frac{1}{2})$ and $(i-\frac{1}{2})$ which have different scattering lengths which we denote as b_+ and b_- respectively. There are $2(i+\frac{1}{2})+1$ states of the spin $(i+\frac{1}{2})$ compound nucleus and $2(i-\frac{1}{2})+1$ states of the spin $(i-\frac{1}{2})$ compound nucleus, giving a total number of $(4i+2)$ intermediate spin states. The fractional probability that a b_+ scattering length occurs is therefore

$$(2i+2)/(4i+2) = (i+1)/(2i+1), \tag{2.1.23}$$

and for the b_- scattering length the probability is

$$2i/(4i+2) = i/(2i+1). \tag{2.1.24}$$

The average coherent scattering length is therefore

$$\bar{b} = \left(\frac{i+1}{2i+1}\right)b_+ + \left(\frac{i}{2i+1}\right)b_- \tag{2.1.25}$$

and the mean value of the square of the scattering length is

$$\overline{|b|^2} = \left(\frac{i+1}{2i+1}\right)b_+^2 + \left(\frac{i}{2i+1}\right)b_-^2. \tag{2.1.26}$$

Following equn (2.1.21), the resulting nuclear spin (NS) incoherence gives a differential cross-section per atom

$$(\mathrm{d}\sigma/\mathrm{d}\Omega)_{\mathrm{NS}} = [(b_+ - b_-)^2 i(i+1)]/(2i+1)^2, \tag{2.1.27}$$

and we shall see later that this form of scattering can in principle be separated from isotopic incoherence by using polarized neutrons.

The isotope with the largest spin incoherent cross-section is the proton, for which $i=\frac{1}{2}$. The two scattering lengths for the $(i+\frac{1}{2})$ and $(i-\frac{1}{2})$ compound nuclear states are approximately

$$b_+ = 1.04 \times 10^{-12}\,\mathrm{cm},$$
$$b_- = 4.74 \times 10^{-12}\,\mathrm{cm},$$

respectively, and from the above expressions we obtain a coherent cross-section

$$\sigma_{\mathrm{coh}} = 4\pi\bar{b}^2 = 2.1 \text{ barns at}^{-1}$$

and a spin incoherent cross-section

$$\sigma_{\mathrm{NS}} = 4\pi(\mathrm{d}\sigma/\mathrm{d}\Omega)_{\mathrm{NS}} = 78.7 \text{ barns at}^{-1}.$$

When the scatterer contains several isotopes with fractional abundances c_ξ then the general expressions for $\bar{b}$ and $\overline{|b|^2}$ are

$$\bar{b} = \sum_\xi [c_\xi/(2i_\xi + 1)][(i_\xi + 1)b_{\xi+} + i_\xi b_{\xi-}] \quad (2.1.28)$$

and

$$\overline{|b|^2} = \sum_\xi [c_\xi/(2i_\xi + 1)][(i_\xi + 1)b_{\xi+}^2 + i_\xi b_{\xi-}^2]. \quad (2.1.29)$$

A compilation of the low-energy neutron coherent scattering lengths b, total scattering cross-sections σ_s, and absorption cross-sections σ_A for most of the naturally occurring isotopes of the elements has been published by Koester and Yelon (1982). Although the absorption cross-section is of little intrinsic interest in scattering experiments, it can be responsible for most of the attenuation of the beam in some samples, and it is an important parameter for the experimentalist.

We reproduce in Table 2.1 some of the nuclear scattering lengths and cross-sections compiled by Koester and Yelon (1982). The examples selected all possess a relatively large spin incoherent cross-section, since these are the isotopes of special interest in polarized neutron experiments. In many instances the individual b_+ and b_- values for these isotopes can in fact be determined using polarized neutrons. The polarized neutron spin precession method in targets containing polarized nuclei (see for example Abragam *et al.* (1973)), is probably the most accurate way to determine $(b_+ - b_-)$ and hence σ_{NS}. This technique, together with results, is described in Chapter 6.

TABLE 2.1 Selected bound coherent scattering lengths and cross-sections for isotopes with a significant spin incoherence (after Koester and Yelon 1982)

Isotope	Abundance (%)	Spin i	$b^{(a)}$	$\sigma_s^{(b)}$	$\sigma_{NS}^{(c)}$	$\sigma_A^{(d)}$
H	100	1/2	−0.3741	81.67	79.9	0.3326
D	0.0149	1	0.6674	7.63	2.03	0.000519
^{7}Li	92.5	3/2	−0.220	1.44	0.83	0.0454
^{23}Na	100	3/2	0.363	3.23	1.57	0.530
^{45}Sc	100	7/2	1.23	23.6	4.6	27.2
^{51}V	99.75	7/2	−0.0414	4.946	4.92	4.9
^{59}Co	100	7/2	0.253	6.15	5.34	37.18
^{65}Cu	30.83	3/2	1.00	13.1	0.53	2.17
^{109}Ag	48.17	1/2	0.419	5.9	3.7	91
^{165}Ho	100	7/2	0.808	9.5	1.3	65
^{197}Au	100	3/2	0.763	7.81	0.49	98.65

[a] Coherent scattering length for bound atoms; units 10^{-12} cm.
[b] Total scattering cross-sections; units 10^{-24} cm^2 = 1 barn.
[c] Nuclear spin incoherent cross-sections calculated from $\sigma_{NS} = \sigma_s - \sigma_{coh}$; units 10^{-24} cm^2 = 1 barn.
[d] Absorption cross-sections for thermal neutrons; units 10^{-24} cm^2 = 1 barn.

We now present specific expressions for the nuclear scattering cross-sections in different types of scattering specimens. The examples chosen are those that have magnetic counterparts which can be studied with polarized neutron beams.

Crystalline samples

Single-crystal and polycrystalline samples contain an ordered array of nuclei, and the neutron–target interaction potential then has the periodicity of the target crystal lattice. The crystal lattice is an assembly of unit cells, and the periodicity is exemplified by a number of lattice vectors $\boldsymbol{l}$, which lead from a point in one unit cell to the corresponding point in another. For a rigid Bravais lattice of N atoms, in which there is only one atom per unit cell, the coherent differential cross-section, following equn (2.1.18), is

$$(\mathrm{d}\sigma/\mathrm{d}\Omega)_{\mathrm{coh}} = |\bar{b}|^2 \left|\sum_{l} \exp(\mathrm{i}\boldsymbol{\kappa}\cdot\boldsymbol{l})\right|^2. \tag{2.1.30}$$

Lovesey (1986) has shown that for a large crystal

$$\left|\sum_{l} \exp(\mathrm{i}\boldsymbol{\kappa}\cdot\boldsymbol{l})\right|^2 = [N(2\pi)^3/v_0] \sum_{\boldsymbol{\tau}} \delta(\boldsymbol{\kappa}-\boldsymbol{\tau}), \tag{2.1.31}$$

where v_0 is the volume of the unit cell, and $\boldsymbol{\tau}$ are reciprocal lattice vectors that satisfy $\exp(\mathrm{i}\boldsymbol{\tau}\cdot\boldsymbol{l}) = 1$. From equns (2.1.30) and (2.1.31) we find,

$$(\mathrm{d}\sigma/\mathrm{d}\Omega)_{\mathrm{coh}} = [N(2\pi)^3/v_0]\, |\bar{b}|^2 \sum_{\boldsymbol{\tau}} (\boldsymbol{\kappa}-\boldsymbol{\tau}). \tag{2.1.32}$$

The scattering described by equn (2.1.32) is Bragg scattering, and, as noted earlier, the scattering length is always the bound scattering length. It is evident from this equation that the scattering is pronouncedly anisotropic, since it only occurs for special orientations of the incident wave vector $\boldsymbol{k}$ and the scattered wave vector $\boldsymbol{k}'$, with $|\boldsymbol{k}| = |\boldsymbol{k}'|$ and also $\boldsymbol{k} - \boldsymbol{k}' = \boldsymbol{\tau}$. The peaks in the scattering along $\boldsymbol{k}'$ which occur when $\boldsymbol{k}$ has the correct orientation with the crystal are called Bragg peaks; there is no scattering outside the peaks. One form of the Bragg equation is $|\boldsymbol{\tau}| = 2k \sin\theta_{\mathrm{B}}$, where θ_{B} is the Bragg angle or half the scattering angle.

In order to extend this to the general case in which there is more than one atom in the unit cell, it is necessary to introduce a nuclear unit cell structure factor $F_{\mathrm{N}}(\boldsymbol{\tau})$ defined by

$$F_{\mathrm{N}}(\boldsymbol{\tau}) = \sum_{R} \bar{b}_R \exp(\mathrm{i}\boldsymbol{\tau}\cdot\boldsymbol{R}) \tag{2.1.33}$$

$$= \sum_{R} \bar{b}_R \exp\{2\pi\mathrm{i}[(hx/a_0) + (ky/b_0) + (lz/c_0)]\}, \tag{2.1.34}$$

where $\boldsymbol{R}$ are now positional vectors within the unit cell with dimensions a_0, b_0 and c_0. Here (x, y, z) are the Cartesian coordinates of the nucleus and $(h\ k\ l)$ are the Miller indices which characterize the reflecting plane and $\boldsymbol{\tau}$. The cross-section becomes

$$(\mathrm{d}\sigma/\mathrm{d}\Omega)_{\mathrm{coh}} = [N(2\pi)^3/v_0] \sum_{\tau} \delta(\boldsymbol{\kappa} - \boldsymbol{\tau})\, |F_{\mathrm{N}}(\boldsymbol{\tau})|^2. \tag{2.1.35}$$

Our discussion up to this point is, however, over-simplified in the sense that in real crystals the nuclei are not rigidly fixed. Atomic vibrations can be accounted for by multiplying the coherent scattering length by a Debye–Waller factor $\exp(-W(\boldsymbol{\kappa}))$, where the function $W(\boldsymbol{\kappa})$ is related to the atomic displacement, hence the unit cell structure factor becomes:

$$F_{\mathrm{N}}(\boldsymbol{\tau}) = \sum_{R} \bar{b}_R \exp(\mathrm{i}\boldsymbol{\tau} \cdot \boldsymbol{R} - W(\boldsymbol{\kappa})). \tag{2.1.36}$$

One of the most convenient methods of displaying the result of an elastic single-crystal experiment is to obtain the inverse Fourier transform of the structure factors, which gives the nuclear density $\rho(\boldsymbol{R})$ in real space:

$$\rho(\boldsymbol{R}) = \sum_{\tau} F_{\mathrm{N}}(\boldsymbol{\tau}) \exp(\mathrm{i}\boldsymbol{\tau} \cdot \boldsymbol{R}). \tag{2.1.37}$$

In principle the sum must contain all the reciprocal lattice vectors $\boldsymbol{\tau}$. There are two practical difficulties with Fourier techniques in neutron scattering. The first is that only the squared modulus of $F_{\mathrm{N}}(\boldsymbol{\tau})$ is obtained from the intensity, and its phase must be found by some other means. Secondly the Fourier series will of necessity be incomplete, since data can only be obtained for $|\boldsymbol{\tau}| < 4\pi/\lambda_{\mathrm{min}}$, where λ_{min} is the shortest wavelength available for the measurements. As a consequence, some method of minimizing termination effects must be introduced.

Liquids and amorphous solids

The range of atomic order in these materials is finite, whereas it is infinite for crystalline solids, and this has a marked effect on the scattering profiles. For liquids it is difficult to define a precise structure, since the atomic arrangement is continually changing. Scattering experiments can nevertheless provide a static structure factor $S(\boldsymbol{\kappa})$, which essentially gives an instantaneous snapshot picture of the atomic distribution in the liquid. Another important feature in the neutron scattering from liquids is the appreciable inelasticity, which cannot be described in a simple way by using a Debye–Waller factor. The strictly elastic scattering measured in a Bragg reflection displays time-averaged properties. Hence, there is a fundamental difference between Bragg scattering and a total scattering experiment which gives $S(\boldsymbol{\kappa})$. This point is taken up in more detail later in this section.

Many of the fundamental aspects of neutron scattering from liquids have been reviewed by Enderby (1968) and by Howe (1978). The scattering properties are most conveniently encapsulated in the so-called scattering law $S(\boldsymbol{\kappa}, \omega)$ given by

$$(\mathrm{d}^2\sigma/\mathrm{d}\Omega\, \mathrm{d}E') = (k'/k)Nb^2 S(\boldsymbol{\kappa}, \omega), \tag{2.1.38}$$

where all the terms are as previously defined. Following our earlier discussion, this scattering cross-section contains two components:

$$(\mathrm{d}^2\sigma/\mathrm{d}\Omega\, \mathrm{d}E')_{\mathrm{inc}} = (k'/k)N[\overline{|b|^2} - |\bar{b}|^2]S_{\mathrm{inc}}(\boldsymbol{\kappa}, \omega) \tag{2.1.39}$$

and

$$(\mathrm{d}^2\sigma/\mathrm{d}\Omega\, \mathrm{d}E')_{\mathrm{coh}} = (k'/k)N\, |\bar{b}|^2\, S_{\mathrm{coh}}(\boldsymbol{\kappa}, \omega). \tag{2.1.40}$$

The scattering laws in $(\boldsymbol{\kappa}, \omega)$ space can be related to space(r)–time(t) correlation functions $G(\boldsymbol{r}, t)$ by taking a double Fourier transform. When classical statistics apply, the double transform of $S_{\mathrm{inc}}(\boldsymbol{\kappa}, \omega)$ gives $G_s(\boldsymbol{r}, t)$, which is a 'self' space–time correlation function that describes the probability of finding an atom which was at the origin at time $t = 0$ at the position $\boldsymbol{r}$ after a time t. $S_{\mathrm{coh}}(\boldsymbol{\kappa}, \omega)$ gives $G(\boldsymbol{r}, t)$ which represents the probability of finding any atom at position $\boldsymbol{r}$ at time t, given that there was an atom at the origin at time $t = 0$. Clearly

$$G(\boldsymbol{r}, t) = G_s(\boldsymbol{r}, t) + G_d(\boldsymbol{r}, t), \tag{2.1.41}$$

where $G_d(\boldsymbol{r}, t)$, which is known as the 'distinct' space–time correlation function, is the probability that an atom is found at $\boldsymbol{r}$ at time t when there was a different atom at the origin at time $t = 0$.

We now restrict the discussion to the coherent part of the scattering cross-section and integrate over all energy transfers:

$$\begin{aligned}(\mathrm{d}\sigma/\mathrm{d}\Omega)_{\mathrm{coh}} &= \int_{E'} (\mathrm{d}^2\sigma/\mathrm{d}\Omega\, \mathrm{d}E')\, \mathrm{d}E' = N\, |\bar{b}|^2 \int_{-\infty}^{+\infty} S(\boldsymbol{\kappa}, \omega)\, \mathrm{d}\omega \\ &= N\, |\bar{b}|^2\, S(\boldsymbol{\kappa}).\end{aligned} \tag{2.1.42}$$

Enderby (1968) has shown how these cross-sections are related to the space–time correlation functions $G(\boldsymbol{r}, t)$ and gives, for example,

$$(\mathrm{d}\sigma/\mathrm{d}\Omega)_{\mathrm{coh}} = N\, |\bar{b}|^2 \int \mathrm{d}\boldsymbol{r}\, \exp(\mathrm{i}\boldsymbol{\kappa} \cdot \boldsymbol{r}) G(\boldsymbol{r}, 0). \tag{2.1.43}$$

It follows that, since

$$\begin{aligned}G(\boldsymbol{r}, 0) &= G_s(\boldsymbol{r}, 0) + G_d(\boldsymbol{r}, 0) \\ &= \delta(r) + \rho[g(r) - 1] + \rho\end{aligned} \tag{2.1.44}$$

where ρ is the number density of atoms, we have

$$(\mathrm{d}\sigma/\mathrm{d}\Omega)_{\mathrm{coh}} = N\, |\bar{b}|^2 \left\{1 + \int \rho[g(r) - 1] \exp(\mathrm{i}\boldsymbol{\kappa} \cdot \boldsymbol{r})\, \mathrm{d}\boldsymbol{r}\right\}. \tag{2.1.45}$$

The quantity $g(r)-1$ occurs here to provide a convergent integral. The constant generates a delta function which is finite for $\kappa=0$, but since this is not scattering, it is not of physical interest. It is worth restating that the pair correlation function $g(r)$ only provides a 'snapshot' function at $t=0$, i.e. it contains no dynamical information. The treatment is often called the 'static' approximation, and is valid when the energy transfers are small in comparison with the incident energy; this follows from the derivation of equn (2.1.43) which assumes that $\boldsymbol{\kappa}$ is kept constant for all energy transfers. This approximation, though excellent for X-ray scattering, is a poor one for neutron scattering, and all data must be corrected for this deficiency using, for example, the method introduced by Placzek (1952).

After averaging the term $\exp(i\boldsymbol{\kappa}\cdot\boldsymbol{r})$ in equn (2.1.45) over all orientations of $\boldsymbol{\kappa}$, and combining the result with equn (2.1.42), we obtain

$$S(\boldsymbol{\kappa}) = 1 + (4\pi\rho/\kappa)\int_0^\infty [g(r)-1]r\sin(\kappa r)\,\mathrm{d}r \qquad (2.1.46)$$

for a monoatomic scattering sample.

Neutrons are particularly powerful in the study of partial structure factors in multicomponent liquids and amorphous alloys, since there is plenty of scope for changing the scattering lengths by isotopic substitution in nuclear scattering samples. By way of example, let us consider a binary alloy with atoms A and B. There are now three interatomic potentials, and correspondingly three partial correlation functions $g_{AA}(r)$, $g_{BB}(r)$ and $g_{AB}(r)$. For example, using our previous definition, $g_{AB}(r)$ is the probability of finding atom B at position r when there is an atom A at the origin and at zero time. The three partial structure factors are obtained by extending equn (2.1.46), and S_{AB} is given by

$$S_{AB}(\boldsymbol{\kappa}) = 1 + (4\pi\rho/\kappa)\int_0^\infty [g_{AB}(r)-1]r\sin(\kappa r)\,\mathrm{d}r, \qquad (2.1.47)$$

where ρ is now the total number of scattering atoms per unit volume. The three partial structure factors are related to the coherent differential cross-section by (see Enderby (1968))

$$(\mathrm{d}\sigma/\mathrm{d}\Omega)_{\mathrm{coh}} \propto N[c_A^2 b_A^2 S_{AA}(\boldsymbol{\kappa}) + 2c_A c_B b_A b_B S_{AB}(\boldsymbol{\kappa}) + c_B^2 b_B^2 S_{BB}(\boldsymbol{\kappa}) - \Delta_1 + \Delta_2], \qquad (2.1.48)$$

where $\Delta_1 = (c_A b_A + c_B b_B)^2$ is a coherent self-scattering term, and $\Delta_2 = c_A b_A^2 + c_B b_B^2$ is a balancing term to give the correct total coherent scattering at large $\boldsymbol{\kappa}$. In equn (2.1.48), c_A and c_B are the atomic fractions of the A and B atoms with coherent scattering lengths b_A and b_B, and N is the total number of atoms in the scatterer. The scattering pattern therefore contains a flat background $(\Delta_2-\Delta_1)$ superimposed by an oscillatory function containing the required structure factors S_{AA}, S_{AB} and S_{BB}. These can be evaluated individually by carrying out three separate experiments with different isotopes of A and B in which the scattering lengths b_A and b_B are varied. An example is the study by Enderby *et al.* (1966) of the alloy Cu_6Sn_5 in which pure tin was alloyed with natural copper, ^{63}Cu and ^{65}Cu.

Diffuse elastic scattering in solids

We have seen that an integration over all neutron energies at fixed $\boldsymbol{\kappa}$ gives us a 'snapshot' view of the atoms in a specimen. By measuring only the purely elastic diffuse scattering all time-dependent effects, like those due to thermal displacements, are eliminated. Bragg scattering is, of course, purely elastic and gives information on the average positions of the atoms in a crystal and, through the relative intensities of the δ-function peaks, the average distribution of the atoms in the unit cell. Such scattering is characteristic of a perfect crystal lattice, however there is frequently some additional scattering between the peaks which arises owing to crystal disorder. This is called the elastic diffuse scattering and can for example provide information on (a) the positions of defects in a lattice, (b) the lattice distortions around defects, and (c) the interactions between defects. A detailed review has been written by Bauer (1979).

The present discussion is again restricted to the case of a binary alloy AB since the treatment can be readily extended to magnetic specimens which can be studied using polarized neutrons. If we assume that the alloy is in the form of a rigid Bravais lattice, then the coherent elastic cross-section, see equn (2.1.30), is

$$(\mathrm{d}\sigma/\mathrm{d}\Omega)_{\mathrm{coh}} = \overline{\left|\sum_l \bar{b}_l \exp(\mathrm{i}\boldsymbol{\kappa}\cdot\boldsymbol{l})\right|^2}, \tag{2.1.49}$$

where the bar denotes an average over all the spatial configurations of the A and B atoms. The configurational average coherent scattering length may be defined as

$$\overline{(\bar{b})} = (1-c)\bar{b}_B + c\bar{b}_A = b, \tag{2.1.50}$$

where c is the atomic fraction of A-type atoms. To obtain the configurational average of the coherent cross-section, we follow the method given by Lovesey (1986) and introduce a site occupation function p_l, where

$$p_l = 1, \text{ if } l \text{ corresponds to a type A atom,} \tag{2.1.51}$$

$$p_l = 0 \text{ otherwise.} \tag{2.1.52}$$

This definition implies that

$$\bar{p}_l = c \tag{2.1.53}$$

and

$$\sum_l p_l = Nc. \tag{2.1.54}$$

Hence it follows that

$$\begin{aligned}\bar{b}_l &= \bar{b}_B + p_l(\bar{b}_A - \bar{b}_B) \\ &= b + (\bar{b}_A - \bar{b}_B)(p_l - c).\end{aligned} \tag{2.1.55}$$

Further, from equn (2.1.53) we obtain

$$\overline{\bar{b}_l \bar{b}_{l'}} = b^2 + (\bar{b}_A - \bar{b}_B)^2\, \overline{(p_{l'} - c)(p_l - c)}. \tag{2.1.56}$$

The second term in this relation can be evaluated by noticing that

$$\begin{aligned}\overline{p_{l'} p_l} &= c, \quad \text{if} \quad l = l' \\ &= c^2, \quad \text{if} \quad l' \neq l,\end{aligned}$$

i.e.

$$p_{l'} p_l = c^2 + c(1-c)\delta_{l,l'} \tag{2.1.57}$$

Substituting (2.1.57) into (2.1.56) gives (see equn (2.1.49)), the elastic coherent cross-section:

$$(\mathrm{d}\sigma/\mathrm{d}\Omega)_{\mathrm{coh}} = (\mathrm{d}\sigma/\mathrm{d}\Omega)_{\mathrm{Bragg}} + N(\bar{b}_A - \bar{b}_B)^2 c(1-c), \tag{2.1.58}$$

where

$$(\mathrm{d}\sigma/\mathrm{d}\Omega)_{\mathrm{Bragg}} = b^2 \left| \sum_l \exp(\mathrm{i}\boldsymbol{\kappa} \cdot \boldsymbol{l}) \right|^2. \tag{2.1.59}$$

The second term in equn (2.1.58) is usually called the Laue monotonic diffuse scattering cross-section, and for a rigid lattice it is $\boldsymbol{\kappa}$-independent. We note that this scattering disappears for a monatomic lattice, and that its magnitude depends strongly on the scattering length difference $(\bar{b}_A - \bar{b}_B)$. The diffuse cross-section (per atom) may be written in the form

$$(\mathrm{d}\sigma/\mathrm{d}\Omega) = c(1-c)(\bar{b}_A - \bar{b}_B)^2 \sum_{j=0}^{\infty} \alpha(\boldsymbol{r}_j) \exp(\mathrm{i}\boldsymbol{\kappa} \cdot \boldsymbol{r}_j), \tag{2.1.60}$$

where $\alpha(\boldsymbol{r}_j)$ are the Cowley short-range order parameters (Cowley 1950), and these give the average distributions of A and B atoms around a particular site in the binary alloy. When describing the defect scattering from impurity atoms A, a positive α denotes clustering, i.e. there is a preference for the defect atom to be surrounded by other defect atoms, whereas a negative α indicates that there is a greater probability of finding host atoms around the defect atom.

2.1.2 Atomic electrons

The unpaired electrons in an atom produce a magnetic field $\boldsymbol{B}$ and the interaction of the neutron dipole moment $\boldsymbol{\mu}_n$ with this field gives rise to magnetic scattering. For a single moving electron, Lovesey (1986) has shown that the magnetic interaction potential may be expressed in terms of the distance $\boldsymbol{R}$ from the electron (see also equn (1.4.9)), as follows

$$\hat{V}_{\mathrm{M}}(\boldsymbol{R}) = -\hat{\boldsymbol{\mu}}_n \cdot \boldsymbol{B} = -\gamma_n \beta_{\mathrm{N}} \hat{\boldsymbol{\sigma}} \cdot \boldsymbol{B}$$
$$= \gamma_n \beta_{\mathrm{N}} \left[2\beta \hat{\boldsymbol{\sigma}} \cdot \mathrm{curl}\left(\frac{\hat{\boldsymbol{s}} \times \boldsymbol{R}}{|\boldsymbol{R}|^3} \right) - \frac{\mathrm{e}}{2m_{\mathrm{e}}c} \left(\hat{\boldsymbol{p}}_{\mathrm{e}} \cdot \frac{\hat{\boldsymbol{\sigma}} \times \boldsymbol{R}}{|\boldsymbol{R}|^3} + \frac{\hat{\boldsymbol{\sigma}} \times \boldsymbol{R}}{|\boldsymbol{R}|^3} \cdot \hat{\boldsymbol{p}}_{\mathrm{e}} \right) \right], \tag{2.1.61}$$

where $\hat{\boldsymbol{\sigma}}$ is the Pauli spin operator for the neutron, and γ_{n} is the magnetic moment of the neutron expressed in units of β_{N}. The factor γ_{n} is dimensionless, and has a magnitude $(\mu_{\mathrm{n}}/\beta_{\mathrm{N}}) = -1.913$. β_{N} and β are the nuclear and Bohr magnetons, $\hat{\boldsymbol{s}}$ is the electron spin operator, and $\hat{\boldsymbol{p}}_{\mathrm{e}}$ is the linear momentum operator for the electron. The first of the terms in parenthesis represents the classical dipole–dipole interaction between the electron and neutron moments, and the second term is the interaction due to the translational (orbital) motion of the electron. The 'spin' scattering and the 'orbital' scattering are more conveniently formulated in reciprocal space, when the matrix element containing the magnetic potential summed over i electrons becomes (Lovesey 1986)

$$\langle \boldsymbol{k}' | \hat{V}_{\mathrm{M}} | \boldsymbol{k} \rangle = -r_0 \hat{\boldsymbol{\sigma}} \cdot \hat{\boldsymbol{Q}}_{\perp}$$

with

$$\hat{\boldsymbol{Q}}_{\perp} = \sum_i \exp(\mathrm{i}\boldsymbol{\kappa} \cdot \boldsymbol{r}_i) \left(\tilde{\boldsymbol{\kappa}} \times (\hat{\boldsymbol{s}}_i \times \tilde{\boldsymbol{\kappa}}) - \frac{i}{\hbar\, |\boldsymbol{\kappa}|} (\tilde{\boldsymbol{\kappa}} \times \hat{\boldsymbol{p}}_i) \right), \tag{2.1.62}$$

where $r_0 = (\gamma_{\mathrm{n}} e^2/m_{\mathrm{e}}c^2) = -0.5391 \times 10^{-12}$ cm is the classical electron radius multiplied by γ_{n}, $\boldsymbol{r}_i$ is the position vector of electron i, and $\hat{\boldsymbol{Q}}_{\perp}$ is an interaction operator. It is particularly helpful to introduce an operator $\hat{\boldsymbol{Q}}$ through a definition suggested by the spin part of equn (2.1.62), i.e.

$$\hat{\boldsymbol{Q}}_{\perp} = \hat{\boldsymbol{Q}} - (\hat{\boldsymbol{Q}} \cdot \tilde{\boldsymbol{\kappa}})\tilde{\boldsymbol{\kappa}}. \tag{2.1.63}$$

Note that $\hat{\boldsymbol{Q}}$ is arbitrary within any function proportional to $\boldsymbol{\kappa}$. $\tilde{\boldsymbol{\kappa}} = \boldsymbol{\kappa}/|\boldsymbol{\kappa}|$ is a unit vector in the direction of the scattering vector $\boldsymbol{\kappa}$, and $\hat{\boldsymbol{Q}}_{\perp}$ represents the component of the interaction which is perpendicular to the scattering vector $\boldsymbol{\kappa}$. The relationship between $\hat{\boldsymbol{Q}}_{\perp}$ and $\hat{\boldsymbol{Q}}$ is illustrated geometrically in Fig. 2.1.

Let us first consider the part of $\hat{\boldsymbol{Q}}_{\perp}$ containing the electron spin operator $\hat{\boldsymbol{s}}_i$, and define a spin magnetization density operator

$$\hat{M}_{\mathrm{s}}(\boldsymbol{r}) = -2\beta \sum_i \hat{\boldsymbol{s}}_i\, \delta(\boldsymbol{r} - \boldsymbol{r}_i). \tag{2.1.64}$$

This gives a spin contribution $\hat{\boldsymbol{Q}}_{\mathrm{s}\perp}$ to $\hat{\boldsymbol{Q}}_{\perp}$ given by

$$\hat{\boldsymbol{Q}}_{\mathrm{s}\perp}(\boldsymbol{\kappa}) = -1/(2\beta) \int \exp(\mathrm{i}\boldsymbol{\kappa} \cdot \boldsymbol{r}) \tilde{\boldsymbol{\kappa}} \times (M_s(\boldsymbol{r}) \times \tilde{\boldsymbol{\kappa}})\, \mathrm{d}\boldsymbol{r}. \tag{2.1.65}$$

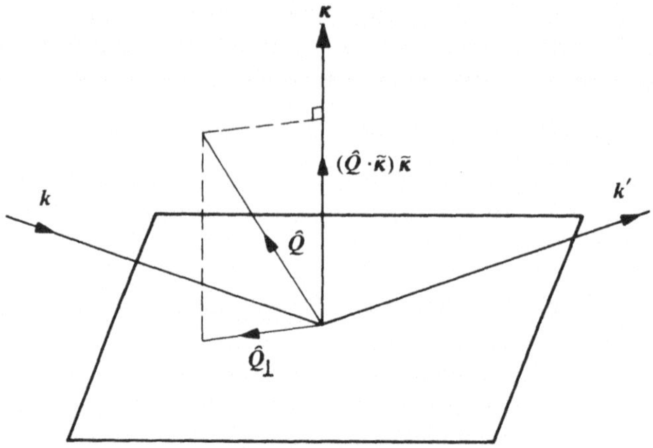

Fig. 2.1 Geometric relation between $\hat{\boldsymbol{Q}}_\perp$ and $\hat{\boldsymbol{Q}}$.

The second part of $\hat{\boldsymbol{Q}}_\perp$ is due to the electron's orbital moment, and this is best treated by introducing an orbital current density operator $\hat{\boldsymbol{j}}(\boldsymbol{r})$ such that

$$\hat{\boldsymbol{j}}(\boldsymbol{r}) = -e/(2m_e)\sum_i (\hat{\boldsymbol{p}}_i\,\delta(\boldsymbol{r}-\boldsymbol{r}_i) + \delta(\boldsymbol{r}-\boldsymbol{r}_i)\hat{\boldsymbol{p}}_i); \qquad (2.1.66)$$

the result, see Lovesey (1986), in terms of the orbital magnetization density $\hat{\boldsymbol{M}}_L(\boldsymbol{r})$ is

$$\hat{\boldsymbol{Q}}_{l\perp}(\boldsymbol{\kappa}) = -1/(2\beta)\int \exp(\mathrm{i}\boldsymbol{\kappa}\cdot\boldsymbol{r})\tilde{\boldsymbol{\kappa}}\times(\hat{\boldsymbol{M}}_l(\boldsymbol{r})\times\tilde{\boldsymbol{\kappa}})\,\mathrm{d}\boldsymbol{r}. \qquad (2.1.67)$$

The similarity with equn (2.1.65) for $\hat{\boldsymbol{Q}}_{s\perp}(\boldsymbol{\kappa})$ is self-evident, and by writing the total magnetization density as

$$\hat{\boldsymbol{M}}(\boldsymbol{r}) = \hat{\boldsymbol{M}}_s(\boldsymbol{r}) + \hat{\boldsymbol{M}}_l(\boldsymbol{r}) \qquad (2.1.68)$$

we see that

$$\hat{\boldsymbol{Q}}_\perp(\boldsymbol{\kappa}) = -1/(2\beta)\int \exp(\mathrm{i}\boldsymbol{\kappa}\cdot\boldsymbol{r})\tilde{\boldsymbol{\kappa}}\times(\hat{\boldsymbol{M}}(\boldsymbol{r})\times\tilde{\boldsymbol{\kappa}})\,\mathrm{d}\boldsymbol{r}, \qquad (2.1.69)$$

i.e. $\hat{\boldsymbol{Q}}_\perp(\boldsymbol{\kappa})$ is proportional to the Fourier transform of the total magnetization density operator. This close association between $\hat{\boldsymbol{Q}}_\perp(\boldsymbol{\kappa})$ and $\hat{\boldsymbol{M}}(\boldsymbol{r})$ is of course the expected result: neutrons are scattered by the net magnetic field generated by the atom's unpaired electrons, and this field in turn depends on the atom's resultant magnetization. The significance of the $\perp$ subscript also now becomes evident, since equn (2.1.69) shows that it is only those magnetization components which are perpendicular to the scattering vector that are effective in scattering neutrons. Much of the remaining discussion of magnetic scattering in this book concentrates on the scattering due to the atom's spin magnetization, it is nevertheless important to emphasize that it is always the total magnetization which is measured in a neutron scattering experiment.

We now derive expressions for the magnetic scattering length, starting with atoms which have only a spin angular momentum. A comparison of the matrix element for nuclear scattering,

$$\langle \boldsymbol{k}' |\, \hat{V}_{\mathrm{N}} \,| \boldsymbol{k} \rangle = b_j \exp(\mathrm{i}\boldsymbol{\kappa} \cdot \boldsymbol{R}_j), \tag{2.1.70}$$

with its magnetic counterpart for a spin-only atom,

$$\langle \boldsymbol{k}' |\, \hat{V}_{\mathrm{M}} \,| \boldsymbol{k} \rangle = -r_0 \hat{\boldsymbol{\sigma}} \cdot [\tilde{\boldsymbol{\kappa}} \times (\hat{\mathbf{s}}_i \times \tilde{\boldsymbol{\kappa}})] \exp(\mathrm{i}\boldsymbol{\kappa} \cdot \boldsymbol{r}), \tag{2.1.71}$$

shows that the magnetic scattering length equivalent to b, which we denote by $\hat{p}$, is

$$\hat{p} = -r_0 \hat{\boldsymbol{\sigma}} \cdot [\tilde{\boldsymbol{\kappa}} \times (\hat{\mathbf{s}}_{\mathrm{i}} \times \tilde{\boldsymbol{\kappa}})]. \tag{2.1.72}$$

This is an operator with respect to the neutron and atomic spins, although it is now in a very unfamiliar form. It is more often expressed by

$$\hat{p} = 2r_0 \langle \hat{s}_i \rangle \langle \hat{\boldsymbol{s}}' \rangle \cdot [\tilde{\boldsymbol{\kappa}}(\tilde{\boldsymbol{\eta}} \cdot \tilde{\boldsymbol{\kappa}}) - \tilde{\boldsymbol{\eta}}], \tag{2.1.73}$$

since

$$\langle \hat{\boldsymbol{\sigma}} \rangle = 2\langle \hat{\boldsymbol{s}}' \rangle, \tag{2.1.74}$$

where $\langle \hat{\boldsymbol{s}}' \rangle$ is the average neutron spin vector. The angular brackets are used to denote the average component of the spin along the quantization direction. Incidentally, see equn (1.4.1), $2\langle \hat{\boldsymbol{s}}' \rangle = \boldsymbol{P}$, where $\boldsymbol{P}$ is the neutron polarization vector. The unit vector $\tilde{\boldsymbol{\eta}}$ is defined as the unit vector in the direction of the atomic spin. Some authors use $\tilde{\boldsymbol{\eta}}$ to denote the unit vector in the direction of the atomic moment, in which case a sign reversal must be applied. Equation (2.1.73) however, is still, incomplete, and a more general formulation requires further modifications. The term in parenthesis, often denoted as the magnetic interaction vector $\boldsymbol{q}$, is a geometric factor of magnitude $\sin \alpha$, where α is the angle between $\tilde{\boldsymbol{\kappa}}$ and $\tilde{\boldsymbol{\eta}}$. The relationship between $\tilde{\boldsymbol{\kappa}}$, $\tilde{\boldsymbol{\eta}}$ and $\boldsymbol{q}$ is shown in Fig. 2.2. $\boldsymbol{q}$ is not a function of the scatterer and is frequently omitted from the scattering length expression. The $\langle \hat{s}_i \rangle$ parameter for one electron can be replaced by $\langle \hat{S} \rangle$, the value of the resultant spin in a spin-only multi-electron atom, or by $(g\langle \hat{J} \rangle/2)$ in an atom with both spin and orbital moments. J is the resultant angular momentum, and g is the Landé splitting factor

$$g = 1 + [J(J+1) + S(S+1) - L(L+1)]/[2J(J+1)]. \tag{2.1.75}$$

Next the term $2\langle \hat{\boldsymbol{s}}' \rangle$ can be replaced by $+1$ or -1, depending on the polarization direction $\boldsymbol{P}$ of the neutron being scattered, and after

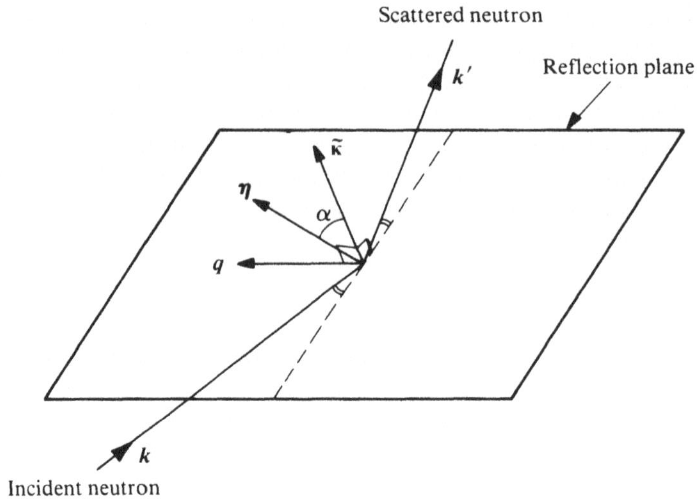

Fig. 2.2 Relationship between the unit vectors $\tilde{\boldsymbol{\kappa}}$, $\tilde{\boldsymbol{\eta}}$ and the magnetic interaction geometric factor $\boldsymbol{q}$.

applying all these changes we obtain

$$p = \pm(r_0/2)2\langle \hat{S} \rangle, \quad \text{or} \quad p = \pm(r_0/2)g\langle \hat{J} \rangle, \tag{2.1.76}$$

where the upper signs give the scattering amplitudes for (+) spin neutrons.

Finally, we must introduce the magnetic form factor $F(\boldsymbol{\kappa})$, which is the Fourier transform of the magnetization density. This is necessary since the electrons responsible for the scattering are, in contrast to the case of nuclear scattering, distributed over a significant volume of space which, for thermal neutron scattering, is comparable with the neutron wavelength. Magnetic form factors are discussed later in some detail, but for the present it suffices to consider these simply as factors which are directly proportional to the magnetic scattering length. We therefore conclude with the expressions

$$p = \pm(r_0/2)2\langle \hat{S} \rangle F(\boldsymbol{\kappa}) \quad \text{or} \quad p = \pm(r_0/2)g\langle \hat{J} \rangle F(\boldsymbol{\kappa}), \tag{2.1.77}$$

and since $2\langle \hat{S} \rangle$ and $g\langle \hat{J} \rangle$ are equal to the spin-only atomic moment $\langle \hat{\mu}_S \rangle$ and the spin-plus-orbital atomic moment $\langle \hat{\mu}_J \rangle$ respectively, we have as alternative expressions for isolated ions

$$p = \mp(r_0/2)\langle \hat{\mu}_S \rangle F(\boldsymbol{\kappa}) \quad \text{and} \quad p = \mp(r_0/2)\langle \hat{\mu}_J \rangle F(\boldsymbol{\kappa}) \tag{2.1.78}$$

with $(r_0/2)$ equal to -0.2695×10^{-12} cm β^{-1}.

The sign reversals are due to the fact that the atomic spins and atomic moments are oppositely directed. We now present some of the more common magnetic cross-section expressions encountered in various elastic neutron scattering experiments.

Fully magnetized scatterers

The coherent elastic magnetic cross-section for a spin-only atom with spin $\langle \hat{S} \rangle$ is obtained by squaring equn (2.1.73), which, after including the form factor $F(\boldsymbol{\kappa})$ and then rearranging becomes

$$(\mathrm{d}\sigma/\mathrm{d}\Omega) = r_0^2 \langle \hat{S}^2 \rangle F(\boldsymbol{\kappa})^2 H^2, \tag{2.1.79}$$

with

$$H = 2(\langle \hat{\boldsymbol{s}}' \rangle \cdot \boldsymbol{q}) \tag{2.1.80}$$

and

$$\boldsymbol{q} = \tilde{\boldsymbol{\kappa}}(\tilde{\boldsymbol{\eta}} \cdot \tilde{\boldsymbol{\kappa}}) - \tilde{\boldsymbol{\eta}}. \tag{2.1.81}$$

For unpolarized neutron scattering, H^2 must be replaced by the average $\langle H^2 \rangle$ and this is given by

$$\langle H^2 \rangle = 4 \sum_{i=1}^{3} q_i^2 \langle \hat{s}_i'^2 \rangle = \tfrac{4}{3} \langle \hat{s}'^2 \rangle \sum_i q_i^2 = \tfrac{4}{3} s'(s'+1) q^2, \tag{2.1.82}$$

and since $s' = \frac{1}{2}$ for neutrons, we obtain $\langle H^2 \rangle = q^2$ and

$$(\mathrm{d}\sigma/\mathrm{d}\Omega) = r_0^2 \langle \hat{S}^2 \rangle F(\boldsymbol{\kappa})^2 [1 - (\tilde{\boldsymbol{\kappa}} \cdot \tilde{\boldsymbol{\eta}})^2]. \tag{2.1.83}$$

This equation provides a relatively simple method for determining the magnetic scattering in the presence of nuclear scattering using unpolarized neutron beams. If the atom is magnetized at right angles to the scattering vector, then the cross-section is a maximum with $q^2 = 1$. However, when the atom is magnetized either parallel or antiparallel to $\boldsymbol{\kappa}$, then there is no magnetic scattering since $q^2 = 0$. An alternative method is to measure the difference scattering between that in the $(\tilde{\boldsymbol{\kappa}} \cdot \tilde{\boldsymbol{\eta}})^2 = 1$ geometry and that for an unmagnetized sample when $(\tilde{\boldsymbol{\kappa}} \cdot \tilde{\boldsymbol{\eta}})^2 = 1/3$, or $q^2 = 2/3$. These unpolarized beam methods, though useful, are far less sensitive than those which use polarized neutron beams.

Paramagnetic scattering

We now consider an unmagnetized paramagnetic scatterer in which the spin-only ions are randomly oriented. Two averagings are necessary, the first being over q^2:

$$\langle q^2 \rangle = \langle \sin^2 \alpha \rangle = 2/3, \tag{2.1.84}$$

where α is the angle between $\tilde{\boldsymbol{\kappa}}$ and $\tilde{\boldsymbol{\eta}}$. Then an averaging must be made over all the orientations of the atomic spins, and this gives

$$\langle \hat{S}^2 \rangle = S(S+1). \tag{2.1.85}$$

The paramagnetic scattering cross-section per atom is identical with that of the free atom, since there is no correlation between the individual spins in the bulk specimen, hence

$$(\mathrm{d}\sigma/\mathrm{d}\Omega)_{\mathrm{para}} = \tfrac{2}{3}r_0^2 S(S+1)F(\boldsymbol{\kappa})^2 \qquad (2.1.86)$$

for unpolarized neutrons. It is instructive to evaluate the magnitude of this cross-section and to compare this with typical values for nuclear scattering. We take as an example Mn^{2+}, ($3d^5$, $S = 5/2$, $L = 0$) and find $(\mathrm{d}\sigma/\mathrm{d}\Omega)_{\mathrm{para}} = 1.69$ barns sterad^{-1} atom^{-1}. The differential nuclear scattering cross-section for Mn is approximately 0.14 barns sterad^{-1} atom^{-1}, so in this example it is ten times weaker than the magnetic scattering. Mn^{2+} is an exceptionally strong magnetic scatterer, and as a general rule we find the magnetic scattering to be weaker than or at best comparable with the strength of the nuclear scattering.

Magnetic diffuse scattering

Magnetic diffuse scattering gives information on the spatial fluctuations of the atomic magnetic moments. At this stage we shall follow the practice adopted with nuclear diffuse scattering, and consider a simple binary magnetic alloy AB where μ_A and μ_B are the atomic moments and c is the concentration of A atoms. The model we describe, which is due to Shull and Wilkinson (1955), assumes that the atoms do not experience local environmental effects, and that changes in the magnetic moment are simply a consequence of concentration fluctuations. The mean magnetic moment is

$$\bar{\mu} = c\mu_A + (1-c)\mu_B \qquad (2.1.87)$$

and the unpolarized beam cross-section is obtained in a straightforward way by substituting equn (2.1.78), together with the orientation parameter q^2, for $\bar{b}_A$ and $\bar{b}_B$ in equn (2.1.58). This gives

$$(\mathrm{d}\sigma/\mathrm{d}\Omega)_{\mathrm{Diff}} = (r_0/2)^2 Nc(1-c)[\mu_A F_A(\boldsymbol{\kappa}) - \mu_B F_B(\boldsymbol{\kappa})]^2[1-(\tilde{\boldsymbol{\kappa}}\cdot\tilde{\boldsymbol{\eta}})^2] \qquad (2.1.88)$$

and it is evident that the field-switching technique, when $(\tilde{\boldsymbol{\kappa}}\cdot\tilde{\boldsymbol{\eta}})^2$ is alternately switched between 0 and 1, can be used to measure this scattering. This method was used by Low and co-workers (see for example Low and Collins (1963) and Holden *et al.* (1967)), to study the structure of moment defects for a range of metallic ion impurities in iron and nickel. The degree of sophistication of the models used to describe binary alloys has been greatly improved since Shull and Wilkinson's early work, see for example Hicks (1979b), and we shall see later that these magnetic diffuse cross-sections are best determined using polarized neutron beams.

Magnetic form factors

One of the major differences between magnetic and nuclear neutron scattering amplitudes is that the former are strongly **κ**-dependent through the magnetic form factor. Having already introduced this concept in the cross-section equations, we now consider the significance of magnetic form factors $F(\boldsymbol{\kappa})$ in greater detail.

First we recall that the atomic form factors obtained by X-ray scattering are determined by the average electron charge distribution for all the electrons in the atom. Only the unpaired electrons, however, contribute to the magnetic scattering observed with neutrons, and the magnetic form factor is associated with only the spatial distributions of these electrons. One of the earliest measurements of a magnetic form factor was that by Shull *et al.* (1951a), who analysed the paramagnetic scattering from MnF_2 using the cross-section equn (2.1.86). Their results, together with estimated uncertainties, are shown in Fig. 2.3. Also shown for comparison is the X-ray form factor for Mn^{2+} calculated by James and Brindley (1932); the faster fall-off of the magnetic form factor with increasing **κ** is consistent with the expected result, since only electrons in the outermost orbitals contribute to the magnetic scattering.

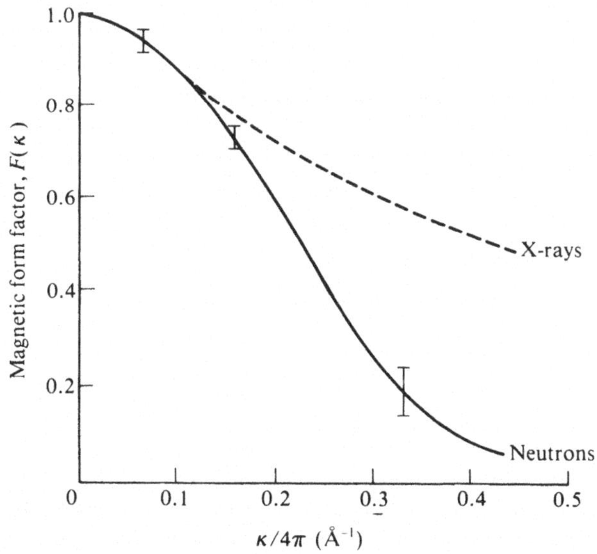

Fig. 2.3 Magnetic form factor $F(\boldsymbol{\kappa})$ for the Mn^{2+} ion determined from measurements of the paramagnetic scattering in MnF_2 by Shull, Strauser and Wollan (1951a). The X-ray scattering form factor is also shown for comparison.

A magnetic form factor may be defined by

$$F(\boldsymbol{\kappa}) = \left[\int \boldsymbol{M}(\boldsymbol{r}) \exp(\mathrm{i}\boldsymbol{\kappa}\cdot\boldsymbol{r})\,\mathrm{d}\boldsymbol{r}^3\right] \Big/ \int \boldsymbol{M}(\boldsymbol{r})\,\mathrm{d}\boldsymbol{r}^3, \qquad (2.1.89)$$

where $\boldsymbol{M}(\boldsymbol{r})$ is the magnetization density associated with the atom. Neutron techniques, particularly polarized neutron scattering from single crystals, allow magnetic structure factors $F_{\mathrm{M}}(\boldsymbol{\kappa})$ to be determined, and these are related to the form factors by

$$F_{\mathrm{M}}(\boldsymbol{\kappa}) = (-r_0/2) \sum_j F_j(\boldsymbol{\kappa})\boldsymbol{\mu}_j \exp\{\mathrm{i}\boldsymbol{\kappa}\cdot\boldsymbol{R}_j\}, \qquad (2.1.90)$$

where the summation is over all j atoms in the crystal at distances $\boldsymbol{R}_j$ from the origin, and $\boldsymbol{\mu}_j$ are the atomic moments. We shall give further details of the experiments later, but it is worthwhile pointing out now that the interest in obtaining accurate $F(\boldsymbol{\kappa})$ values follows largely because these can be calculated theoretically, and a comparison of experiment and theory can provide detailed information on unpaired electron wavefunctions in the solid state. A review of what has been achieved has been made by Moon (1982). By way of illustration, we now present the results of two theoretical predictions for the magnetic form factor: first the difference between spin and orbital form factors, followed by the effect of ligands on magnetic ion form factors in dilute salts.

We have seen that the spin and orbital angular momenta contribute to the atomic magnetization in different ways. Spin magnetism is a simple consequence of the magnetic moment of the electron, whereas orbital magnetism arises from the circulating current density due to the electron's motion, and gives an orbital magnetization density which is much more localized than the spin density. One result of this is that the integrals, as for example in equn (2.1.89), which give the spin and orbital form factors, are intrinsically different. Brown (1979) has shown that for spherical symmetry, and in the dipole approximation, the radial form factor may be written as

$$F(\boldsymbol{\kappa}) = \langle j_0(\boldsymbol{\kappa})\rangle + \{\mu_L/(\mu_S + \mu_L)\} j_2(\boldsymbol{\kappa}) \qquad (2.1.91)$$

where

$$\langle j_n(\boldsymbol{\kappa})\rangle = \int_0^\infty j_n(\kappa r)\, U^2(r)\, r^2\,\mathrm{d}r \qquad (2.1.92)$$

and $j_n(\kappa r)$ are spherical Bessel functions, $U(r)$ is the electron radial wave function, and μ_S and μ_L are the spin and orbital magnetic moments. Equation (2.1.91) can therefore be used, with experimental data, to deduce the spin and orbital contributions to the resultant moment. The separate spin and orbital form factors for the Fe^{2+} ion given by Brown

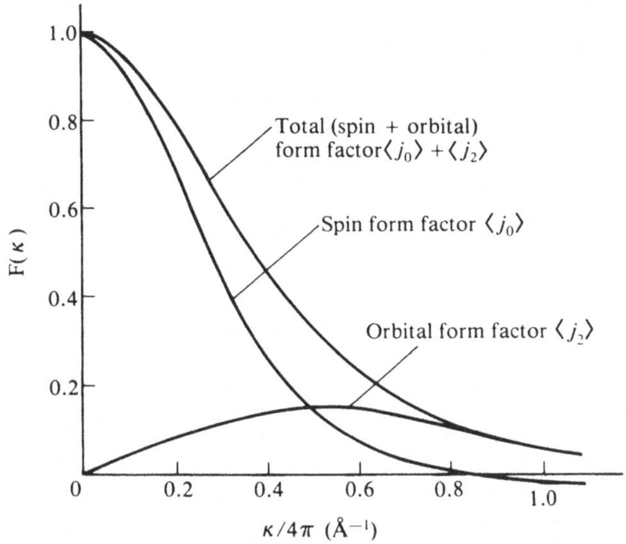

Fig. 2.4 Spin and orbital form factors for the Fe^{2+} ion from calculations by Watson and Freeman (1961); see Brown (1979). Note that for $\kappa \to 0$, $\langle j_0 \rangle = 1$ and $\langle j_2 \rangle = 0$.

(1979), using calculations by Watson and Freeman (1961), are shown in Fig. 2.4. The resultant form factor is wider than the spin-only form factor $\langle j_0(\kappa) \rangle$, since the orbital form factor decreases less rapidly with $\boldsymbol{\kappa}$.

As our second example we reproduce the theoretical form factor predictions made by Rimmer (1970), which show the effects of two nearest-neighbour ligands on the form factor of a 3d ion. There are three contributions to the resultant covalent form factor:

(a) A 3d form factor, which is slightly reduced compared with the free ion form factor, and is governed by the 3d wavefunction of extent ~0.5 Å.
(b) An overlap term, with negative amplitude; since the overlap charge is typically centred at ~1 Å from the metal nucleus, this contribution falls off more rapidly with $\boldsymbol{\kappa}$ than in (a).
(c) A ligand term, which is positive, and since this extends to ~2 Å from the atomic nucleus, it is only effective at very low values of $\boldsymbol{\kappa}$.

Calculated values of these contributions, together with the summed covalent form factor are shown in Fig. 2.5. We shall see in Section 4.2.6 that a careful analysis of polarized beam diffraction data from single crystals containing 3d complex ions has provided experimental evidence for the existence of ligand unpaired electron spin densities.

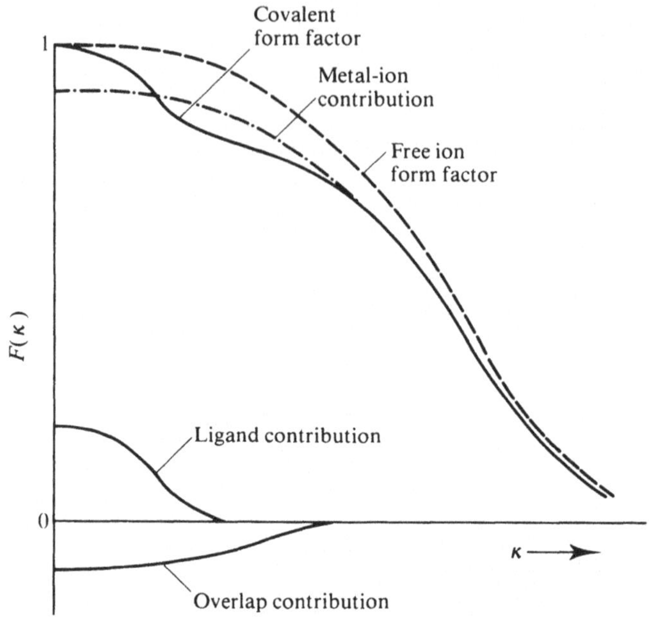

Fig. 2.5 Magnetic form factors for a 3d metal ion showing the effect of covalent ligand bonding (from Rimmer 1970).

2.2 POLARIZED NEUTRON SCATTERING

In the previous section it was shown how the neutron's magnetic moment allows neutron beams to probe electronic magnetism on an atomic scale. Owing to the vector nature of the magnetic interaction potential and also the **κ**-dependence of the scattering through the magnetic form factor, it is possible to distinguish between magnetic and nuclear scattering even with unpolarized neutrons. It has, however, been clear following Halpern and Johnson's (1939) publication that a more sophisticated investigation of atomic magnetism becomes available if both the scattering atoms and the incident neutrons are polarized. The scattering of polarized neutron beams is treated in this section. We shall see that this technique also allows us to study the nature of nuclear magnetism in samples in which the nuclear spins can be polarized.

In order to orient our discussion of the theory it is advantageous to start by giving a short description of the classical polarized neutron diffractometer. Essentially all polarized neutron experiments up to about 1970 were carried out on this type of instrument. Figure 2.6 shows a schematic diagram of the diffractometer originated by Nathans *et al.* (1959). The polarizing monochromator provides a monochromatic beam (polarized say in the (+) or spin 'up' state) at a sample with some

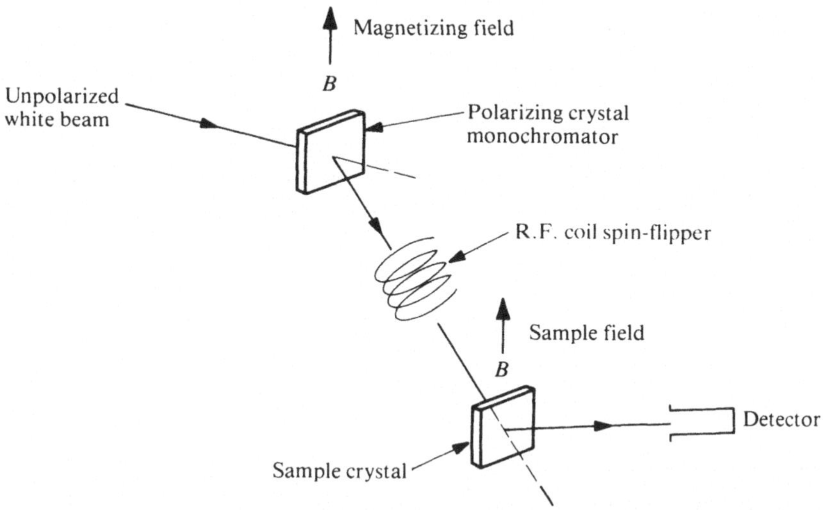

Fig. 2.6 Schematic diagram of a polarized neutron diffractometer.

preferred orientation axis, and a scattered neutron intensity I_+ is detected which, for a perfect polarizer, is proportional to a spin-dependent cross-section which we designate as $(d\sigma/d\Omega)_+$. During part of the experiment the spin-flipper after the polarizer is activated, and the inverted neutron spins give a detected intensity I_- which in an idealized instrument is proportional to $(d\sigma/d\Omega)_-$. The spin-dependence of the scattering is usually analysed from the flipping ratio $R = (I_+/I_-)$, which can be determined with high precision. The origin of the polarization-dependent scattering from oriented nuclear and magnetic systems will now be described.

2.2.1 Oriented nuclei

We have already seen that the neutron–nucleus interaction depends on their spins, and this can be generalized by introducing a spin-dependent scattering amplitude operator $\hat{b}$. For a single isotope

$$\hat{b} = \mathscr{A} + \tfrac{1}{2}\mathscr{B}\hat{\boldsymbol{\sigma}} \cdot \hat{\boldsymbol{i}}, \tag{2.2.1}$$

where $\frac{1}{2}\hat{\boldsymbol{\sigma}} \equiv \hat{\boldsymbol{s}}'$ is the neutron spin operator, and $\hat{\boldsymbol{i}}$ is that for the nuclear spin. Lovesey (1986) showed that the two scattering lengths b_+ and b_- are related to the coefficients $\mathscr{A}$ and $\mathscr{B}$ as follows:

$$b_+ = \mathscr{A} + \tfrac{1}{2}\mathscr{B}[(i+\tfrac{1}{2})(i+\tfrac{3}{2}) - i(i+1) - \tfrac{3}{4}] = \mathscr{A} + i\mathscr{B}/2 \tag{2.2.2}$$

and

$$b_- = \mathscr{A} + \tfrac{1}{2}\mathscr{B}[(i-\tfrac{1}{2})(i+\tfrac{1}{2}) - i(i+1) - \tfrac{3}{4}] = \mathscr{A} - (i+1)\mathscr{B}/2. \tag{2.2.3}$$

Thus

$$\mathscr{A} = [1/(2i+1)][(i+1)b_+ + ib_-] \quad (2.2.4)$$

and

$$\mathscr{B} = [2/(2i+1)][b_+ - b_-]. \quad (2.2.5)$$

If the spin directions of either the neutron or the nucleus are completely random, i.e. when either of the spins averages to zero, we have

$$\bar{\bar{b}} = \mathscr{A}, \quad (2.2.6)$$

i.e. equn (2.2.4) agrees with equn (2.1.25). A further result is that

$$\begin{aligned}\overline{|\bar{b}|^2} &= |\mathscr{A}|^2 + |\mathscr{B}|^2\, i(i+1)/4 \\ &= [1/(2i+1)][(i+1)b_+^2 + ib_-^2] \quad (2.2.7)\end{aligned}$$

in agreement with equn (2.1.26). $\mathscr{B}$ is therefore related to the nuclear spin (NS) incoherent scattering cross-section, and if we write the incoherent scattering length b_{NS} as

$$b_{\mathrm{NS}} = [\overline{|b|^2} - |\bar{b}|^2]^{1/2} \quad (2.2.8)$$

we obtain

$$\begin{aligned}b_{\mathrm{NS}} &= [i(i+1)]^{1/2}(b_+ - b_-)/(2i+1) \\ &= \mathscr{B}[i(i+1)]^{1/2}/2. \quad (2.2.9)\end{aligned}$$

The first comprehensive treatment giving the cross-section equations for the scattering of polarized neutrons by polarized nuclei was that of Schermer and Blume (1968). This was formulated in terms of a new symbol $\boldsymbol{d}_{\mathrm{N}}$ for each nucleus N, and this is related to our nuclear spin parameter $\mathscr{B}_{\mathrm{N}}$ as follows

$$\boldsymbol{d}_{\mathrm{N}} = \mathscr{B}_{\mathrm{N}} i_{\mathrm{N}} \boldsymbol{P}_{\mathrm{N}}, \quad (2.2.10)$$

where i_{N} is the nuclear spin quantum number, $\boldsymbol{P}_{\mathrm{N}}$ is the vector nuclear polarization, and by analogy with Section 1.4 we have

$$\boldsymbol{P}_{\mathrm{N}} = \langle \hat{\boldsymbol{i}}_{\mathrm{N}} \rangle / i_{\mathrm{N}}, \quad (2.2.11)$$

where $\langle \hat{\boldsymbol{i}}_{\mathrm{N}} \rangle$ is the expectation value of the nuclear spin operator. It was shown that the vector $\frac{1}{2}\boldsymbol{d}_{\mathrm{N}}$ can in fact be considered as a coherent nuclear spin scattering amplitude, and plays the same role as b_{N} and $p(\boldsymbol{\kappa})\boldsymbol{q}$ in nuclear and magnetic coherent scattering experiments:

$$\tfrac{1}{2}\boldsymbol{d}_{\mathrm{N}} \equiv b_{\mathrm{N}} \equiv p(\boldsymbol{\kappa})\boldsymbol{q} = r_0\langle \hat{S} \rangle\, F(\boldsymbol{\kappa})\boldsymbol{q}. \quad (2.2.12)$$

Although Schermer and Blume (1968) give the complete expressions for systems with both nuclear and magnetic scattering, we restrict this discussion to purely nuclear scatterers. The polarized neutron coherent elastic cross-section, where the nucleus N is polarized, is

$$(\mathrm{d}\sigma/\mathrm{d}\Omega)_{\mathrm{coh}} = \left|\sum_{N} \exp(\mathrm{i}\boldsymbol{\kappa}\cdot\boldsymbol{R}_{\mathrm{N}})\langle\{b_{\mathrm{N}}\}\rangle\right|^2 + \tfrac{1}{4}\left|\sum_{\mathrm{N}} \exp(\mathrm{i}\boldsymbol{\kappa}\cdot\boldsymbol{R}_{\mathrm{N}})\langle\boldsymbol{d}_{\mathrm{N}}\rangle\right|^2$$
$$+ \mathrm{Re}\left\{\sum_{\mathrm{NN}'} \exp[\mathrm{i}\boldsymbol{\kappa}\cdot(\boldsymbol{R}_{\mathrm{N}} - \boldsymbol{R}_{\mathrm{N}'})]\langle\{b^*_{\mathrm{N}'}\}\rangle\langle\boldsymbol{d}_{\mathrm{N}}\cdot\boldsymbol{P}\rangle\right\}, \tag{2.2.13}$$

where $\{b_{\mathrm{N}}\}$ is the familiar coherent scattering length for the nucleus N at position $\boldsymbol{R}_{\mathrm{N}}$ with spin i_{N}, the brackets $\langle\ \rangle$ denote isotropic averages, and $\boldsymbol{P}$ is the neutron polarization vector. The final term in this expression is the interference term, and is the one of prime interest in polarized beam experiments. All experiments with polarized nuclei require a large applied field $\boldsymbol{B}$ in order to generate sufficient nuclear polarization. If we define $\tilde{\boldsymbol{B}}$ as the unit vector in the direction of this field, and the neutron spin polarization states as being either (+) or (−) with respect to this unit vector, then equn (2.2.13) may be written as

$$(\mathrm{d}\sigma/\mathrm{d}\Omega)_{\pm} = \left|\sum_{\mathrm{N}} (\mathrm{i}\boldsymbol{\kappa}\cdot\boldsymbol{R}_{\mathrm{N}})\langle\{b_{\mathrm{N}}\} \pm \tfrac{1}{2}\boldsymbol{d}_{\mathrm{N}}\cdot\tilde{\boldsymbol{B}}\rangle\right|^2. \tag{2.2.14}$$

The polarized beam coherent scattering cross-sections per nucleus for a sample containing a single polarizable isotope then become

$$(\mathrm{d}\sigma/\mathrm{d}\Omega)_{\pm} = [b_{\mathrm{N}} \pm (\tfrac{1}{2})d_{\mathrm{N}}]^2 = [b_{\mathrm{N}} \pm (\tfrac{1}{2})\mathscr{B}_{\mathrm{N}} i_{\mathrm{N}} P_{\mathrm{N}}]^2, \tag{2.2.15}$$

or, after substituting for b_{N} and $\mathscr{B}_{\mathrm{N}}$ (see equns (2.2.4) and (2.2.5)),

$$(\mathrm{d}\sigma/\mathrm{d}\Omega)_{\pm} = \left[\left(\frac{i+1}{2i+1}\right)b_+\left(1 \pm \frac{iP_{\mathrm{N}}}{(i+1)}\right) + \left(\frac{i}{2i+1}\right)b_-(1 \mp P_{\mathrm{N}})\right]^2. \tag{2.2.16}$$

In the limit $P_{\mathrm{N}} = 1$ we find $(\mathrm{d}\sigma/\mathrm{d}\Omega)_+ = b_+^2$, due solely to a compound nucleus with spin $(i + \frac{1}{2})$, and there is therefore no spin incoherent scattering when both the neutron and nucleus have 'up' spins. The (−) spin neutrons experience an effective coherent scattering amplitude

$$b_{\mathrm{eff}} = [1/(2i+1)](b_+ - 2ib_-), \tag{2.2.17}$$

and this contains a mixture of the two spin-dependent scattering amplitudes, since the nucleus 'up' and neutron 'down' combined initial state does not produce an eigenstate of the compound system.

Next we consider the specific example of the coherently scattered intensities in a single-crystal Bragg peak. These are proportional to the square of the total coherent nuclear structure factor $F = F_{\mathrm{N}} + F_{\mathrm{NS}}$, therefore for one contributing nucleus:

$$|F|^2 = b^2 + (\mathscr{B}/2)^2 i^2 P_{\mathrm{N}}^2 + b\mathscr{B}i\boldsymbol{P}_{\mathrm{N}}\cdot\boldsymbol{P}, \tag{2.2.18}$$

where we have reintroduced the individual nuclear and neutron polarization vectors.

The corresponding nuclear spin (NS) incoherent scattering cross-section for a single isotope was given by Schermer and Blume (1968) to be

$$(\mathrm{d}\sigma/\mathrm{d}\Omega)_{\mathrm{NS}} = b_{\mathrm{NS}}^2\{1 - [i/(i+1)]P_{\mathrm{N}}^2 - [1/(i+1)]\boldsymbol{P}_{\mathrm{N}}\cdot\boldsymbol{P}\}, \quad (2.2.19)$$

and by using equn (2.2.9) we see that this may be rewritten as follows:

$$(\mathrm{d}\sigma/\mathrm{d}\Omega)_{\mathrm{NS}} = (\mathscr{B}/2)^2\{i(i+1) - i^2P_{\mathrm{N}}^2 - i\boldsymbol{P}_{\mathrm{N}}\cdot\boldsymbol{P}\} \qquad (2.2.20)$$

It is instructive to use these two expressions for $|F|^2$ and $(\mathrm{d}\sigma/\mathrm{d}\Omega)_{\mathrm{NS}}$ to calculate the intensities of the nuclear Bragg peaks and the accompanying spin incoherent scattering for various combinations of the nuclear and neutron polarizations. The results are shown in Table 2.2.

The principal conclusion is that the creation of nuclear polarization changes the scattering amplitude and hence the scattered intensities of both the Bragg peaks as well as the incoherent 'background'. Useful experiments are possible with nuclear polarized samples and unpolarized beams, though, in a directly amalogous way to the scattering by (electron) magnetized samples, the most sensitive experiments require polarized nuclei and polarized neutrons. A further advantage of using polarized neutrons is that the sign of $\mathscr{B}$, or $(b_+ - b_-)$, can be obtained provided that the sign of $\boldsymbol{P}_{\mathrm{N}}$ is known. Further, it is evident that such methods are only useful for isotopes with a large $\mathscr{B}$, i.e. those with large spin incoherent scattering cross-sections. The diffraction technique with polarized nuclei is analogous to the isotopic substitution method which is widely used in neutron scattering to create a contrast in the scattering pattern. An important advantage of the polarization method is that it

TABLE 2.2 Dependences of the Bragg peak and spin incoherent intensities on the nuclear and neutron polarizations for one polarizable nucleus (see equns (2.2.18) and (2.2.20)

Nuclear polarization	Neutron polarization	Bragg peak intensities	Spin-incoherent cross-section $(\mathrm{d}\sigma/\mathrm{d}\Omega)_{\mathrm{NS}}$
0	0	b^2	$\left(\frac{\mathscr{B}^2}{4}\right)i(i+1)$
0	$\boldsymbol{P}$	b^2	$\left(\frac{\mathscr{B}^2}{4}\right)i(i+1)$
$\boldsymbol{P}_{\mathrm{N}}$	0	$b^2+\left(\frac{\mathscr{B}^2}{4}\right)i^2P_{\mathrm{N}}^2$	$\left(\frac{\mathscr{B}^2}{4}\right)i(i+1)-\left(\frac{\mathscr{B}^2}{4}\right)i^2P_{\mathrm{N}}^2$
$\boldsymbol{P}_{\mathrm{N}}$	$\boldsymbol{P}$	$b^2+\left(\frac{\mathscr{B}^2}{4}\right)i^2P_{\mathrm{N}}^2+b\mathscr{B}i\boldsymbol{P}_{\mathrm{N}}\cdot\boldsymbol{P}$	$\left(\frac{\mathscr{B}^2}{4}\right)i(i+1)-\left(\frac{\mathscr{B}^2}{4}\right)i^2P_{\mathrm{N}}^2 - \left(\frac{\mathscr{B}^2}{4}\right)i\boldsymbol{P}_{\mathrm{N}}\cdot\boldsymbol{P}$

does not require the composition of the scatterer to be changed, i.e. only one sample is needed.

2.2.2 Magnetic materials

The major application of polarized neutron beams at reactor sources has been the determination of form factors and spin densities in magnetic materials. The measurements make use of the polarization-dependent interference term between the coherent nuclear and magnetic scattering amplitudes, and experimentally this involves determining the ratio R of the Bragg peak intensities from a magnetized single crystal with both (+) and (−) incident neutron spin polarizations. In addition to providing great sensitivity, this technique also supplies the sign of the magnetic scattering amplitude. The method is not necessarily restricted to single-crystal specimens, and in this section we also include the application of polarized neutron beams to the study of (a) diffraction from amorphous magnetic solids, and (b) magnetic diffuse scattering in binary alloys.

Magnetic single-crystal diffraction

We start by deriving the expressions for the coherent elastic scattering cross-sections in polarized beam diffraction experiments. The scattering always contains a nuclear as well as a magnetic component, and for simplicity we shall restrict this discussion to examples in which the nuclei have only a coherent scattering amplitude, and in which the atomic magnetism is due solely to the unpaired electron spin angular momentum S. We shall also assume that this magnetism is saturated, as for example in a strongly magnetized ferromagnet, or in a paramagnetic ion at low temperatures in a large applied magnetic field.

Before we determine the cross-sections, it is useful to restate our definitions for the unit cell structure factors. The unit cell structure factor for coherent nuclear scattering (see equn (2.1.36)) is

$$F_N(\boldsymbol{\kappa}) = \sum_{R} \bar{b} \exp(i\boldsymbol{\kappa} \cdot \boldsymbol{R}) \exp(-W), \qquad (2.2.21)$$

where $\boldsymbol{R}$ is a position vector in the unit cell, $\bar{b}$ is the mean coherent scattering length, and $\exp(-W)$ is a temperature factor, the Debye–Waller factor, which allows for thermal vibrations.

There is also a nuclear spin structure factor $F_{NS}(\boldsymbol{\kappa})$ in which the term $\frac{1}{2}\boldsymbol{d}_N$ in equn (2.2.12) replaces $\bar{b}$:

$$\begin{aligned} F_{NS}(\boldsymbol{\kappa}) &= \sum_{R,N} (\mathscr{B}_N \mid 2) i\boldsymbol{P}_N \exp(i\boldsymbol{\kappa} \cdot \boldsymbol{R}) \exp(-W) \\ &= \sum_{R,N} (\boldsymbol{d}_N/2) \exp(i\boldsymbol{\kappa} \cdot \boldsymbol{R}) \exp(-W), \qquad (2.2.22) \end{aligned}$$

where P_N is the nuclear polarization of nucleus N, and $\mathscr{B}_N$ is the now-familiar nuclear spin scattering parameter. This nuclear polarization contribution can sometimes be important, as we shall see in Section 4.5. It is, however, negligible for the majority of magnetic specimens, and is ignored in the remainder of this section.

Thirdly, there is a unit cell magnetic structure factor $F_M(\boldsymbol{\kappa})$, which for isolated spin-only ions may be written, following equn (2.1.77), as

$$F_M(\boldsymbol{\kappa}) = \sum_R (r_0/2)gSF(\boldsymbol{\kappa})\boldsymbol{P}\cdot\boldsymbol{q}\exp(i\boldsymbol{\kappa}\cdot\boldsymbol{R})\exp(-W), \qquad (2.2.23)$$

where $\boldsymbol{P}$ is the neutron polarization vector, the magnetic interaction geometric factor $\boldsymbol{q}$ has been reintroduced, and g is set equal to 2.

The coherent elastic scattering cross-section for neutrons with polarization $\boldsymbol{P}$ in a centrosymmetric ferromagnetic crystal becomes

$$(d\sigma/d\Omega) = \{F_N(\boldsymbol{\kappa})^2 + 2F_N(\boldsymbol{\kappa})F_M(\boldsymbol{\kappa})\boldsymbol{P}\cdot[\tilde{\boldsymbol{\kappa}}(\tilde{\boldsymbol{\eta}}\cdot\tilde{\boldsymbol{\kappa}}) - \tilde{\boldsymbol{\eta}}] + F_M(\boldsymbol{\kappa})^2[1-(\tilde{\boldsymbol{\eta}}\cdot\tilde{\boldsymbol{\kappa}})^2]\} \qquad (2.2.24)$$

and in the usual experimental geometry for polarized beam diffraction when the applied field and atomic spins are perpendicular to the scattering vector

$$\begin{aligned}(d\sigma/d\Omega) &= \{F_N(\boldsymbol{\kappa})^2 + 2F_N(\boldsymbol{\kappa})F_M(\boldsymbol{\kappa})(\boldsymbol{P}\cdot -\tilde{\boldsymbol{\eta}}) + F_M(\boldsymbol{\kappa})^2\}\\ &= \left|\sum_R \exp(i\boldsymbol{\kappa}\cdot\boldsymbol{R})\right|^2 [\bar{b}^2 + r_0 g\bar{b}F(\boldsymbol{\kappa})\langle\hat{S}\rangle(\boldsymbol{P}\cdot -\tilde{\boldsymbol{\eta}})\\ &\quad + (r_0 g/2)^2F(\boldsymbol{\kappa})^2\langle\hat{S}\rangle^2]\exp(-2W). \end{aligned} \qquad (2.2.25)$$

For positively polarized neutrons the neutron and atomic spins are oppositely directed, i.e. $\boldsymbol{P}\cdot\tilde{\boldsymbol{\eta}} = -1$, and

$$(d\sigma/d\Omega)_+ = \left|\sum_R \exp(i\boldsymbol{\kappa}\cdot\boldsymbol{R})\right|^2 [\bar{b} + (r_0 g/2)F(\boldsymbol{\kappa})\langle\hat{S}\rangle]^2\exp(-2W). \qquad (2.2.26)$$

Similarly for negatively polarized neutrons, $\boldsymbol{P}\cdot\tilde{\boldsymbol{\eta}} = +1$, and

$$(d\sigma/d\Omega)_- = \left|\sum_R \exp(i\boldsymbol{\kappa}\cdot\boldsymbol{R})\right|^2 [\bar{b} - (r_0 g/2)F(\boldsymbol{\kappa})\langle\hat{S}\rangle]^2\exp(-2W). \qquad (2.2.27)$$

In a perfectly efficient polarized beam instrument, for which the incident beam polarization and flipping efficiency are both 100%, the measured quantities are the flipping ratios $R(\boldsymbol{\kappa})$ given by

$$\begin{aligned}R(\boldsymbol{\kappa}) &= (d\sigma/d\Omega)_+/(d\sigma/d\Omega)_-\\ &= \frac{[1 + (r_0 g/2\bar{b})F(\boldsymbol{\kappa})\langle\hat{S}\rangle]^2}{[1 - (r_0 g/2\bar{b})F(\boldsymbol{\kappa})\langle\hat{S}\rangle]^2}\\ &= [(1+\gamma(\boldsymbol{\kappa}))/(1-\gamma(\boldsymbol{\kappa}))]^2 \end{aligned} \qquad (2.2.28)$$

where $\gamma(\boldsymbol{\kappa}) = F_M(\boldsymbol{\kappa})/F_N(\boldsymbol{\kappa})$ are the ratios of the magnetic to the nuclear structure factors. This relationship demonstrates the sensitivity of the polarized beam technique in measuring weak magnetic scattering, i.e. where $\gamma \ll 1$. An unpolarized beam experiment gives a peak intensity which is proportional to $b^2(1+\gamma^2)$, where b is the nuclear scattering amplitude, whereas the polarized beam experiment gives an intensity proportional to $b^2(1+2\gamma)$ for (+) spin neutrons, and a flipping ratio $R \cong 1+4\gamma$. A good example of the ultimate sensitivity achievable by using the polarized beam method is the search for the neutron electric dipole moment published by Shull and Nathans (1967). These authors reported a $\gamma = 0$ result to an accuracy of 1 part in 10^4; this accuracy being dictated by their counts statistics.

The equations we have just derived also define the main criterion required of a good polarizing crystal monochromator. We saw in Chapter 1 that one of the best polarizing crystals currently in use is $Co_{0.92}Fe_{0.08}$, for which $g = 2$, $\overline{r_0\langle \hat{S}\rangle} = \overline{(-r_0\langle \hat{\mu}_s\rangle)/2} = 0.48 \times 10^{-12}$ cm, and $\bar{b} = 0.37 \times 10^{-12}$ cm. The magnetic form factors $F(\boldsymbol{\kappa})$ for the (111) and (200) Bragg reflections in $Co_{0.92}Fe_{0.08}$ have been given by Nathans and Paoletti (1959) to be

$$F(\boldsymbol{\kappa})_{(111)} = 0.781 \quad \text{and} \quad F(\boldsymbol{\kappa})_{(200)} = 0.652.$$

Hence we calculate

$$R_{(111)} = 100.9 \quad \text{and} \quad R_{(200)} = 10820$$

for this crystal. The maximum polarizing efficiencies P_{Max} attainable using these reflections is given by

$$P_{\text{Max}} = (I_+ - I_-)/(I_+ + I_-) = (R-1)/(R+1), \tag{2.2.29}$$

and this is 99.98% for the (200) reflection, but significantly less at 98.03% for the (111) reflection.

The polarized beam diffraction experiment with single crystals provides a number of γ values for a series of Bragg reflections characterized by their (hkl) values, and with $\boldsymbol{\kappa} = \boldsymbol{\tau}$. The first stage in the data handling is to use accurately known nuclear structure factors $F_N(hkl)$ to give the magnetic structure factors $F_M(hkl)$, and hence the magnetic scattering amplitudes $\mu F(hkl)$. Normalized form factors can then be obtained by using magnetization (μ) data to calculate the zero scattering vector form factor $F(\boldsymbol{\kappa} = 0)$ which equals $(-r_0\mu/2) = (0.2695 \times 10^{-12}\mu)\ \text{cm}^{-1}$. The most revealing method of analysing the flipping ratio data is to evaluate the inverse Fourier transforms of the magnetic structure factors in precisely the same way as shown in equn (2.1.37) for crystals in which there is only nuclear scattering. This gives the unpaired electron spin density distribution $\rho(x, y, z)$ which may be expressed by

$$\rho(x, y, z) = (1/v_0) \sum_{h,k,l} F_M(hkl) \cos 2\pi[(hx/a_0) + (ky/b_0) + (lz/c_0)] \tag{2.2.30}$$

where a_0, b_0, c_0, are the unit cell dimensions, and the summation is over all the (hkl) reflections; v_0 is the unit cell volume and typical units for $\rho(x, y, z)$ are Bohr magnetons (β) per cubic angström. It is common practice to build up three-dimensional Fourier maps of the unpaired electron distribution using exactly the same techniques as used in X-ray single-crystal structural determinations.

A good illustration of the power of the polarized beam diffraction technique is provided by showing the form factor and magnetic moment distributions observed by Mook (1966) for nickel. A comparison of the measured and calculated form factors is presented in Fig. 2.7, and the magnetic moment distribution measured in the (001) plane is shown in Fig. 2.8. An important deduction from these and similar Fourier plots was that there is a region between neighbouring nickel atoms where there appears to be a constant negative spin density of about $-0.0085\beta/\text{Å}^3$; when integrated this gives a moment corresponding to 0.09 electrons per

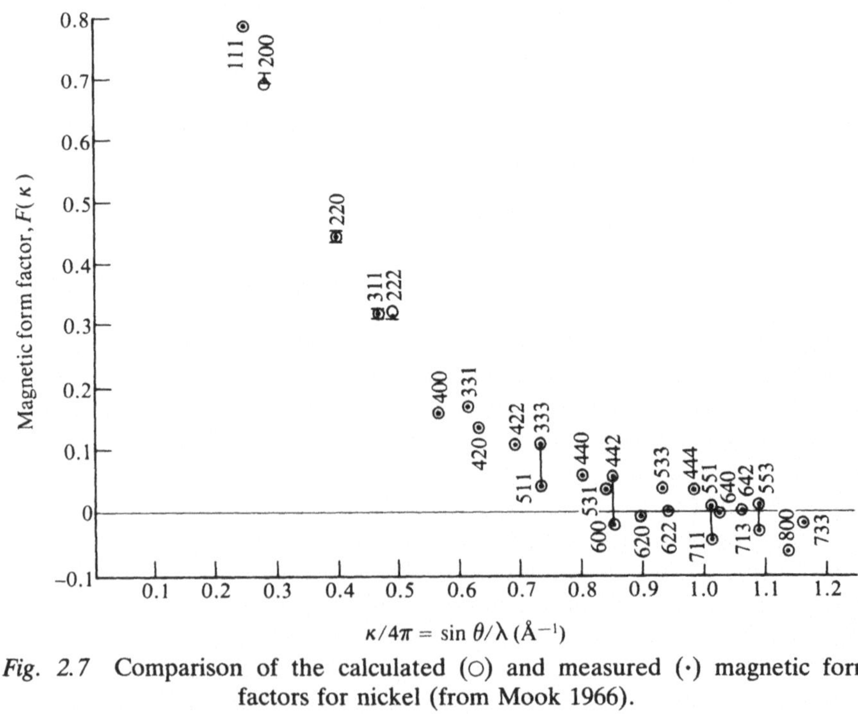

Fig. 2.7 Comparison of the calculated (○) and measured (·) magnetic form factors for nickel (from Mook 1966).

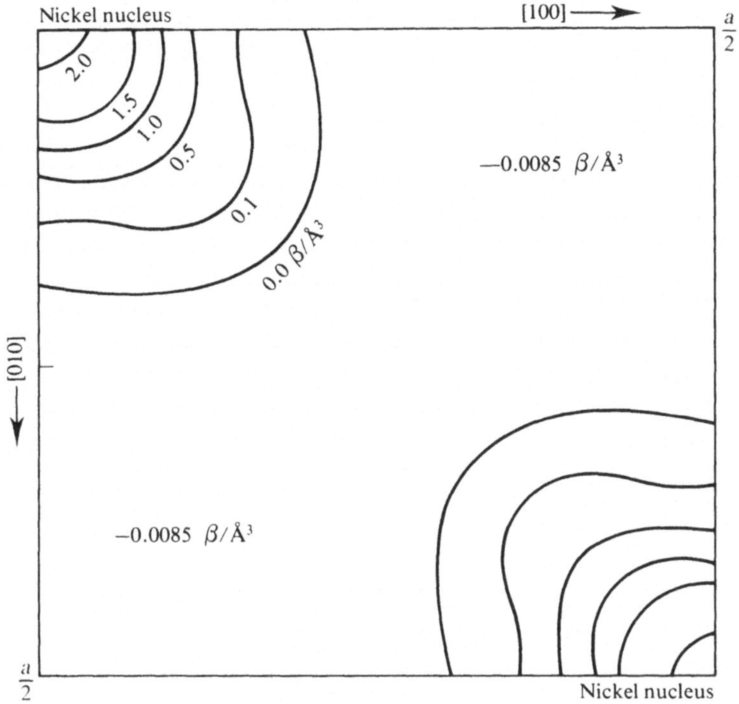

Fig. 2.8 Magnetic moment distribution in the (001) plane of the unit cell of nickel (from Mook 1966).

Ni atom. The contours are clearly not circular, and this illustrates in a very convincing way that the spin density is distributed unsymmetrically along the different crystal axes.

Before completing this section, it should be noted that the polarized beam technique is only directly applicable in crystals for which the nuclear structure is centrosymmetric. In acentric structures it can be shown (see for example Forsyth (1979)), that the measured flipping ratio does not determine the (F_M/F_N) ratio. Both F_M and F_N are complex quantities in such structures and in order to explain the measured flipping ratios it is necessary to postulate a model for the spin density, and then compare calculation with experiment. An example of a study of an acentric structure is the determination of the spin density distribution in a crystal of DPPH:C_6H_6 by Boucherle *et al.* (1982).

Some of the materials which can be investigated by the polarized beam method include ferromagnets, ferrimagnets, and paramagnets in which the atomic spins are highly oriented by an applied magnetic field. It is also being increasingly applied in structural studies of single crystals containing polarized nuclei. Examples of the results obtained with all these types of scattering samples will be presented in Chapter 4.

Amorphous magnetic solids

The determination of the partial structure factors in magnetized amorphous ferromagnets uses the fact that measurements of the cross-sections with both (+) and (−) spin neutrons give two contrasting scattering patterns; experiments analogous to those which use isotopic substitution with nuclear scatterers can therefore be performed. Further, by carrying out measurements with the applied field ($B \parallel P$) both perpendicular and parallel to the scattering vector $\boldsymbol{\kappa}$, there is sufficient information to obtain the three partial structure factors in a binary alloy without recourse to isotopic substitution. The coherent differential cross-sections for the nuclear scattering from a binary alloy, as given in equn (2.1.48), are as follows (Guoan *et al.* 1982):

$\boldsymbol{P} \perp \boldsymbol{\kappa}$:

$$(\mathrm{d}\sigma/\mathrm{d}\Omega)^{\perp}_{\pm} = c_A^2(b_A \pm p_A)^2 S_{AA}(\boldsymbol{\kappa}) + 2c_A c_B b_B (b_A \pm p_A) S_{AB}(\boldsymbol{\kappa}) + c_B^2 b_B^2 S_{BB}(\boldsymbol{\kappa}), \tag{2.2.31}$$

$\boldsymbol{P} \parallel \boldsymbol{\kappa}$:

$$(\mathrm{d}\sigma/\mathrm{d}\Omega)^{\parallel}_{+} = (\mathrm{d}\sigma/\mathrm{d}\Omega)^{\parallel}_{-} = c_A^2 b_A^2 S_{AA}(\boldsymbol{\kappa}) + 2c_A c_B b_A b_B S_{AB}(\boldsymbol{\kappa}) + c_B^2 b_B^2 S_{BB}(\boldsymbol{\kappa}). \tag{2.2.32}$$

All the terms, other than the magnetic scattering length for the magnetic atom A, where

$$p_A = -(r_0/2)\mu_A F_A(\boldsymbol{\kappa}) \tag{2.2.33}$$

are as given previously. The method was pioneered by Blétry and Sadoc (1975), and representative results are described in Section 4.3.

Magnetic diffuse scattering in magnetized alloys

We again adopt as our example the binary magnetic alloy discussed in Section 2.1.2, and we also assume that the diffuse scattering follows the simple Shull and Wilkinson model. The diffuse cross-section in the polarized beam experiment contains the same term as given by equn (2.1.88), but there is also, see Williams and Rainford (1974), an additional polarization-dependent term

$$(\mathrm{d}\sigma/\mathrm{d}\Omega)^{\mathrm{Pol}}_{\pm} = \mp r_0 c(1-c)[\tilde{\boldsymbol{\kappa}}(\tilde{\boldsymbol{\eta}} \cdot \tilde{\boldsymbol{\kappa}}) - \tilde{\boldsymbol{\eta}}](b_A - b_B)[F_A(\boldsymbol{\kappa})\mu_A - F_B(\boldsymbol{\kappa})\mu_B], \tag{2.2.34}$$

where c is the fractional concentration of the A atoms.

In the conventional geometry, when the magnetic field is applied perpendicular to the scattering vector, this additional cross-section becomes

$$(\mathrm{d}\sigma/\mathrm{d}\Omega)^{\mathrm{Pol}}_{\pm} = \mp r_0 c(1-c)(b_{\mathrm{A}}-b_{\mathrm{B}})[F_{\mathrm{A}}(\boldsymbol{\kappa})\mu_{\mathrm{A}} - F_{\mathrm{B}}(\boldsymbol{\kappa})\mu_{\mathrm{B}}]. \quad (2.2.35)$$

There are many more elaborate expressions for this diffuse cross-section, some of which are described with experimental results in Chapter 4. The main conclusions from equn (2.2.35), however, are general to them all and are: (i) that the polarized beam method, in contrast to the unpolarized beam experiment, allows the sign of $(\mu_{\mathrm{A}}-\mu_{\mathrm{B}})$ to be determined, and (ii) that the essential requirement to obtain a good diffuse scattering difference signal $[(\mathrm{d}\sigma/\mathrm{d}\Omega)_{+} - (\mathrm{d}\sigma/\mathrm{d}\Omega)_{-}]$ is that there should be a good contrast between the two coherent nuclear scattering amplitudes b_{A} and b_{B}.

2.3 POLARIZATION ANALYSIS

The previous sections have amply demonstrated how polarizing the incident beam in a neutron scattering experiment can produce new and detailed information on the scattering system in comparison with unpolarized beam methods. A further extension allows us to use a neutron polarizer in the scattered beam, and such experiments provide information on how the scattering process changes the neutron spin; these measurements are commonly referred to as polarization analysis. We should, however, start by repeating the warning of Moon *et al.* (1969a) who pioneered this experimental technique. In Halpern and Johnson's remarkably prophetic 1939 paper the neutron spin polarization was described as a vector, and its change on scattering required a vector–vector relationship. Strictly then, the term 'polarization analysis' implies that a polarization vector is measured. 'Three-dimensional' polarization analysis experiments can now be performed, though this was not the case when Moon *et al.* (1969a) introduced their technique. To quote from their original paper these authors stated, 'We do not measure polarization at all. Perhaps a more descriptive phrase might be neutron spin spectroscopy.' They in fact measured a scalar quantity, the single component of the beam polarization which was parallel to the incident beam polarization, the magnetic field applied to the sample, and the field applied at the analyser; their technique is more accurately described as 'one-dimensional polarization analysis'. Despite this restriction, the method has provided a powerful tool for identifying various forms of magnetic and nuclear scattering, and its description as simply 'polarization analysis' now seems to be established.

The main aim of this section is to describe the theoretical basis for understanding polarization analysis experiments, and to combine this with selective and unambiguous illustrative examples; a fuller presentation of experimental results is delayed until Chapter 4. However, in order to provide a greater appreciation of the significance of the derived spin-dependent cross-sections, it will again first be fruitful to describe the essential features of the polarization analysis experiment.

As previously, we shall use the sign (+) to denote the neutron spin-up state with eigenvalue +1 for the operator $\hat{\sigma}_z$, and (−) for the spin-down state with eigenvalue −1. In the most general experiment on a steady-state source, a monochromatic polarized neutron beam is scattered by the sample, and the scattered neutrons are analysed in energy, wave vector, and spin. The spin (or polarization) analysis enables the four cross-sections which connect all possible neutron spin states before and after the scatterer to be evaluated. The four spin-dependent or partial cross-sections are related to the cross-section measured in the unpolarized beam experiment as follows:

$$(\mathrm{d}^2\sigma/\mathrm{d}\Omega\,\mathrm{d}E') = \tfrac{1}{2}[(\mathrm{d}^2\sigma/\mathrm{d}\Omega\,\mathrm{d}E')_{++} + (\mathrm{d}^2\sigma/\mathrm{d}\Omega\,\mathrm{d}E')_{--} + (\mathrm{d}^2\sigma/\mathrm{d}\Omega\,\mathrm{d}E')_{+-} + (\mathrm{d}^2\sigma/\mathrm{d}\Omega\,\mathrm{d}E')_{-+}], \quad (2.3.1)$$

and the factor $\frac{1}{2}$ denotes that in the unpolarized experiment the (+) and (−) spin neutrons occur with equal probabilities of $\frac{1}{2}$. The (+ +) and (− −) cross-sections are commonly referred to as non-spin-flip (NSF) events, and the (+ −) and (− +) cross-sections as spin-flip (SF) events. The partial cross-sections equate to the 'polarized beam' cross-sections as follows:

$$(\mathrm{d}^2\sigma/\mathrm{d}\Omega\,\mathrm{d}E')_{+} = (\mathrm{d}^2\sigma/\mathrm{d}\Omega\,\mathrm{d}E')_{++} + (\mathrm{d}^2\sigma/\mathrm{d}\Omega\,\mathrm{d}E')_{+-} \quad (2.3.2)$$

and

$$(\mathrm{d}^2\sigma/\mathrm{d}\Omega\,\mathrm{d}E')_{-} = (\mathrm{d}^2\sigma/\mathrm{d}\Omega\,\mathrm{d}E')_{--} + (\mathrm{d}^2\sigma/\mathrm{d}\Omega\,\mathrm{d}E')_{-+}. \quad (2.3.3)$$

Figure 2.9 shows a diagram of the basic experimental arrangement, as used by Moon *et al.* (1969a). In the original instrument, magnetized cobalt–iron monochromating/polarizing crystals were used as the polarizer and spin analyser. The scattering sample was held in a magnetic field which defines the polarization direction of the incident and scattered neutrons. In order to measure the four partial cross-sections, neutron spin-flippers, which reverse the polarization direction, are required both before (efficiency f_1) and after (efficiency f_2) the scattering sample. The polarizer (efficiency P) and spin analyser (efficiency P_A) may be considered as devices which reflect predominantly either (+) or (−) spin neutrons, and the experiment involves counting the following intensities $I_{\sigma\sigma'}$ in the four possible instrument configurations:

I_{++}: Spin-flipper 1 'off', spin-flipper 2 'off'.
I_{-+}: Spin-flipper 1 'on', spin-flipper 2 'off'.
I_{+-}: Spin-flipper 1 'off', spin-flipper 2 'on'.
I_{--}: Spin-flipper 1 'on', spin-flipper 2 'on'.

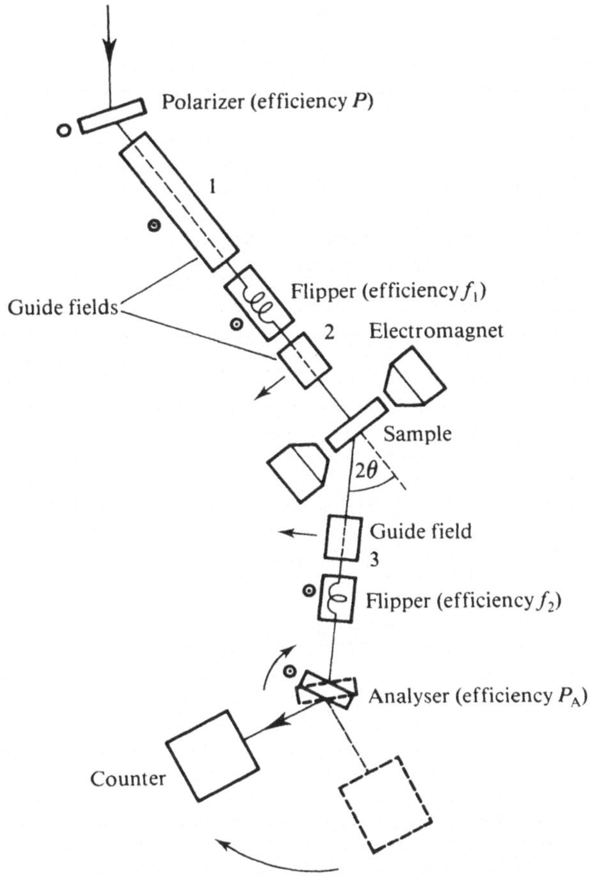

Fig. 2.9 Basic layout of the polarization analysis triple-axis spectrometer originated by Moon *et al.* 1969a).

For perfect polarizers and spin flippers, the four intensities are directly proportional to the corresponding partial cross-sections. In practice, component inefficiencies do occur, and can be allowed for using methods to be described in Section 5.1.1. Many experiments only require the spin-flip and non-spin-flip scattering processes to be separated, and for these only one spin-flipper, often though not necessarily the one in the incident beam, is required.

To take account of the polarization of both the incident and scattered beams, we need to define a set of scattering amplitudes U in terms of the Pauli spin matrices for the neutron spin. For a single spin-only magnetic atom containing a nucleus with spin i, the most general expression for this amplitude is

$$u = (u'\,d')\,\{b - (|r_0|\,g/2)F(\boldsymbol{\kappa})[\tilde{\boldsymbol{\eta}} - (\tilde{\boldsymbol{\eta}}\cdot\tilde{\boldsymbol{\kappa}})\tilde{\boldsymbol{\kappa}}]\langle\hat{S}\rangle\cdot\hat{\boldsymbol{\sigma}} + (\mathscr{B}/2)\mathbf{i}\cdot\hat{\boldsymbol{\sigma}}\}\begin{pmatrix}u\\d\end{pmatrix}$$

$$= (u'\,d')\,\{b + [-m\mathbf{S}_\perp + (\mathscr{B}/2)i\tilde{\boldsymbol{i}}]\cdot\hat{\boldsymbol{\sigma}}\}\begin{pmatrix}u\\d\end{pmatrix} \qquad (2.3.4)$$

where $\hat{\boldsymbol{\sigma}}$ is the Pauli spin operator for the neutron spin and the matrix elements u, d and u', d' refer to the spin states of the incident and scattered beams, i.e. either as 'up' or 'down'. Note that we now use the magnitude of the constant r_0, rather than explicitly writing its negative sign. In the condensed notation we have used $\boldsymbol{S}_\perp$ to denote the atomic spin component perpendicular to $\boldsymbol{\kappa}$, i.e. the active component in magnetic scattering, and m is defined by

$$m = (|r_0|/2)g\,F(\boldsymbol{\kappa}). \qquad (2.3.5)$$

$\tilde{\boldsymbol{i}}$ is a unit vector in the direction of the nuclear spin. We further define a vector $\boldsymbol{A}$ by

$$\boldsymbol{A} = -m\boldsymbol{S}_\perp + (\mathscr{B}/2)i\tilde{\boldsymbol{i}} \qquad (2.3.6)$$

to obtain

$$U = (u'\,d')(b + \boldsymbol{A}\cdot\hat{\boldsymbol{\sigma}})\begin{pmatrix}u\\d\end{pmatrix}. \qquad (2.3.7)$$

This may be rewritten in terms of the matrix components of the Pauli spin operator as follows:

$$U = (u'\,d')\left\{b + A_X\begin{pmatrix}0 & 1\\1 & 0\end{pmatrix} + A_Y\begin{pmatrix}0 & \sqrt{(-1)}\\\sqrt{(-1)} & 0\end{pmatrix} + A_Z\begin{pmatrix}1 & 0\\0 & -1\end{pmatrix}\right\}\begin{pmatrix}u\\d\end{pmatrix} \qquad (2.3.8)$$

or

$$\begin{aligned}U = {} & b(u'u + d'd) + A_X(u'd + ud')\\ & + \sqrt{(-1)}A_Y(-u'd + ud') + A_Z(u'u - d'd),\end{aligned} \qquad (2.3.9)$$

where the reference direction Z is parallel to that of the neutron polarization. The four spin-dependent scattering amplitudes are then given by

$$\begin{aligned}U^{++} &= b + A_Z && (u = u' = 1;\ d = d' = 0)\\ U^{--} &= b - A_Z && (d = d' = 1;\ u = u' = 0)\\ U^{+-} &= A_X + \sqrt{(-1)}A_Y && (u = d' = 1;\ d = u' = 0)\\ U^{-+} &= A_X - \sqrt{(-1)}A_Y && (d = u' = 1;\ u = d' = 0)\end{aligned}, \qquad (2.3.10)$$

or, after resubstituting equn (2.3.6),

$$\begin{aligned} U^{++} &= b - mS_{\perp Z} + (\mathcal{B}/2)i_Z \\ U^{--} &= b + mS_{\perp Z} - (\mathcal{B}/2)i_Z \\ U^{+-} &= -m(S_{\perp X} + \surd(-1)S_{\perp Y}) + (\mathcal{B}/2)(i_X + \surd(-1)i_Y) \\ U^{-+} &= -m(S_{\perp X} - \surd(-1)S_{\perp Y}) + (\mathcal{B}/2)(i_X - \surd(-1)i_Y). \end{aligned} \tag{2.3.11}$$

We now apply these equations to determine the spin-dependences of the scattering for a variety of important practical examples.

2.3.1 Coherent and incoherent nuclear scattering

The coherent cross-sections are proportional to the squares of the average values of the amplitudes $U_j^{\sigma\sigma'}$, summed over all j. The (+ +) cross-section, for example, is

$$(\overline{U_j^{++}})^2 = [\langle(b_j + (\mathcal{B}_j/2)i_{jZ})_{\text{sp}}\rangle_{\text{iso}}]^2 \tag{2.3.12}$$

where $(\ )_{\text{sp}}$ denotes an average over the spin states of the nuclei in the scatterer, and $\langle\ \rangle_{\text{iso}}$ denotes an average over isotopes. For most scattering samples the nuclear spins are randomly oriented, that is

$$(i_{jX})_{\text{sp}} = (i_{jY})_{\text{sp}} = (i_{jZ})_{\text{sp}} = 0 \tag{2.3.13}$$

and we find

$$|\overline{U^{++}}|^2 = |\overline{U^{--}}|^2 = \langle b_j\rangle^2_{\text{iso}}, \tag{2.3.14}$$

$$|\overline{U^{+-}}|^2 = |\overline{U^{-+}}|^2 = 0. \tag{2.3.15}$$

The nuclear coherent scattering is therefore always non-spin-flip, and the same result also holds for the isotopic incoherent scattering.

Next we evaluate the incoherent scattering cross-sections which are proportional to $\overline{(U^{\sigma\sigma'})^2} - (\overline{U^{\sigma\sigma'}})^2$. As an example we again consider the (+ +) partial cross-section

$$\begin{aligned} \overline{(U^{++})^2} - (\overline{U^{++}})^2 &= \langle([b_j + (\mathcal{B}/2)i_{jZ}]^2)_{\text{sp}}\rangle_{\text{iso}} - \langle b_j\rangle^2_{\text{iso}} && (2.3.16a) \\ &= \langle b_j^2\rangle_{\text{iso}} - \langle b_j\rangle^2_{\text{iso}} + \tfrac{1}{4}\langle\mathcal{B}_j^2(i_{jZ})^2_{\text{sp}}\rangle_{\text{iso}} \\ &\quad + \langle b_j\mathcal{B}_j(i_{jZ})_{\text{sp}}\rangle_{\text{iso}} && (2.3.16b) \\ &= \langle b_j^2\rangle_{\text{iso}} - \langle b_j\rangle^2_{\text{iso}} + \tfrac{1}{12}\langle\mathcal{B}_j^2 i(i+1)\rangle_{\text{iso}}. && (2.3.16c) \end{aligned}$$

The latter expression follows from equn (2.3.13) and also

$$(i_{jX})^2_{\text{sp}} = (i_{jY})^2_{\text{sp}} = (i_{jZ})^2_{\text{sp}} = i(i+1)/3 \tag{2.3.17}$$

for randomly oriented nuclear spins. The other non-spin-flip cross-section $\overline{(U^{--})^2} - (\overline{U^{--}})^2$ gives an identical result, whereas the spin-flip cross-sections are

$$\overline{(U^{+-})^2} - (\overline{U^{+-}})^2 = \overline{(U^{-+})^2} - (\overline{U^{-+}})^2 = \langle\mathcal{B}_j^2 i(i+1)\rangle_{\text{iso}}/6 \tag{2.3.18}$$

We note that the total spin incoherent differential cross-section per atom is $\mathscr{B}^2 i(i+1)/4$ for a scatterer containing a single isotope, and that this agrees with equns (2.1.27) and (2.2.5). The spin-dependent elastic cross-sections in this case are

$$(\mathrm{d}\sigma/\mathrm{d}\Omega)_{++} = (\mathrm{d}\sigma/\mathrm{d}\Omega)_{--} = \mathscr{B}^2\, i(i+1)/12 \qquad (2.3.19)$$

and

$$(\mathrm{d}\sigma/\mathrm{d}\Omega)_{+-} = (\mathrm{d}\sigma/\mathrm{d}\Omega)_{-+} = \mathscr{B}^2\, i(i+1)/6. \qquad (2.3.20)$$

If the incident beam polarization is unity, then the scattered beam polarization is

$$P' = \frac{(\mathscr{B}^2 i(i+1)/12) - (\mathscr{B}^2 i(i+1)/6)}{(\mathscr{B}^2 i(i+1)/12) + (\mathscr{B}^2 i(i+1)/6)} = -\frac{1}{3} \qquad (2.3.21)$$

and it is readily shown that the more general result for an incident beam polarization $\boldsymbol{P}$ is

$$\boldsymbol{P}' = -\boldsymbol{P}/3. \qquad (2.3.22)$$

The different spin dependences of the nuclear coherent and spin incoherent scattering means that these two types of scattering can be separated by polarization analysis. This is useful for instance in structural studies of hydrogenous and deuterated liquids, when the potentially simpler experiment with a polarized beam and a polarized nuclear sample is not possible because high nuclear polarizations can only be produced at low temperatures ($\ll$1 K) in solids. Multiple scattering of course complicates the analysis of data, since, for example, twice-scattered neutrons from an incoherent scatterer have a final polarization $+P/9$. The dependence of the scattered-beam polarization on multiple scattering has been applied in an experimental test of multiple scattering calculations in ^{51}V (Freeman *et al.* 1978).

Figure 2.10 shows the results of the elastic polarization analysis experiment reported by Moon *et al.* (1969a) on ^{51}V, which is nearly 100% naturally abundant, and is to a good approximation a pure spin incoherent scatterer. The flipper 'on' or spin-flip intensity is twice as large as the flipper 'off' or non spin-flip intensity, as predicted by the above equations. This result is independent of the polarization direction with respect to the scattering vector, showing that the magnetic scattering in vanadium is negligibly small.

2.3.2 Paramagnetic scattering

First we state a general conclusion from equn (2.3.11) which is valid for all forms of magnetic scattering. The equation for the magnetic interaction vector $\boldsymbol{q}$ shows that magnetic scattering is only possible for those atomic spin components which are perpendicular to the scattering vector

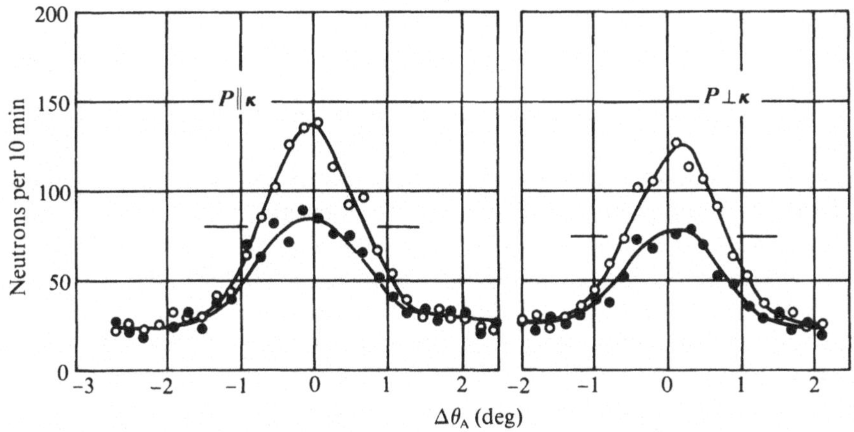

Fig. 2.10 Polarization analysis of vanadium scattering showing that the spin-flip scattering observed with the spin-flipper 'on' (○) is twice the non-spin-flip (●) scattering (from Moon *et al.* 1969a).

$\boldsymbol{\kappa}$. Thus, if the neutron polarization is along the scattering vector, $S_{\perp z} = 0$, and all the magnetic scattering is spin-flip scattering. This is true for all types of magnetic scattering: coherent, incoherent, elastic and inelastic.

We now consider the elastic scattering from the unpaired electrons of spin-only paramagnetic ions ($g = 2$) which are localized in a crystal lattice. If we denote the $(+\,+)$ and $(-\,-)$ partial cross-sections as NSF or non-spin-flip cross-sections, and the $(+\,-)$ and $(-\,+)$ cross-sections as SF or spin-flip cross-sections, the total cross-section equn (2.1.86) is, for neutron polarization $\boldsymbol{P} = 0$, given by

$$(\mathrm{d}\sigma/\mathrm{d}\Omega)_{\mathrm{NSF}} + (\mathrm{d}\sigma/\mathrm{d}\Omega)_{\mathrm{SF}} = \tfrac{2}{3} r_0^2 (g/2)^2\, F(\boldsymbol{\kappa})^2 S(S+1). \quad (2.3.23)$$

The partial cross-sections in terms of the relative orientations of $\tilde{\boldsymbol{\kappa}}$ and $\tilde{\boldsymbol{P}}$ are

$$(\mathrm{d}\sigma/\mathrm{d}\Omega)_{\mathrm{NSF}} = \tfrac{1}{3} r_0^2 (g/2)^2 F(\boldsymbol{\kappa})^2 S(S+1)[1 - (\tilde{\boldsymbol{P}} \cdot \tilde{\boldsymbol{\kappa}})^2] \quad (2.3.24)$$

and

$$(\mathrm{d}\sigma/\mathrm{d}\Omega)_{\mathrm{SF}} = \tfrac{1}{3} r_0^2 (g/2)^2 F(\boldsymbol{\kappa})^2 S(S+1)[1 + (\tilde{\boldsymbol{P}} \cdot \tilde{\boldsymbol{\kappa}})^2] \quad (2.3.25)$$

where $\tilde{\boldsymbol{P}}$ and $\tilde{\boldsymbol{\kappa}}$ are unit vectors in the neutron polarization and scattering vector directions.

A further result is that for an incident beam polarization $P = 1$, the scattered beam polarization P' is

$$P' = \left[\frac{(1 - (\tilde{\boldsymbol{P}} \cdot \tilde{\boldsymbol{\kappa}})^2) - (1 + (\tilde{\boldsymbol{P}} \cdot \tilde{\boldsymbol{\kappa}})^2)}{(1 - (\tilde{\boldsymbol{P}} \cdot \tilde{\boldsymbol{\kappa}})^2) + (1 + (\tilde{\boldsymbol{P}} \cdot \tilde{\boldsymbol{\kappa}})^2)}\right]$$
$$= -(\tilde{\boldsymbol{P}} \cdot \tilde{\boldsymbol{\kappa}})^2 \quad (2.3.26)$$

and for incident polarization $\boldsymbol{P}$ this becomes

$$\boldsymbol{P}' = -(\tilde{\boldsymbol{P}} \cdot \tilde{\boldsymbol{\kappa}})^2 \boldsymbol{P} \tag{2.3.27}$$

This expression was first derived by Halpern and Johnson in their 1939 paper.

The $\tilde{\boldsymbol{P}} \cdot \tilde{\boldsymbol{\kappa}}$ dependence of the magnetic scattering provides a convenient method of using polarization analysis to measure paramagnetic cross-sections in unmagnetized samples. When the applied field or $\boldsymbol{P}$ direction at the scatterer is directed along the scattering vector, i.e. $\tilde{\boldsymbol{P}} \cdot \tilde{\boldsymbol{\kappa}} = 1$, the spin-flip cross-section equals the total paramagnetic cross-section as observed in the unpolarized beam experiment. In the $(\tilde{\boldsymbol{P}} \cdot \tilde{\boldsymbol{\kappa}}) = 0$ geometry, both the spin-flip and non-spin-flip cross-sections are equal, i.e. each are half the value of the total paramagnetic cross-section.

The isolation of paramagnetic scattering by polarization analysis is an important application of the technique, since this cannot readily be separated from isotopic disorder, multiple Bragg, thermal diffuse, Bragg or spin incoherent scattering by other neutron scattering methods. A full separation is possible by performing experiments in both the $\tilde{\boldsymbol{P}} \cdot \tilde{\boldsymbol{\kappa}} = 1$ and $\tilde{\boldsymbol{P}} \cdot \tilde{\boldsymbol{\kappa}} = 0$ geometries. For a paramagnetic specimen which contains a nucleus with a spin incoherent scattering cross-section (NS), the experiment in the $\tilde{\boldsymbol{P}} \cdot \tilde{\boldsymbol{\kappa}} = 1$ geometry gives

$$(\mathrm{d}\sigma/\mathrm{d}\Omega)_{\mathrm{SF}} = (\mathrm{d}\sigma/\mathrm{d}\Omega)_{\mathrm{para}} + \tfrac{2}{3}(\mathrm{d}\sigma/\mathrm{d}\Omega)_{\mathrm{NS}}, \tag{2.3.28}$$

whereas, in the $\tilde{\boldsymbol{P}} \cdot \tilde{\boldsymbol{\kappa}} = 0$ geometry

$$(\mathrm{d}\sigma/\mathrm{d}\Omega)_{\mathrm{SF}} = \tfrac{1}{2}(\mathrm{d}\sigma/\mathrm{d}\Omega)_{\mathrm{para}} + \tfrac{2}{3}(\mathrm{d}\sigma/\mathrm{d}\Omega)_{\mathrm{NS}}. \tag{2.3.29}$$

Measurements taken in both of these geometries therefore allow an unambiguous separation of these cross-sections. In many samples the spin incoherent scattering cross-section is well known, or may in fact be zero, and in these instances only one experiment is needed to determine $(\mathrm{d}\sigma/\mathrm{d}\Omega)_{\mathrm{para}}$.

Figure 2.11 shows the paramagnetic scattering measurements of Moon *et al.* (1969a) for MnF_2. The results were obtained by scanning over the elastic peak using the polarization analysis triple-axis spectrometer described earlier. In the $\boldsymbol{P} \parallel \boldsymbol{\kappa}$ geometry, the full paramagnetic scattering cross-section is observed with the spin-flipper activated. The non-spin-flip or flipper 'off' paramagnetic cross-section in this geometry is zero. In the $\boldsymbol{P} \perp \boldsymbol{\kappa}$ geometry, the spin-flip scattering represents half the paramagnetic cross-section, as does the non-spin-flip or flipper 'off' intensity. The small signal observed in the $\boldsymbol{P} \parallel \boldsymbol{\kappa}$ experiment with the spin-flipper 'on' could have been caused by multiple Bragg scattering in the sample, or might simply have been due to the imperfect efficiencies of the instrument's polarizers and spin-flipper.

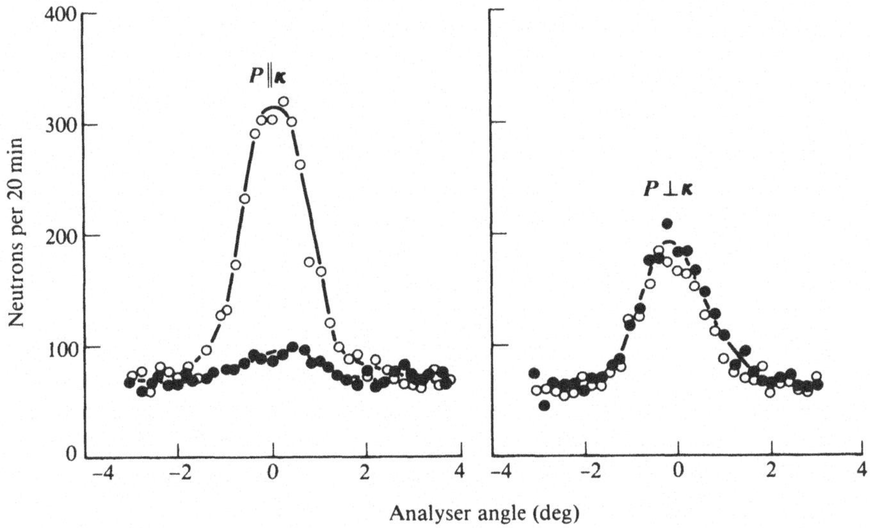

Fig. 2.11 Polarization analysis of the paramagnetic scattering from MnF_2; ●, flipper off; ○, flipper on (from Moon *et al.* 1969a).

2.3.3 Nuclear and magnetic Bragg scattering

Before proceeding further it will be convenient to express the generalized vector

$$\boldsymbol{A} = -m\boldsymbol{S}_{\perp} + (\mathscr{B}/2)\boldsymbol{i} \tag{2.3.30}$$

introduced in equn (2.3.6) into a form which more readily reveals the cross-sections for Bragg scattering. We define in a new condensed notation, first the matrix elements

$$\langle b_i \rangle = \langle \lambda' | \hat{b}_i | \lambda \rangle, \tag{2.3.31}$$

$$\langle A_{i\alpha} \rangle = \langle \lambda' | \hat{A}_{i\alpha} | \lambda \rangle, \tag{2.3.32}$$

and the scattering probabilities

$$W^{\sigma\sigma'} = \left| \langle \lambda' | \sum_i \exp(\mathrm{i}\boldsymbol{\kappa} \cdot \boldsymbol{r}_i) U_i^{\sigma\sigma'} | \lambda \rangle \right|^2, \tag{2.3.33}$$

where these parameters characterize a scattering system which changes from state λ to λ' after scattering the neutron with a double differential cross-section

$$(\mathrm{d}^2\sigma/\mathrm{d}\Omega\, \mathrm{d}E') = \sum_{\lambda} p_{\lambda} \sum_{\lambda'} (k'/k) W^{\sigma\sigma'} \delta(\hbar\omega + E_{\lambda} - E_{\lambda'}). \tag{2.3.34}$$

The delta function merely expresses the energy conservation condition.

By expanding the double sum in equn (2.3.33), we obtain

$$W^{\pm\pm} = \sum_{i,j} \exp[\mathrm{i}\boldsymbol{\kappa}\cdot(\boldsymbol{r}_i - \boldsymbol{r}_j)](\langle b_i\rangle\langle b_j^*\rangle \pm \langle b_i\rangle\langle A_{jZ}^*\rangle$$
$$\pm \langle A_{iZ}\rangle\langle b_j^*\rangle + \langle A_{iZ}\rangle\langle A_{jZ}^*\rangle) \qquad (2.3.35)$$

and

$$W^{\pm\mp} = \sum_{i,j} \exp[\mathrm{i}\boldsymbol{\kappa}\cdot(\boldsymbol{r}_i - \boldsymbol{r}_j)](\langle A_{iX}\rangle\langle A_{jX}^*\rangle + \langle A_{iY}\rangle\langle A_{iY}^*\rangle$$
$$\mp \sqrt{(-1)}\tilde{Z}\cdot(\langle \boldsymbol{A}_i\rangle \times \langle \boldsymbol{A}_j^*\rangle)). \qquad (2.3.36)$$

It is now possible to evaluate all the partial cross-sections, $(\mathrm{d}^2\sigma/\mathrm{d}\Omega\,\mathrm{d}E')_{++}$, etc. These can then be combined to give general expressions for the total double differential cross-section, and also the scattered beam polarization P'. For clarity, we abbreviate $(\mathrm{d}^2\sigma/\mathrm{d}\Omega\,\mathrm{d}E')$ by σ. We recall from Section 1.4 that if ϕ_+ and ϕ_- are the fractions of spin 'up' and spin 'down' neutrons in a beam of polarization P, which is now the beam incident on the scatterer, then

$$\phi_+ + \phi_- = 1, \qquad (2.3.37)$$
$$\phi_+ - \phi_- = P. \qquad (2.3.38)$$

Hence

$$\phi_+ = \tfrac{1}{2}(1 + P) \qquad (2.3.39)$$

and

$$\phi_- = \tfrac{1}{2}(1 - P). \qquad (2.3.40)$$

The total cross-section in this notation is

$$\sigma = \phi_+(\sigma_{++} + \sigma_{+-}) + \phi_-(\sigma_{-+} + \sigma_{--}) \qquad (2.3.41)$$

and the scattered beam polarization is

$$P' = \frac{(\phi_+\sigma_{++} + \phi_-\sigma_{-+}) - (\phi_-\sigma_{--} + \phi_+\sigma_{+-})}{(\phi_+\sigma_{++} + \phi_-\sigma_{-+}) + (\phi_-\sigma_{--} + \phi_+\sigma_{+-})}, \qquad (2.3.42)$$

i.e.

$$\sigma P' = \phi_+\sigma_{++} + \phi_-\sigma_{-+} - \phi_-\sigma_{--} - \phi_+\sigma_{+-}. \qquad (2.3.43)$$

After substituting into equn (2.3.34), we find

$$(\mathrm{d}^2\sigma/\mathrm{d}\Omega\,\mathrm{d}E') = \sum_{\lambda\lambda'} p_\lambda(k'/k)\Big\{\sum_{i,j} \exp[\mathrm{i}\boldsymbol{\kappa}\cdot(\boldsymbol{r}_i - \boldsymbol{r}_j)][\langle b_i\rangle\langle b_j^*\rangle + \langle \boldsymbol{A}_i\rangle\cdot\langle \boldsymbol{A}_j\rangle$$
$$+\boldsymbol{P}\cdot(\langle b_i\rangle\langle \boldsymbol{A}_j^*\rangle + \langle b_j^*\rangle\langle \boldsymbol{A}_i\rangle - \sqrt{(-1)}\langle \boldsymbol{A}_i\rangle \times \langle \boldsymbol{A}_j\rangle)]$$
$$\times\, \delta(\hbar\omega + E_\lambda - E_{\lambda'}), \qquad (2.3.44)$$

where $\boldsymbol{P} = P\tilde{Z}$, and the scattered beam polarization P' in the Z direction is

$$P'(\mathrm{d}^2\sigma/\mathrm{d}\Omega\,\mathrm{d}E') = \sum_{\lambda\lambda'} p_\lambda(k'/k)\Big\{\sum_{i,j} \exp[\mathrm{i}\boldsymbol{\kappa}\cdot(\boldsymbol{r}_i - \boldsymbol{r}_j)]$$
$$\times [\tilde{Z}\cdot(\langle b_i\rangle\langle \boldsymbol{A}_j^*\rangle + \langle b_j^*\rangle\langle \boldsymbol{A}_i\rangle$$
$$+ \sqrt{(-1)}\langle \boldsymbol{A}_i\rangle \times \langle \boldsymbol{A}_j^*\rangle) + P(\langle b_i\rangle\langle b_j^*\rangle + \langle A_{iZ}\rangle\langle A_{jZ}^*\rangle$$
$$- \langle A_{iX}\rangle\langle A_{jX}^*\rangle - \langle A_{iY}\rangle\langle A_{jY}^*\rangle)] \cdot \delta(\hbar\omega + E_\lambda - E_{\lambda'}).$$
$$(2.3.45)$$

In this last equation the first two terms represent the familiar interference between the nuclear and magnetic scattering, and we have seen several examples of how this can be utilized to produce polarized neutron beams. The third term also creates polarization, as, for example, in inelastic magnon scattering, or in elastic scattering from spiral magnetic structures. The effect of all three terms can be conveniently summarized by stating that, in order to produce polarization, the scattering system must provide its own directional axis. This occurs in a ferromagnetic, and also in more complex spiral atomic spin arrangements. The final four terms in equn (2.3.45) demonstrate again that the initial polarization P is conserved in coherent nuclear scattering and in magnetic scattering for those atomic spin components which are along the polarization direction, but that transverse atomic spin components flip the neutron spin.

Equations (2.3.44) and (2.3.45) also illustrate the fundamental difference between the conventional polarized beam method in which the total cross-section is measured with a single polarizer, and the polarization analysis experiment (albeit one-dimensional) in which the cross-section and the spin changes on scattering are measured. The first experiment is only useful for oriented systems such as ferromagnets, certain antiferromagnets, or samples containing polarized nuclei, whereas polarization analysis can usefully be employed to investigate scattering systems which have a random collection of atomic or nuclear spins.

We now apply these last two equations to illustrate the value of polarization analysis in studying coherent Bragg scattering either in polycrystals or single crystals. The discussion is simplified by restricting ourselves to examples in which there is no nuclear polarization. Reference to the amplitude equns (2.3.11) shows that nuclear peaks contain only non-spin-flip scattering, whereas magnetic peaks may contain a mixture of spin-flip and non-spin-flip scattering, depending on the orientations of the neutron polarization and the atomic moments relative to the scattering vector.

As a first example we consider the geometry in which the neutron polarization and scattering vector directions are parallel, as shown in Fig. 2.12(a). Only S_X and S_Y contribute to the magnetic scattering, which is

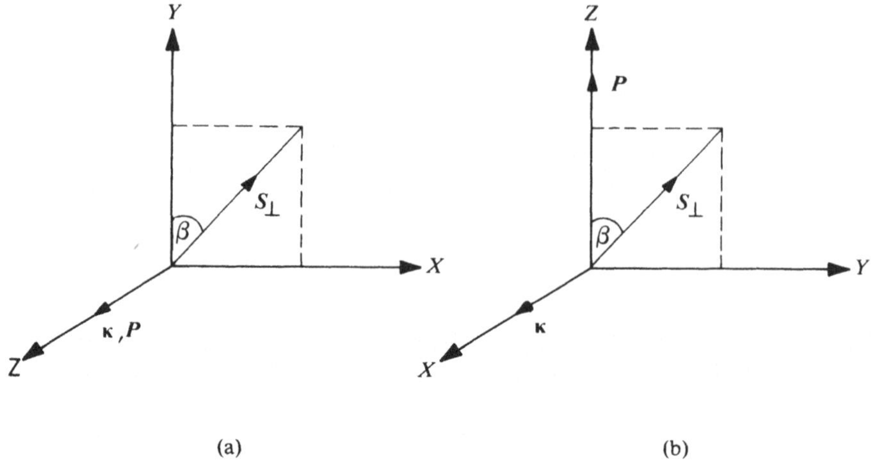

Fig. 2.12 Vector directions for elastic magnetic scattering: (a) $\boldsymbol{P}\cdot\boldsymbol{\kappa}=1$, and (b) $\boldsymbol{P}\cdot\boldsymbol{\kappa}=0$.

therefore entirely of the spin-flip type, whereas the nuclear scattering is non-spin-flip. The spin-dependent cross-sections are obtained from equns (2.3.35) and (2.3.36) and are

$$(\mathrm{d}\sigma/\mathrm{d}\Omega)_{\pm\pm}=\sum_{i,j}\exp[\mathrm{i}\boldsymbol{\kappa}\cdot(\boldsymbol{r}_i-\boldsymbol{r}_j)]\langle b_i\rangle\langle b_j^*\rangle \tag{2.3.46}$$

and

$$(\mathrm{d}\sigma/\mathrm{d}\Omega)_{\pm\mp}=\sum_{i,j}\exp[\mathrm{i}\boldsymbol{\kappa}\cdot(\boldsymbol{r}_i-\boldsymbol{r}_j)]m_i m_j^*[\boldsymbol{S}_{\perp i}\cdot\boldsymbol{S}_{\perp j}\mp\sqrt{(-1)}\tilde{Z}\cdot(\boldsymbol{S}_{\perp i}\times\boldsymbol{S}_{\perp j}^*)]. \tag{2.3.47}$$

This method is a particularly powerful one for studying antiferromagnetic structures, since it allows a complete separation of the nuclear and magnetic scattering at a single temperature, and with only a modest magnetic field ($\sim 10^{-2}$ T) at the scattering sample to guide the neutron polarization. For many antiferromagnets some of the magnetic Bragg peaks overlap nuclear peaks, and in the usual unpolarized beam experiment the nuclear peaks are measured at a temperature above the Néel point. The sample is then cooled below the Néel point and the combined nuclear-plus-magnetic intensities measured. The magnetic intensities are found by subtraction. This procedure often involves large Debye–Waller corrections, and in some cases the crystal structure may change at or below the Néel point. An isothermal method for separating the nuclear and magnetic intensities is therefore of considerable value. This technique was used, for example, by Jacobson and Fender (1975) to separate the nuclear and magnetic Bragg peaks in antiferromagnetic $BiFeO_3$.

Next we discuss the $\tilde{\boldsymbol{P}} \cdot \tilde{\boldsymbol{\kappa}} = 0$ geometry shown in Fig. 2.12(b), where we have $S_{\perp X} = 0$. The magnetic scattering can now be a mixture of the non-spin-flip and spin-flip types, each with contributions ranging from 100% to zero, depending on the direction of the atomic moments. The spin-dependent cross-sections are

$$(\mathrm{d}\sigma/\mathrm{d}\Omega)_{\pm\pm} = \sum_{i,j} \exp[\mathrm{i}\boldsymbol{\kappa} \cdot (\boldsymbol{r}_i - \boldsymbol{r}_j)][\langle b_i \rangle \langle b_j^* \rangle \mp (\langle b_i \rangle m_j^* S_{\perp Z_j} + \langle b_j^* \rangle m_i S_{\perp Zi}) + m_i m_j^* S_{\perp Zi} S_{\perp Zj}^*] \qquad (2.3.48)$$

and

$$(\mathrm{d}\sigma/\mathrm{d}\Omega)_{\pm\mp} = \sum_{i,j} \exp[\mathrm{i}\boldsymbol{\kappa} \cdot (\boldsymbol{r}_i - \boldsymbol{r}_j)] m_i m_j^* S_{\perp Yi} S_{\perp Yj}^*. \qquad (2.3.49)$$

The terms in the non-spin-flip cross-section are, in the order in which they appear, the pure nuclear scattering, the nuclear–magnetic interference, and the pure magnetic scattering. The contributing atomic spin components are all perpendicular to $\boldsymbol{\kappa}$ and parallel to $\boldsymbol{P}$. Spin-flip scattering arises from components which are perpendicular to both $\boldsymbol{\kappa}$ and $\boldsymbol{P}$. Since the atomic moments in collinear ferromagnets are all oriented along $\boldsymbol{P}$, there is zero spin-flip scattering, and polarization analysis is not generally helpful with such scatterers. In some instances it may, however, be profitable to use polarization analysis to investigate deviations from collinearity.

The $\tilde{\boldsymbol{P}} \cdot \tilde{\boldsymbol{\kappa}} = 0$ geometry is again much more beneficial in the study of antiferromagnets. Here the nuclear–magnetic interference terms are often zero, and the magnetic intensity from a single-domain collinear antiferromagnet has the following neutron spin and directional dependences:

$$I_{\pm\pm}^{(\mathrm{M})} = I_0 \cos^2 \beta, \qquad (2.3.50)$$

$$I_{\pm\mp}^{(\mathrm{M})} = I_0 \sin^2 \beta, \qquad (2.3.51)$$

where β is the angle between $\boldsymbol{S}_\perp$ and $\boldsymbol{P}$. Thus, by rotating an antiferromagnetic single crystal about the scattering vector, the magnetic intensity should change from pure non-spin-flip scattering (when $\boldsymbol{S}_\perp \parallel \boldsymbol{P}$) to pure spin-flip scattering (when $\boldsymbol{S}_\perp \perp \boldsymbol{P}$). An experiment using this principle, but with a multidomain chromium single crystal, has been carried out by Werner *et al.* (1969), and is described in Section 5.3.1.

2.3.4 Coherent excitations

The polarization analysis of coherent excitations is primarily of interest in the scattering from magnetic single-crystal specimens. Pure phonon scattering is always non-spin-flip and the spin-flip scattering therefore identifies the magnon scattering. However, there are some ordered ferromagnets, particularly those with a large average moment per atom, in which the phonons couple with the magnetic ions and the phonon peaks then exhibit a polarization dependence. The neutron spin-dependent effects expected due to magnetovibrational coupling are well understood, and we shall return to describe these later. We start this discussion by considering the magnon scattering observed in ferromagnets and antiferromagnets. The detailed derivations of the cross-section relationships are most conveniently performed using magnetic correlation functions, and are fully described by Lovesey (1986). Other major papers on this subject include those of Sáenz (1960), and Izyumov and Maleev (1962). In order to understand the essential aspects of the use of polarization analysis to investigate coherent magnetic inelastic scattering, it is sufficient to examine the geometric factors in the cross-section equations.

We first deal with ferromagnets and use the cross-section formulae given by Lovesey (1986). In a Heisenberg ferromagnet the spin wave creation (upper sign) and spin wave annihilation (lower sign) cross-sections depend on the neutron polarization $\boldsymbol{P}$ as follows

$$(\mathrm{d}^2\sigma/\mathrm{d}\Omega\,\mathrm{d}E')^{\pm} = r_0^2(k'/k)[(g/2)F(\boldsymbol{\kappa})]^2\tfrac{1}{2}S[(2\pi)^3/v_0]\sum_{\boldsymbol{q},\boldsymbol{\tau}}(n_{\boldsymbol{q}}+\tfrac{1}{2}\pm\tfrac{1}{2})$$
$$\times\,\delta(\hbar\omega\mp\hbar\omega_{\boldsymbol{q}})\,\delta(\boldsymbol{\kappa}\mp\boldsymbol{q}-\boldsymbol{\tau})[1+(\tilde{\boldsymbol{\kappa}}\cdot\tilde{\boldsymbol{\eta}})^2\mp 2(\boldsymbol{P}\cdot\tilde{\boldsymbol{\kappa}})(\tilde{\boldsymbol{\kappa}}\cdot\tilde{\boldsymbol{\eta}})], \quad (2.3.52)$$

where $\hbar\omega_{\boldsymbol{q}}$ is the magnon energy with wave vector $\boldsymbol{q}$, and $\tilde{\boldsymbol{\eta}}$ is a unit vector in the direction of the atomic spin. The behaviour of this scattering with polarized beams is most easily described if we substitute the unit vector in the direction of the magnetization, $\tilde{\boldsymbol{M}}$, for $-\tilde{\boldsymbol{\eta}}$, since it frequently happens in practice that $\tilde{\boldsymbol{M}}$ is parallel to the neutron polarization vector. The geometric factor becomes

$$(\mathrm{d}^2\sigma/\mathrm{d}\Omega\,\mathrm{d}E')^{\pm}\propto[1+(\tilde{\boldsymbol{\kappa}}\cdot\tilde{\boldsymbol{M}})^2\pm 2(\boldsymbol{P}\cdot\tilde{\boldsymbol{\kappa}})(\tilde{\boldsymbol{\kappa}}\cdot\tilde{\boldsymbol{M}})]. \quad (2.3.53)$$

The cross-section for (+) spin neutrons is then obtained with a neutron polarization vector $\boldsymbol{P}=1$, and correspondingly $\boldsymbol{P}=-1$ for (−) spin neutrons. This is the formulation used by Sáenz (1960).

Let us now examine some typical examples to which this cross-section expression can be applied. The first point to note is that with ferromagnets it is essential to magnetize the specimen to saturation so as to minimize any depolarization of the neutron beam as it traverses the magnetic domains in the sample; thus $\boldsymbol{P}\|\boldsymbol{M}$ is a prerequisite. One important consequence of this is that, since spin waves are a result of correlations between S_X and S_Y spin components, the magnetic spin-dependent amplitude equns (2.3.11) tell us that all magnon scattering in saturated ferromagnets must be spin-flip processes. This result is independent of the direction of the scattering vector $\boldsymbol{\kappa}$.

We now show that the most useful geometry for polarization analysis magnon experiments is when the scattering vector is also collinear with $\tilde{\boldsymbol{P}}$

and $\tilde{\boldsymbol{M}}$. From equn (2.3.53) we see that the spin wave creation cross-section in the one-dimensional polarization analysis experiment is given by

$$(\mathrm{d}^2\sigma/\mathrm{d}\Omega\,\mathrm{d}E')^{+} \propto [1 + (\tilde{\boldsymbol{\kappa}}\cdot\tilde{\boldsymbol{M}})^2 \pm 2P_Z(\tilde{\boldsymbol{P}}\cdot\tilde{\boldsymbol{\kappa}})(\tilde{\boldsymbol{\kappa}}\cdot\tilde{\boldsymbol{M}})] \propto 2(1 + P_Z), \quad (2.3.54)$$

where P_Z is now a scalar incident beam polarization, and the corresponding spin wave annihilation cross-section is

$$(\mathrm{d}^2\sigma/\mathrm{d}\Omega\,\mathrm{d}E')^{-} \propto 2(1 - P_Z). \quad (2.3.55)$$

Both these cross-sections are of the spin-flip type, and for fully polarized beams, where $|P_Z| = 1$, we see that finite cross-section values are only possible if P_Z is $+1$ in equn (2.3.54) and -1 in equn (2.3.55). It follows that the creation cross-section is a $(+, -)$ neutron spin-flip cross-section, whereas the annihilation cross-section is a $(-, +)$ spin-flip scattering process.

We see immediately that two types of experiment are possible in this $\tilde{\boldsymbol{P}}\cdot\tilde{\boldsymbol{\kappa}} = 1$ or $\tilde{\boldsymbol{\kappa}}\cdot\tilde{\boldsymbol{\eta}} = -1$ geometry. The first, which is the inelastic analogue of the polarized beam diffraction method, requires only one polarizer and is usually called partial polarization analysis. The method was first demonstrated for the ferrite scatterer $Li_{0.5}Fe_{2.5}O_4$ by Riste *et al.* (1968). The results of their experiment, which used an unpolarized incident beam and a positive polarization analyser ($Co_{0.92}Fe_{0.08}$) are shown in Fig. 2.13. Spin wave annihilation $(-\ +)$ is observed when the spin-flipper between

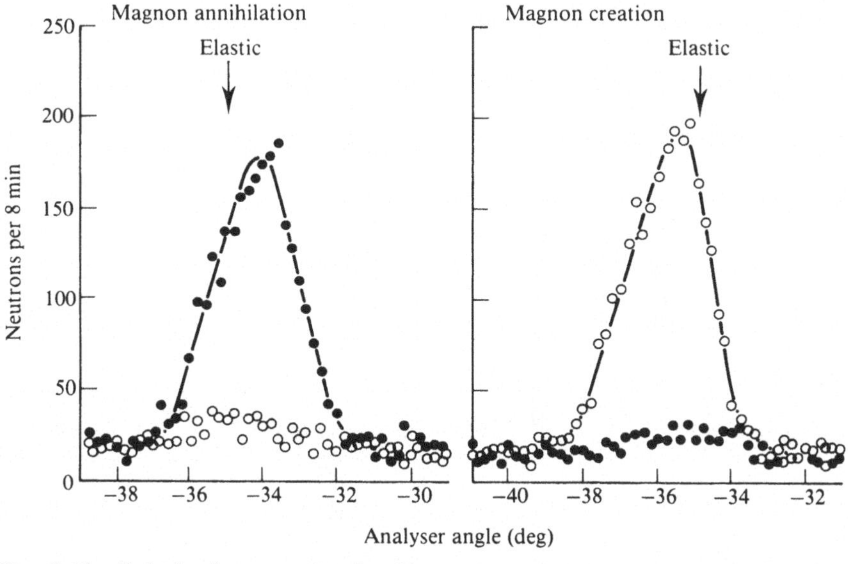

Fig. 2.13 Polarization creation in the magnon scattering from lithium ferrite. The incident beam is unpolarized, the spin-flipper is after the scatterer, and the geometry is $\tilde{\boldsymbol{\kappa}}\cdot\tilde{\boldsymbol{\eta}} = -1$; ●, flipper off; ○, flipper on (from Riste *et al.* 1968).

the scatterer and analyser is 'off', whereas spin wave creation (+ −) is observed when the flipper is activated. The same kind of experiment is possible with a polarized incident beam and a non-polarizing crystal analyser; this arrangement was used by Holden and Stirling (1977) in their study of ferromagnetic Pd_3Fe.

The second type of experiment uses two polarizers, i.e. full polarization analysis. It is evident that if the polarizers are the same (or select the same neutron spins) and have efficiencies approaching 100%, then magnon peaks from ferromagnets can only be observed by activating a spin-flipper either before or after the scatterer. For positive polarizers like $Co_{0.92}Fe_{0.08}$ magnon creation is observed with a spin-flipper 'on' behind the scatterer, and magnon annihilation is observed with a spin-flipper 'on' in front of the scatterer. Figure 2.14(a) shows the data of Moon *et al.* (1969a) for lithium ferrite with the former instrumental arrangement.

A somewhat less useful geometry for studying ferromagnets is that in which the polarization and magnetization directions are perpendicular to the scattering vector, i.e. the $\tilde{\boldsymbol{P}} \cdot \tilde{\boldsymbol{\kappa}}$ or $\tilde{\boldsymbol{M}} \cdot \tilde{\boldsymbol{\kappa}} = 0$ geometry. If we choose a coordinate system where $\tilde{\boldsymbol{\kappa}}$ is along the Y direction, then magnons can

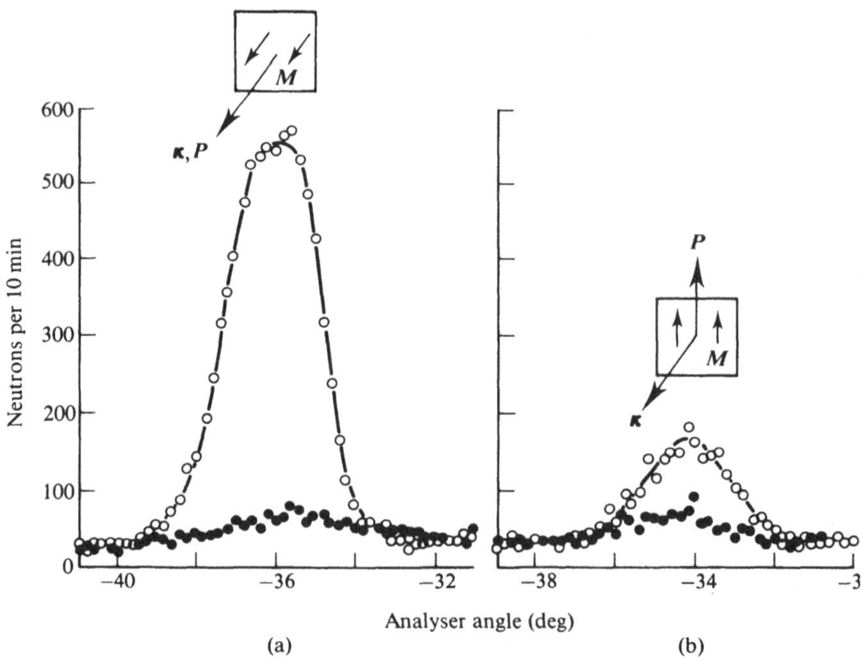

Fig. 2.14 Polarization analysis of magnon scattering from lithium ferrite in two experimental arrangements as described in the text; ●, flipper off; ○, flipper on (from Moon *et al.* 1969a).

only occur due to $S_X S_X$-type correlations, though these are still entirely spin-flip in character. Equation (2.3.53) shows that the creation and annihilation cross-sections are now equal and that their peaks are expected to have a quarter of the intensity observed in the $\tilde{\boldsymbol{P}} \cdot \tilde{\boldsymbol{\kappa}} = \tilde{\boldsymbol{M}} \cdot \tilde{\boldsymbol{\kappa}} = 1$ geometry, i.e. that shown in Fig. 2.14(a). However, in a full polarization analysis experiment with identical polarizers, they can only be observed by activating a spin-flipper. Figure 2.14(b) shows the data of Moon *et al.* for $Li_{0.5}Fe_{2.5}O_4$ with positive polarizers and the spin flipper located before the scatterer; the flipper 'on' signal is due to the $(-\,+)$ magnon annihilation cross-section. The non-zero flipper 'off' signals in Figs 2.14(a) and (b) can be attributed to slight imperfections in the efficiencies of the polarizers and the spin-flipper.

The polarization analysis of magnon scattering in antiferromagnets is more complicated than that in ferromagnets, since the scattering can now involve non-spin-flip events. This is because the neutron polarization and sublattice magnetization directions can be different, so that the spin components involved with the spin waves need not necessarily be perpendicular to the neutron polarization direction. An incident polarized beam on an antiferromagnet can therefore be scattered with zero spin change, partial depolarization or complete spin-flip, depending on the experimental arrangement.

All three possibilities have been demonstrated by Moon *et al.* (1969a) for α-Fe_2O_3. Figure 2.15 shows their results for the polarization analysis of magnon creation with an energy $\hbar\omega \sim 5$ meV near the (1,1,1) position under a variety of conditions. In Fig. 2.14(a) a 0.9 T vertical (Z direction) field in the (111) plane induces a unique horizontal antiferromagnetic axis. Referring again to the amplitude equations for magnetic scattering, and arbitrarily setting a coordinate system with $S_{\perp X} = 0$, the spins then lie along the Y-axis and spin waves can only arise by correlations in the S_Z components. The result is therefore that only non-spin-flip scattering is possible, and Fig. 2.15(a) only shows an intensity in the polarization analysis experiment when the spin-flipper is 'off'. In Fig. 2.15(b) only a weak field, sufficient to maintain the neutron beam polarization, is applied vertically. The orientation of the spins in the (111) plane becomes randomized, and roughly equal contributions to the spin wave scattering are expected from the S_Y (spin-flip) and S_Z (non-spin-flip) components. The magnon creation peaks observed in the polarization analysis experiment therefore have the same intensities. Finally, in Fig. 2.15(c) the neutron polarization direction is arranged to be along the scattering vector ($S_{\perp Z} = 0$), and only spin-flip scattering is observed in an analogous way to the situation with ferromagnets.

We now return briefly to discuss polarization effects resulting from magnetovibrational coupling. This scattering is only inelastic in the phonon system, and appears in the neutron spectrum at the same peak

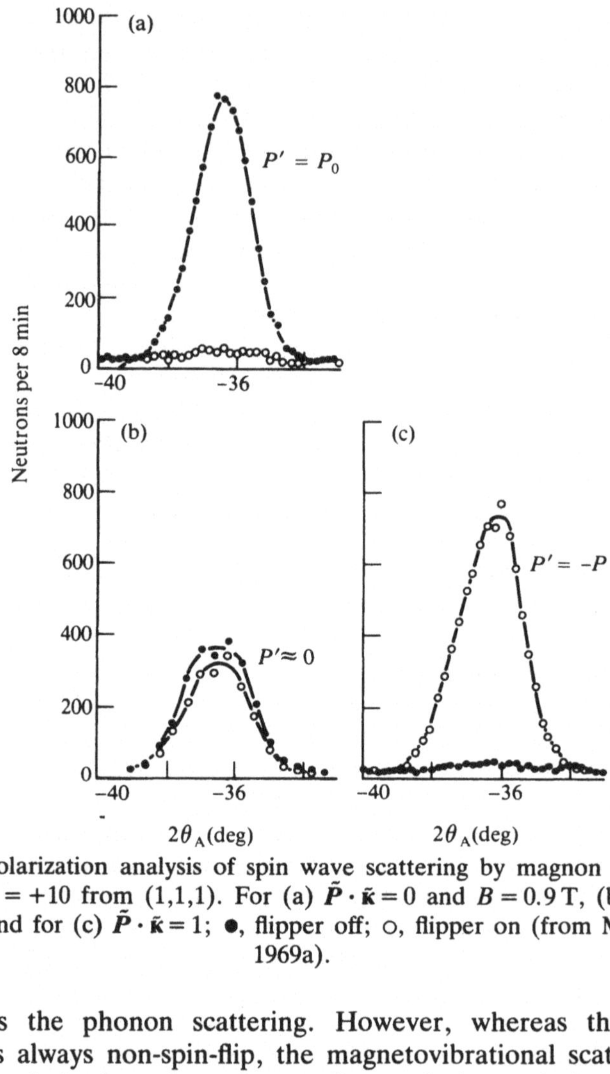

Fig. 2.15 Polarization analysis of spin wave scattering by magnon creation in α-Fe_2O_3, $\Delta\theta = +10$ from (1,1,1). For (a) $\tilde{\boldsymbol{P}} \cdot \tilde{\boldsymbol{\kappa}} = 0$ and $B = 0.9\,\mathrm{T}$, (b) $\tilde{\boldsymbol{P}} \cdot \tilde{\boldsymbol{\kappa}} = 0$ and $B = 0$, and for (c) $\tilde{\boldsymbol{P}} \cdot \tilde{\boldsymbol{\kappa}} = 1$; ●, flipper off; ○, flipper on (from Moon *et al.* 1969a).

positions as the phonon scattering. However, whereas the phonon scattering is always non-spin-flip, the magnetovibrational scattering has the same polarization dependence discussed for polarized neutron diffraction. Although full polarization analysis is potentially the best way of identifying magnetovibrational scattering, one of the clearest demonstrations of its existence was made in a series of partial polarization analysis experiments on iron and nickel by Steinsvoll *et al.* (1981). The measurements were performed on a triple-axis spectrometer with a $Co_{0.92}Fe_{0.08}$ polarizing monochromator and a non-polarizing graphite energy analyser; the flipper 'off' data thus corresponds to the combined $(++)$ and $(+-)$ cross-sections. In the experimental geometry used the applied field was normal to the scattering vector, i.e. $\tilde{\boldsymbol{M}} \cdot \tilde{\boldsymbol{\kappa}} = 0$, where, as is evident from equn (2.3.53), the magnon peak intensities (M) are expected to be insensitive to the incident beam polarization. The results of a typical constant-energy scan with energy transfer $\hbar\omega = 17\,\text{meV}$ in nickel are shown in Fig. 2.16. The flipping ratios for the longitudinal acoustic (LA) modes were analysed in a similar way to that for elastic Bragg scattering in magnetized single crystals, and yielded a set of inelastic magnetic form factors for both Fe and Ni. One of the merits of this type of experiment is that it can provide form factor results at scattering vectors other than the reciprocal lattice vectors as observed by elastic diffraction. In fact, the inelastic form factors measured for Ni were the same as the elastic form factors, though significant differences were measured in the case of Fe. Two explanations were suggested for the Fe result: (i) that there could be a dynamic distortion of the 3d electron distribution as neighbouring atoms oscillate with respect to each other, or (ii) that the phenomenon is a static one associated with the nature of the band wave functions. It was not possible to distinguish between these two possibilities, though further polarized-beam magnetovibrational scattering experiments may help to elucidate this matter.

Before completing this section on inelastic polarization analysis, a brief reference should be made to one of its more recent applications,

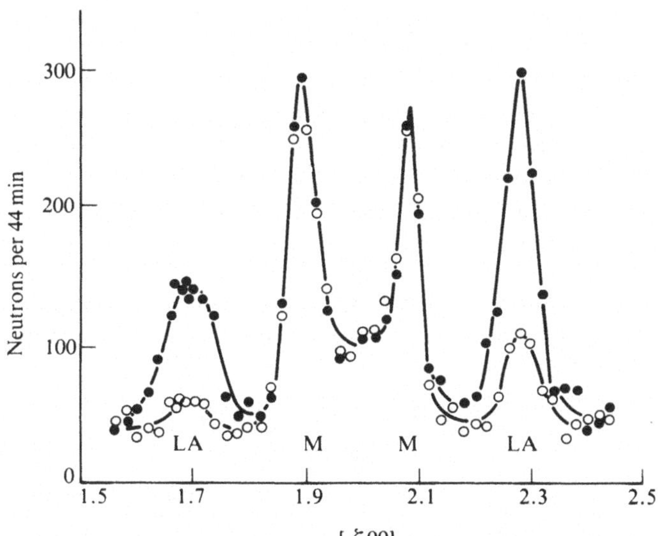

Fig. 2.16 Typical constant-energy scans of inelastic scattering from a ^{60}Ni single crystal showing the existence of magnetovibrational scattering in the longitudinal acoustic (LA) modes. The scattering due to the magnon modes is neutron-polarization-independent in the $\tilde{\boldsymbol{M}} \cdot \tilde{\boldsymbol{\kappa}} = 0$ geometry used in this experiment; $\hbar\omega = 17\,\text{meV}$; •, (+) incident spins; ○, (−) incident spins (from Steinsvoll *et al.* 1981).

primarily by the Brookhaven and Oak Ridge National Laboratories, in studying paramagnetic scattering. Experiments on EuO, Fe and Ni, in particular at and above their critical temperatures, were used to elucidate the form of the critical scattering. Polarization analysis is essential when investigating high-T_c magnetic materials like Fe and Ni, since the vibrational scattering is strong at T_c. Some of these experiments are described in Section 5.4.2.

2.4 NEUTRON OPTICS

Many optical phenomena that were first observed with visible light have now been shown to have their neutron counterparts. These include refraction, total reflection, slit and ruled grating diffraction, thin-film interference, etc. All these effects are the result of the interference between the forward scattered waves and incident beam waves in various optical devices. The important parameter which determines their behaviour is the neutron refractive index. As the neutron wave passes through one medium (1) into another (2) its wave vector changes from $\boldsymbol{k}_1$ to $\boldsymbol{k}_2$ and we define the refractive index for the boundary as

$$n_{1,2} = k_2/k_1. \tag{2.4.1}$$

Since the scattering is always in the forward direction ($\boldsymbol{\kappa} = 0$), neutron optical phenomena do not depend on either the structure or dynamics of the media involved, and are in fact usually observed more readily at wavelengths which are much larger than normal interatomic distances. Another consequence of the forward scattering is that we can neglect recoil, so the refractive index depends on the bound coherent scattering amplitude b, even in the case of gases. The relationship between n and b, including the situation in magnetic media, can be deduced from simple physical arguments.

Let us assume that the neutron kinetic energy in medium 1 is E_1, and that it experiences a mean change of potential $\langle V \rangle$ when it enters medium 2. We then have

$$\begin{aligned} n_{1,2} = k_2/k_1 &= [(E_1 - \langle V \rangle)/E_1]^{1/2} \\ &\simeq 1 - \langle V \rangle / 2E_1. \end{aligned} \tag{2.4.2}$$

The approximation follows since $\langle V \rangle \ll E_1$, and further we have $n \to 1$. Typical values of $|n - 1|$ for thermal neutrons are in the region of 10^{-6}.

We now discuss an example of importance in the production of polarizing mirrors. The first medium is assumed to be a vacuum and the energy E_1 is, in terms of the de Broglie wavelength λ,

$$E_1 = (2\pi^2 \hbar^2)/m\lambda^2. \tag{2.4.3}$$

The second medium is magnetic and provides both a nuclear and magnetic part to $\langle V \rangle$. The nuclear part $\langle V \rangle_N$ is given by an averaged optical potential which is derived from the Fermi pseudopotential (see equn (2.1.12)), i.e.

$$\langle V \rangle_N = h^2 N\bar{b}/(2\pi m), \qquad (2.4.4)$$

where $\bar{b}$ is the mean bound coherent scattering amplitude in medium 2, and N is the number density of the scattering nuclei. The magnetic part of the potential is given by equn (2.1.61):

$$\langle V \rangle_M = -\boldsymbol{\mu}_n \cdot \boldsymbol{B} = \pm \mu_n B, \qquad (2.4.5)$$

where the $\pm$ signs refer to the neutron spins which are parallel and antiparallel respectively to the $\boldsymbol{B}$ field direction.

Substituting into equn (2.4.2), we obtain

$$n_\pm = 1 - \lambda^2 (N\bar{b}/(2\pi) \pm m\mu_n B/h^2), \qquad (2.4.6a)$$

$$n_\pm = 1 - (\lambda^2 N/(2\pi))(\bar{b} \pm (B/B_s)p), \qquad (2.4.6b)$$

where $p = 2\pi m\mu_n B_s/h^2 N$ is the magnetic coherent scattering amplitude at $\boldsymbol{\kappa} = 0$ expressed in an analogous way to $\bar{b}$, and B_s is the magnetic induction of the magnetic medium at saturation. The magnetic contribution to the refractive index in a ferromagnet depends on the mean induction B, and in general this is less than the saturation value. The most significant feature of this expression from the aspect of polarized neutrons is that the refractive index in magnetized materials is neutron-spin-dependent, and this property is exploited in all optical neutron polarizers.

2.4.1 Mirror reflection

For most media, $\bar{b} > 0$ and $n < 1$, so that neutrons entering these media from a vacuum (or air) pass into a lower-density region, and can undergo total reflection if the glancing angle is sufficiently small. The critical reflection angle below which mirror reflection takes place can be derived by reference to Fig. 2.17. In Fig. 2.17(a), Snell's law for refraction states

$$n_{1,2} = \cos\theta_1/\cos\theta_2. \qquad (2.4.7)$$

Thus the critical glancing angle shown in Fig. 2.17(b) is given by

$$n_{1,2} = \cos\theta_c \qquad (2.4.8)$$

and total reflection occurs in Fig. 2.17(c) when $\theta < \theta_c$.

For small values of θ_c, as are appropriate for thermal and even slow neutrons, we can use the following approximation when medium 1 is

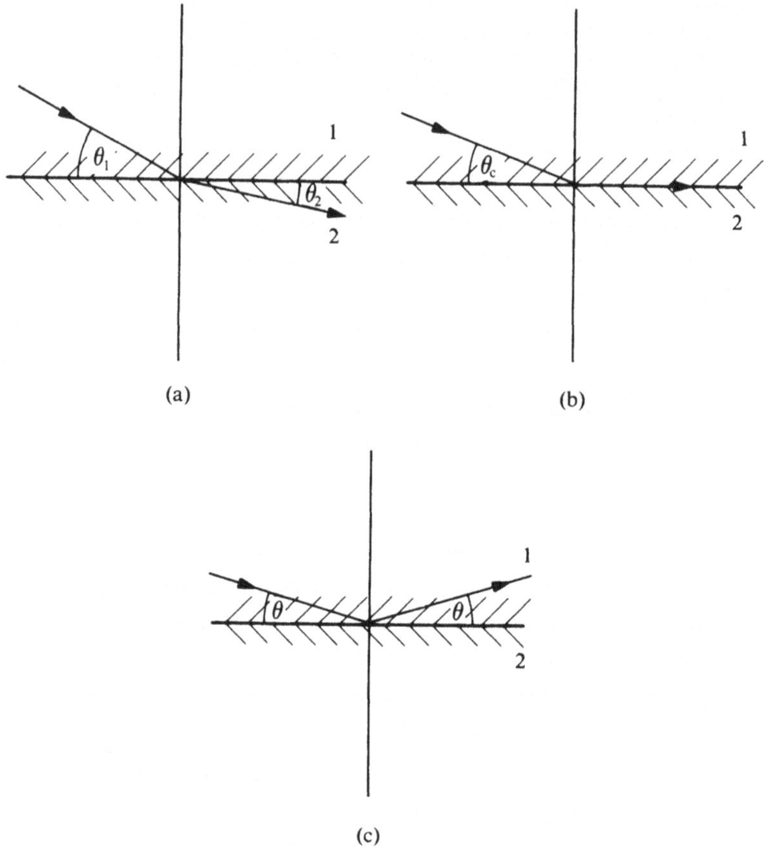

Fig. 2.17 Neutron optical phenomena at an interface between media 1 and 2: (a) refraction, (b) critical glancing angle, and (c) total reflection.

either a vacuum or air:

$$n = \cos\theta_c \simeq 1 - (\theta_c^2/2) \tag{2.4.9}$$

$$\theta_c = [2(1-n)]^{1/2}$$
$$= \lambda(N/\pi)^{1/2}[\bar{b} \pm (B/B_s)p]^{1/2} \tag{2.4.10}$$

for magnetic mirrors.

The phenomenon of total reflection from non-magnetic mirror surfaces has been exploited in the construction of neutron guide tubes, which can conduct thermal and cold neutron beams over long distances with very little intensity loss. The four flat walls of the tube are constructed from highly polished glass which are coated with a film ~1000 Å thick of a high-$\bar{b}$ material, normally nickel. The tubes are usually assembled in a curved geometry and this prevents fast neutrons and γ-rays produced in the source from reaching the neutron instruments. The utilization of neutron guide tubes has been reviewed by Jacrot (1970).

Magnetic mirrors are characterized by having two critical glancing angles, and this property has been extensively applied in the production of polarizing mirrors and polarizing guides. These devices are discussed in Chapter 3.

A partial mirror reflection remains at glancing angles greater than θ_c, and the mirror reflectivity in this region was shown by Goldberger and Sietz (1947) to be given by

$$\mathscr{R} = \left(\frac{(n^2 - \cos^2\theta)^{1/2} - \sin\theta}{(n^2 - \cos^2\theta)^{1/2} + \sin\theta}\right)^2$$

$$\simeq \left(\frac{1 - [1 - (\theta_c/\theta)^2]^{1/2}}{1 + [1 - (\theta_c/\theta)^2]^{1/2}}\right)^2 \tag{2.4.11}$$

These equations show that as θ increases beyond θ_c there is a rapid decrease in the reflectivity, for example when $(\theta/\theta_c) = 1.1$ we have $\mathscr{R} \sim 15\%$. The theory we have presented was also extended by Goldberger and Seitz (1947) to include the effects of neutron absorption and incoherent scattering in the reflectivity medium. These are not treated here since they are of only minor importance in optical polarizing devices. Two other phenomena, however, namely refraction and interference, are exploited in multilayer thin-film polarizers, and we now proceed to discuss these.

2.4.2 Thin-film devices

During the mid-1970s several workers, notably Schoenborn *et al.* (1974) and Mezei (1976a), showed that, using mirrors composed of multilayers of thin metallic films, it was possible to effect large increases in the (reflection angle/wavelength) range over which bulk material mirrors are reflective to neutrons. This enhancement arises due to the interference between the neutron waves as they traverse the different devices, and with magnetic layers the phenomenon can be applied to produce polarized neutrons. Figure 2.18 shows the three types of optical systems which can be used in this way: (a) a single thin film, (b) a periodic bilayer with a repeat spacing d, and (c) a generalized multilayer structure.

The most versatile method of calculating the optical characteristics of thin-film multilayers is, following Goldberger and Seitz (1947), to adopt the techniques used in conventional optics. We may assign a refractive index n_j and a thickness d_j to each successive layer and use the standard methods (see for example Born and Woolf (1970)) to determine the reflection and transmission coefficients at each interface. This procedure was first used for neutrons by Gukasov *et al.* (1977), who defined the

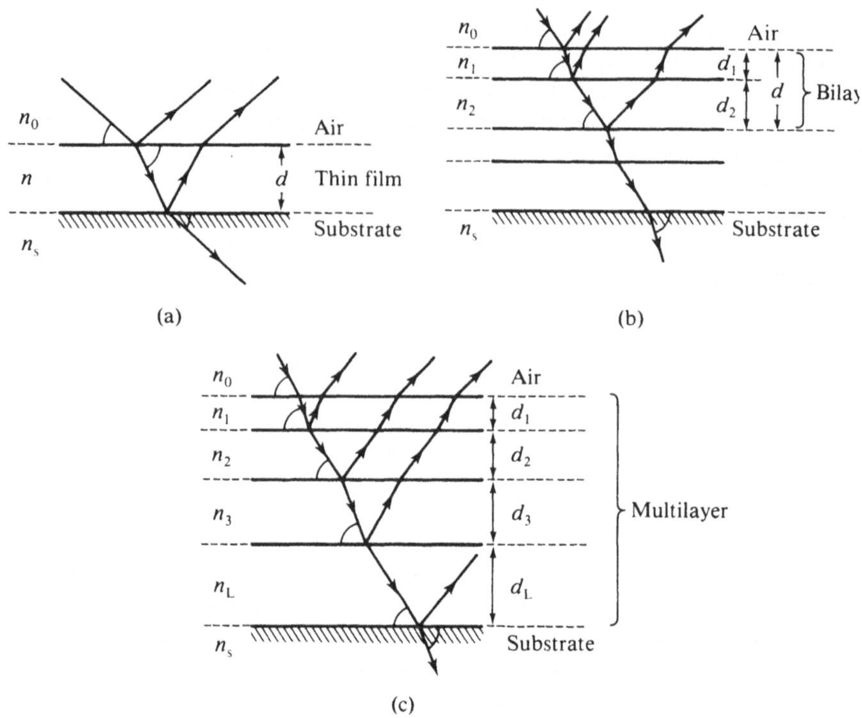

Fig. 2.18 Thin-film devices: (a) a single thin film, (b) a periodic bilayer which acts as a monochromator, and (c) a generalized supermirror.

following characteristic matrix for each jth layer

$$[\hat{M}_j] = \begin{bmatrix} \cos\beta_j & -\mathrm{i}\sin\beta_j/\rho_j \\ -\mathrm{i}\rho_j\sin\beta_j & \cos\beta_j \end{bmatrix}, \tag{2.4.12}$$

where

$$\rho_j = n_j \sin\theta_j \tag{2.4.13}$$

and

$$\beta_j = (2\pi/\lambda)n_j d_j \sin\theta_j. \tag{2.4.14}$$

θ_j is the incident glancing angle at the jth layer and is determined by Snell's law:

$$n_1\cos\theta_1 = n_2\cos\theta_2 + \cdots = n_L\cos\theta_L. \tag{2.4.15}$$

The resultant multilayer matrix $[\hat{M}]$ is obtained by multiplying the matrices of all the layers j up to the final layer L

$$[\hat{M}] = [\hat{M}_1][\hat{M}_2]\cdots[\hat{M}_j]\cdots[\hat{M}_L] \tag{2.4.16}$$

$$= \begin{bmatrix} M_{11} & M_{12} \\ M_{21} & M_{22} \end{bmatrix}. \tag{2.4.17}$$

The reflection (r) and transmission (t) coefficients for the multilayer assembly are then related to the matrix elements M_{lm} and ρ parameters for the media before and after the multilayer structure by

$$r = \frac{(M_{11} + M_{12}\rho_s)\rho_0 - (M_{21} + M_{22}\rho_s)}{(M_{11} + M_{12}\rho_s)\rho_0 + (M_{21} + M_{22}\rho_s)} \tag{2.4.18}$$

and

$$t = \frac{2\rho_0}{(M_{11} + M_{12}\rho_s)\rho_0 + (M_{21} + M_{22}\rho_s)}, \tag{2.4.19}$$

where the subscripts s and 0 refer to the multilayer substrate and air (or vacuum) layers. The resultant reflectivity $\mathcal{R}$ and transmittivity T are given by

$$\mathcal{R} = rr^* \tag{2.4.20}$$

and

$$T = tt^* \tag{2.4.21}$$

In polarizing devices, it is arranged that the interference which gives rise to a reflectivity above the mirror critical glancing angle occurs for only one of the neutron spin states. In practice this involves matching the refractive index for the unreflected spins with that for the non-magnetic component in the multilayer structure; the unwanted spin then experiences a uniform refractive index, so that interference is not possible. It was shown in Section 1.5 how a periodic Fe–Ge bilayer can act as a long-wavelength monochromator (see Fig. 1.15). In the supermirror concept introduced by Mezei (1976a), the multilayer film is prepared with a gradient in the 'lattice' spacing, and its effect is to provide a high reflectivity up to (θ/λ) values approximately three times greater than encountered with normal mirror reflections. All these devices are described in greater detail in Chapter 3. Figure 3.16(a) shows the performance characteristics of a typical polarizing mirror, and the polarizing supermirror performance that can now be achieved following development work by Schärpf (1981) is illustrated in Fig. 3.16(c).

The method of calculating neutron reflectivity profiles described in this section has not only been used in aiding the design of optical devices, but has also been applied in the study of surface phenomena. Of particular interest to this book is the method pioneered by Felcher (1981), which measures polarized beam (+) and (−) neutron spin reflectivities at reflection angles above the critical angle. This provides a very sensitive method for studying the depth profile of surface magnetism. A discussion of the principles and instrumentation needed for this type of experiment is given in Section 3.3.5, and some experimental results are described in Section 4.6.

2.5 NUCLEAR RESONANCES

A neutron reacts with a target nucleus to form a relatively long-lived ($\tau \sim 10^{-16}$ s) compound nucleus with an excitation energy equal to the neutron's kinetic energy plus the binding energy (typically 6–10 MeV) of the captured neutron. This gives a series of resonance reactions with probabilities which peak at neutron energies E_R, the resonance energies, corresponding to excited nuclear levels with energies E_R plus the binding energy. The total cross-sections for neutron–nuclear interactions are therefore composed of a slowly varying component due to direct elastic scattering, as well as cross-sections for compound nucleus reactions which have sharp maxima at the resonance energies E_R. Figure 2.19(a) shows a typical energy scheme for a compound nucleus of mass $(N+1)$ formed by the absorption of a neutron by a nucleus with mass N.

The lifetime τ of a compound nucleus state gives the resonance, through the uncertainty principle, an energy width

$$\Gamma = \hbar/\tau, \tag{2.5.1}$$

where τ^{-1} is the probability per second that the compound state decays. For thermal and epithermal neutrons, there are two main mechanisms for this decay: neutron emission as shown in Fig. 2.19(b), and γ-emission as shown in Fig. 2.19(c). These processes are characterized by a neutron

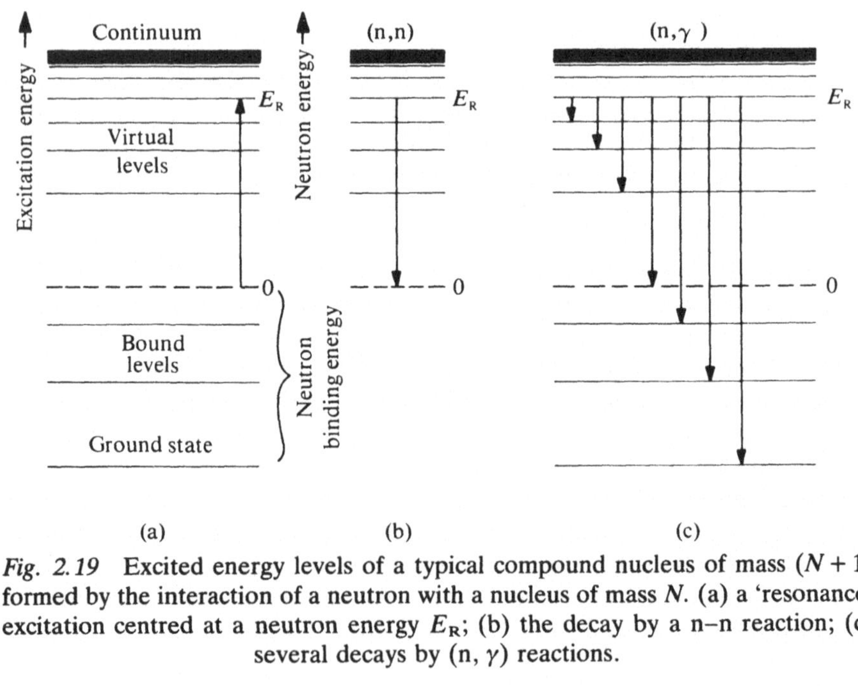

Fig. 2.19 Excited energy levels of a typical compound nucleus of mass $(N+1)$ formed by the interaction of a neutron with a nucleus of mass N. (a) a 'resonance' excitation centred at a neutron energy E_R; (b) the decay by a n–n reaction; (c) several decays by (n, γ) reactions.

width Γ_n and a radiation width Γ_γ respectively, and in the absence of other decay mechanisms

$$\Gamma = \Gamma_\gamma + \Gamma_n. \tag{2.5.2}$$

The probability that the compound nucleus decays through the emission of γ-radiation is Γ_γ/Γ, and that through re-emission of a neutron is Γ_n/Γ. Γ_γ varies only slightly from resonance to resonance, and is found to have a value of ~0.1–0.2 eV for intermediate mass nuclei decreasing to ~30 meV for heavy nuclei. Γ_n, on the other hand, steadily increases with the neutron energy, and is also largest for light nuclei. Typical magnitudes for Γ_n are keV for light nuclei, eV for intermediate mass nuclei, and meV for heavy nuclei.

The energy-dependences of nuclear scattering lengths in the region of a resonance follow the well known Breit–Wigner formula, and is given for example by Bacon (1975). These lengths have real and imaginary components, and their variations with the neutron energy over the width of a resonance can in principle be used to solve the phase problem in structural determinations using neutron diffraction. Possible candidates for this type of experiment are materials containing ^{113}Cd and ^{149}Sm, which have strong thermal neutron resonances.

For heavy nuclei and low neutron energies, $\Gamma_\gamma \gg \Gamma_n$, and resonance capture dominates over resonance scattering. We shall see that if the capturing nucleus can be polarized, then the neutron absorption around the resonance energy depends strongly on the neutron spin, and can provide a basis for producing a neutron polarizing filter. First we describe the neutron energy dependence of the (n, γ) cross-section.

For an isolated resonance, the variation of the absorption cross-section σ_A with the neutron energy E follows closely the Breit–Wigner relationship,

$$\sigma_A(E) = \pi \lambda\!\!\!^-{}^2 \frac{g\Gamma_n\Gamma_\gamma}{[(E-E_R)^2 + (\Gamma/2)^2]}, \tag{2.5.3}$$

where

$$\lambda\!\!\!^- = \lambda/(2\pi) = \hbar/(mv) = \hbar/(2mE)^{1/2}. \tag{2.5.4}$$

The cross-section maximum occurs at $E = E_R$ and has a value

$$\sigma_A^0 = 4\pi \lambda\!\!\!^-{}_R^2 g\Gamma_n\Gamma_\gamma/\Gamma^2. \tag{2.5.5}$$

$\sigma_A(E)$ may therefore be written as

$$\sigma_A(E) = \sigma_A^0/[1 + [(E-E_R)/(\Gamma/2)]^2]. \tag{2.5.6}$$

The factor g is a statistical weighting factor which is unity for target nuclei with zero spin but is otherwise given by

$$g = (2J+1)/[2(2i+1)], \tag{2.5.7}$$

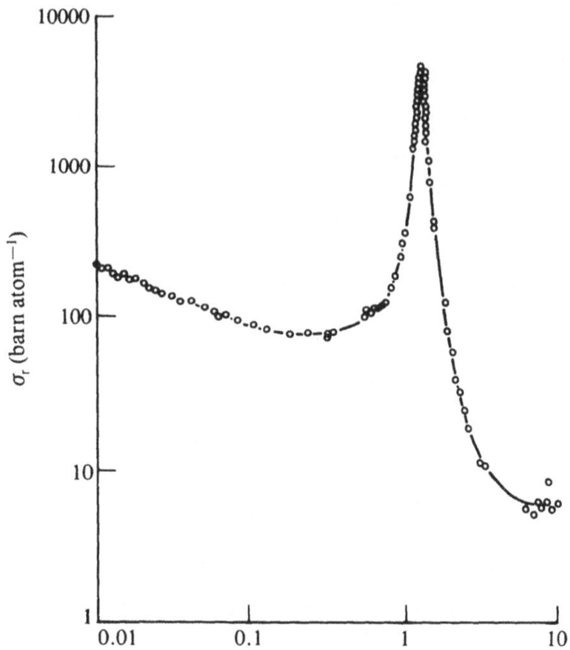

Fig. 2.20 The total cross-section in rhodium as a function of the neutron energy. The curve is a Breit–Wigner fit to experimental data (from Beckurts and Wirtz 1964).

where i is the spin of the target nucleus before capture, and J is the spin of the compound nucleus. Most slow and intermediate energy neutron resonances are s-wave resonances, for which the captured neutron does not impart any orbital angular momentum to the compound nucleus; their angular momentum values are either $J_+ = i + \frac{1}{2}$ or $J_- = i - \frac{1}{2}$. Following Beckurts and Wirtz (1964) it is possible to assign a reduced neutron width Γ_{n0} to each level which is related to the real neutron width Γ_n as follows:

$$\Gamma_n = \Gamma_{n0}\sqrt{E}. \tag{2.5.8}$$

Figure 2.20 shows the energy dependence of the cross-section for a typical isolated resonance, the 1.23 eV nuclear resonance in rhodium. The curve in the figure was calculated using the Breit–Wigner expression. At very low neutron energies when $E \rightarrow 0$ and $\Gamma \ll E_R$, $\sigma_A(E)$ reduces to

$$\begin{aligned}\sigma_A(E) &= \pi \lambda\!\!\!^{-2} g \Gamma_n \Gamma_\gamma / E_R^2 \\ &\simeq \pi \lambda\!\!\!^{-2}_R g \Gamma_{n0} \Gamma_\gamma / (E_R \sqrt{E}) \propto E^{-1/2} \propto v^{-1}.\end{aligned} \tag{2.5.9}$$

This famous $1/v$ law for low-energy neutron capture has a simple interpretation, which is that the capture probability depends on the time the neutron spends in the vicinity of the nucleus. It is often possible to calculate the $1/v$ cross-section from the resonance parameters observed at higher energies.

We now derive expressions for the resonance capture cross-sections for polarized nuclei and polarized neutrons. The general form for the cross-section may be written (see for example Daniels (1965)) as

$$\sigma = A + B\boldsymbol{P}_{\mathrm{N}} \cdot \boldsymbol{P}, \tag{2.5.10}$$

where A and B are scalar quantities, $\boldsymbol{P}_{\mathrm{N}}$ is the nuclear polarization vector, and $\boldsymbol{P}$ is the neutron polarization vector. For randomly oriented nuclei and neutrons, the second term in this equation vanishes, and the cross-section corresponds to the normal unpolarized beam cross-section, hence $A = \sigma_{\mathrm{A}}(E)$.

Next we consider the $J_- = i - \frac{1}{2}$ cross-section, and assume that the neutrons and nuclei are fully polarized parallel to each other, i.e. $\boldsymbol{P}_{\mathrm{N}} \cdot \boldsymbol{P} = 1$. However, we also know that the $i - \frac{1}{2}$ state cannot be formed for parallel neutron and nuclear spins, in which case this cross-section is zero. Mathematically,

$$0 = A + B,$$

i.e.

$$B = -A = -\sigma_{\mathrm{A}}(E),$$

or

$$\begin{aligned}\sigma_{i-1/2} &= \sigma_{\mathrm{A}}(E) - \sigma_{\mathrm{A}}(E)(\boldsymbol{P}_{\mathrm{N}} \cdot \boldsymbol{P}) \\ &= \sigma_{\mathrm{A}}(E)(1 - \boldsymbol{P}_{\mathrm{N}} \cdot \boldsymbol{P}).\end{aligned} \tag{2.5.11}$$

For the $J_+ = i + \frac{1}{2}$ cross-section we first examine the case when the spins are randomly oriented. The total number of combined states for a neutron with spin $\frac{1}{2}$ and a nucleus with spin i is $2(2i + \frac{1}{2})$; these are subdivided into $2(i + \frac{1}{2}) + 1 = 2(i + 1)$ states with $J_+ = i + \frac{1}{2}$, and $2(i - \frac{1}{2}) + 1 = 2i$ states with $J_- = i - \frac{1}{2}$. The probability of a J_+ capture is therefore $2(i + 1)/2(2i + 1) = (i + 1)/(2i + 1)$, and this in turn is proportional to A. Next we again consider parallel neutron and nuclear spins, when the J_+ capture probability is 1, and which is proportional to $A + B$. We therefore have

$$\left(\frac{\text{Fully polarized } \sigma_{i+1/2}}{\text{Unpolarized } \sigma_{i+1/2}}\right) = 1\bigg/\left(\frac{i+1}{2i+1}\right) = \frac{A+B}{A}, \tag{2.5.12}$$

from which we obtain

$$B = Ai/(i + 1) \tag{2.5.13}$$

and

$$\begin{aligned}\sigma_{i+1/2} &= \sigma_{\mathrm{A}}(E) + \sigma_{\mathrm{A}}(E)\frac{i}{(i+1)}\boldsymbol{P}_{\mathrm{N}} \cdot \boldsymbol{P} \\ &= \sigma_{\mathrm{A}}(E)\left(1 + \frac{i}{(i+1)}\boldsymbol{P}_{\mathrm{N}} \cdot \boldsymbol{P}\right).\end{aligned} \tag{2.5.14}$$

These equations are often abbreviated in the following form for positively ($P = +1$) and negatively ($P = -1$) polarized neutrons:

$$\sigma_{A\pm}(E) = \sigma_A(E)(1 \pm \rho P_N), \qquad (2.5.15)$$

where $\rho = -1$ for $(i - \frac{1}{2})$ resonances, and $\rho = i/(i+1)$ for $(i + \frac{1}{2})$ resonances. This expression was first derived by Rose (1949).

It frequently happens, particularly for thermal neutron capture, that the total cross-section contains contributions from more than one resonant state. An example with which we shall be concerned later is the thermal neutron capture in ^{149}Sm, which proceeds principally through a $(i + \frac{1}{2})$ spin state, but to which there is also a contribution from a $(i - \frac{1}{2})$ spin bound state resonance, where the resonance energy E_R in Fig. 2.18 is negative. The cross-section for the $(i - \frac{1}{2})$ capture in ^{149}Sm has been estimated by Höhne (1961). For mixed resonances, the spin-dependent cross-sections are still given by equn (2.5.15), but with a statistical weighting factor (see Freeman and Williams (1978)):

$$\rho = [i(1 - 2x) - x]/(i + 1) \qquad (2.5.16)$$

where

$$x = \sigma_{i-1/2}/(\sigma_{i+1/2} + \sigma_{i-1/2}) \qquad (2.5.17)$$

is the fractional capture into the $(i - \frac{1}{2})$ spin state.

Reference was made in the introductory chapter to one important application of equn (2.5.15), that is the use of polarized neutrons and polarized nuclei to determine the J states of nuclear resonances. In the experiment, the transmittances of (+) and (−) spin neutron beams through a target in which the sign of the nuclear polarization is known are measured at the nuclear resonance energy. The sense of the transmittance difference $(T_+ - T_-)$ then gives the J state of the resonance. An alternative method for determining the J values of nuclear resonance states is to use Breit–Wigner curve-fittings of measured cross-sections to give the product $g\Gamma_n$, and to eliminate the lower g value if $\Gamma_n > \Gamma$; J can then be found with the higher g value substituted into equn (2.5.7). All unpolarized beam methods, however, are prone to much greater uncertainties than the polarized beam transmission measurement. Table 2.3 shows representative experimental results obtained using polarized neutrons. The nuclear polarizations needed for these experiments were created by cooling the targets to low temperatures, often $T \ll 1$ K, and either applying a strong external magnetic field ('brute force' polarization), or more conveniently by using the large nuclear hyperfine fields which occur in many atoms which have unpaired electrons (hyperfine polarization).

More recently the difference in the capture cross-sections for (+) and (−) spin neutrons, as given by equn (2.5.15), has been used to produce a

TABLE 2.3 Spins of nuclear resonances determined by the transmission of polarized neutrons through targets containing polarized nuclei

Isotope	Nuclear spin i	Nuclear polarization P_N; method and magnitude where given	Energies (eV) of resonances below 10 eV with their spins: $J_+ = i + \frac{1}{2}$	$J_- = i - \frac{1}{2}$	Reference
^{59}Co	7/2	Hyperfine field $P_N \simeq -0.14$	~0.001–0.1 eV (78.3%)	~0.001–0.1 eV (21.7%)	(a)
^{115}In	9/2	Brute force $P_N \simeq 0.027$	1.46 9.10	3.86	(b)
^{149}Sm	7/2	Hyperfine field $P_N = -0.13$	0.0976 0.873 4.93 8.9		(c)
^{151}Eu	5/2	Hyperfine field $P_N = -0.016$	0.32 0.46 1.06	3.37	(d)
^{153}Eu	5/2	Hyperfine field P_N not given	2.46		(d)
^{155}Gd	3/2	Hyperfine field $P_N = 0.028$	0.0268 2.57	2.01	(d)
^{157}Gd	3/2	Hyperfine field $P_N = 0.036$	0.0314		(d)
^{165}Ho	7/2	Hyperfine field $P_N = 0.176$	3.92		(e)
^{168}Er	7/2	Hyperfine field $P_N \simeq 0.45$	0.46	0.58 6.1 6.9	(f)

(a) Schermer, R. I. (1963). *Phys. Rev.*, **130,** 1907.
(b) Stolovy, A. (1960). *Phys. Rev.*, **118,** 211.
(c) Marshak, H., Postma, H., Sailor, V. L., Shore, F. J., and Reynolds, C. A. (1962). *Phys. Rev.*, **128,** 1287.
(d) Stolovy, A. (1964). *Phys. Rev.*, **134B,** 68.
(e) Brunhardt, G., Postma, H., and Sailor, V. L. (1965). *Phys. Rev.*, **137B,** 1484.
(f) Schermer, R. I., Passell, L., Brunhart, G., Reynolds, C. A., Sailor, L. V., and Shore, F. J. (1968). *Phys. Rev.*, **167,** 1121.

neutron polarizing filter. The method was first proposed by Williams (1973), who suggested using a filter containing highly polarized ^{149}Sm nuclei as a broad-band polarizer for thermal neutrons. Further details, together with experimental measurements, are provided in Chapter 3. A polarizing filter which uses the spin-dependent capture principle can only be effective if three conditions apply: (a) it must be possible to obtain a high nuclear polarization ($|P_N| \gtrsim 0.8$); (b) $|\rho|$ should approach unity, which means preferably using a $(i - \frac{1}{2})$ resonance or possibly a $(i + \frac{1}{2})$ resonance where i is large; and (c) σ_A should be much larger than the total cross-sections of the non-absorbing nuclei in the filter.

3

Neutron polarizers and instrumentation

This chapter describes many of the techniques, devices and instruments which have either been built or proposed to exploit the unique characteristics of polarized neutron beams. The emphasis is placed on the operation of the spin-sensitive components in the instruments, and how their performances depend on the neutron energy. We shall cover methods that are applicable in the cold, thermal and epithermal spectral regions. Instruments which rely on using precessing polarization have been omitted, since these are treated in Chapter 6.

3.1 NEUTRON POLARIZERS

We saw in Chapter 1 that the number of methods now available for polarizing neutron beams are many and varied, and that the choice of a polarizer for an instrument should be determined primarily by the intended application. The properties of the different types of neutron polarizers are discussed in this section. One of the important parameters which is used to assess the effectiveness of a polarizer is its polarizing efficiency P. This is defined as the polarization of the outgoing beam after the polarizer has acted on an unpolarized beam. Following equn (1.4.10) the polarizing efficiency may be expressed as

$$P = (N_+ - N_-)/(N_+ + N_-), \tag{3.1.1}$$

where N_+ and N_- are the numbers of (+) and (−) spin neutrons in the outgoing beam.

The second factor of importance is the intensity loss caused by the polarizing process. This depends on terms like the angular acceptance of the polarizer together with its reflectivity in the case of mosaic crystals and mirrors, or its transmittance for polarizing filters.

3.1.1 Polarizing crystals

Polarizing crystal monochromators are well suited to the needs of polarized beam diffractometry since, as their name implies, they simultaneously polarize and monochromate. They are also being used increasingly in polarized-beam triple-axis spectrometers. The principle by which they operate has been described in Section 2.2, see especially equns (2.2.26) and (2.2.27). However, it is worthwhile reiterating the fundamental cross-section equation, and to clarify the geometrical requirements.

The geometrical arrangement used with all polarizing crystals is shown in Fig. 3.1. Here a vertical magnetic field $\boldsymbol{B}$ is applied to a ferromagnetic single crystal which is oriented so that a Bragg reflection takes place from an (hkl) plane with suitably matched nuclear and magnetic structure factors $F_N(\boldsymbol{\kappa})$ and $F_M(\boldsymbol{\kappa})$. The field saturates the atomic moments which point along $\boldsymbol{B}$, whereas the atomic spins $\boldsymbol{S}$ with unit vectors $\tilde{\boldsymbol{\eta}}$ are directed antiparallel to $\boldsymbol{B}$. The unpolarized neutron beam may be regarded as the sum of two polarized beams with either (+) spins ($\boldsymbol{P}=1$) or (−) spins ($\boldsymbol{P}=-1$) respectively with respect to $\boldsymbol{B}$, and the cross-section for Bragg reflection in this geometry is given by equn (2.2.25):

$$(\mathrm{d}\sigma/\mathrm{d}\Omega) = F_N(\boldsymbol{\kappa})^2 + 2F_N(\boldsymbol{\kappa})F_M(\boldsymbol{\kappa})(\boldsymbol{P}\cdot -\tilde{\boldsymbol{\eta}}) + F_M(\boldsymbol{\kappa})^2.$$

For $\boldsymbol{P}\cdot\tilde{\boldsymbol{\eta}} = -1$ this becomes $(F_N(\boldsymbol{\kappa}) + F_M(\boldsymbol{\kappa}))^2$, whereas for $\boldsymbol{P}\cdot\tilde{\boldsymbol{\eta}} = +1$ the cross-section is $(F_N(\boldsymbol{\kappa}) - F_M(\boldsymbol{\kappa}))^2$, and the reflected intensities are proportional to these terms. Clearly, a polarizing reflection requires $|F_N(\boldsymbol{\kappa})| = |F_M(\boldsymbol{\kappa})|$; these quantities may be either positive or negative, and can produce both positively or negatively polarized neutron beams, see Fig. 3.1. By combining equn (2.2.25) with equn (3.1.1), the polarizing efficiency of the Bragg reflected beam is found to be

$$P = 2F_N(\boldsymbol{\kappa})F_M(\boldsymbol{\kappa})/[F_N(\boldsymbol{\kappa})^2 + F_M(\boldsymbol{\kappa})^2]. \tag{3.1.2}$$

The properties of polarizing monochromator crystals have been summarized by Freund and Forsyth (1979). As stated in Chapter 1, the first successful and extensively used polarizing reflection was the (200) reflection from f.c.c. $Co_{0.92}Fe_{0.08}$. This is still commonly used today, though its main drawback is the relatively high absorption due to cobalt (21 barns atom^{-1} at $\lambda = 1.08$ Å). Crystal slices of 2–3 mm thickness are used in a symmetrical transmission geometry so as to provide usefully large beam areas with relatively small crystal areas. The structure factors $F_N(\boldsymbol{\kappa})$ and $F_M(\boldsymbol{\kappa})$ are both positive for the (200) reflection, hence the polarizing efficiency is positive. Moon *et al.* (1969a) estimate that the intensity available from a $Co_{0.92}Fe_{0.08}$ crystal is approximately four times less than that from a good unpolarized beam monochromator at ~1 Å. This loss compounds to a loss factor ~15 in a triple-axis polarization analysis spectrometer, with the result that these instruments only become a practical proposition at a high-flux reactor.

$Co_{0.92}Fe_{0.08}$ has been largely superseded by the Heusler alloy Cu_2MnAl, especially at lower thermal and cold neutron energies. The (111) reflection gives a higher reflectivity than $Co_{0.92}Fe_{0.08}$ (200) and has $F_N \simeq -F_M$, thus producing a negatively polarized beam; it was developed

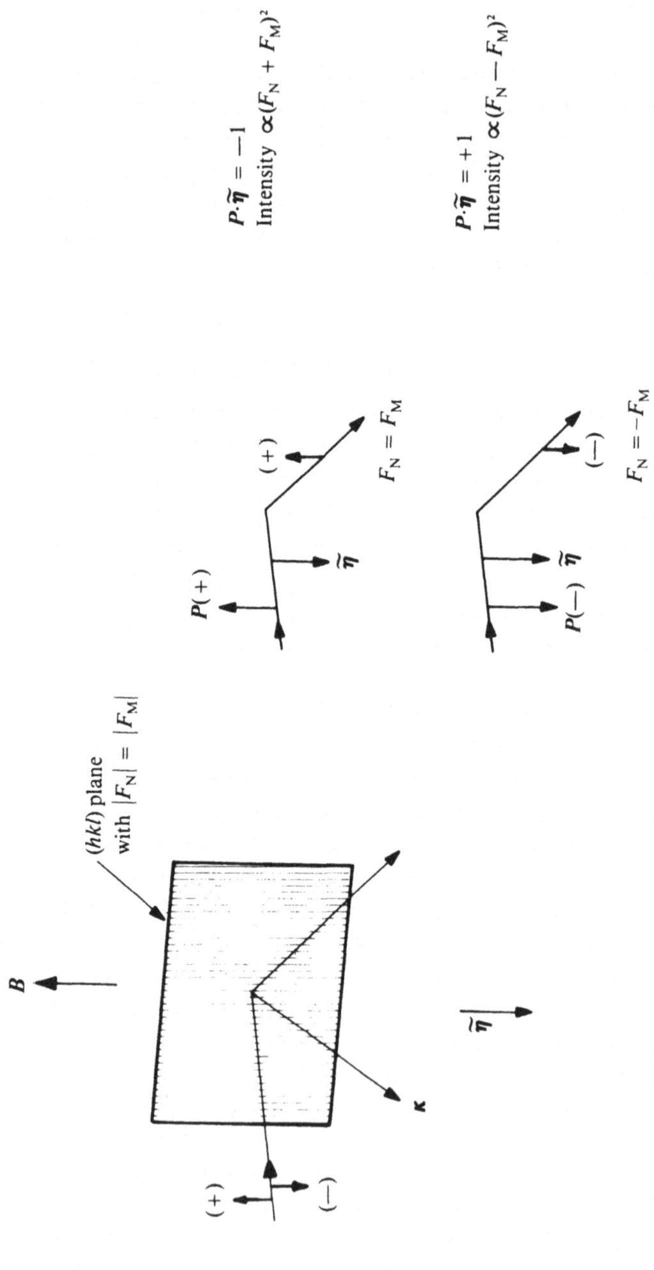

Fig. 3.1 Principle and geometry used for producing polarized neutron beams by Bragg reflection from magnetized ferromagnetic single-crystal monochromators.

largely by Delapalme *et al.* (1971). Unfortunately the Heusler alloy (222) reflection has a much higher nuclear structure factor than the (111) reflection, and second-order contamination can be problematical; this is often removed by using λ-selective resonance absorption filters. A similar material developed by Bednarski *et al.* (1980) is $Fe_{2.95}Mn_{0.05}Si$; these authors claim that the (111) reflection has a polarizing efficiency ~95%, and produces an intensity gain between twice and three times that for $Co_{0.92}Fe_{0.08}$.

The peak reflectivities and widths obtained for Heusler alloy crystals with a mosaic spread of 0.186°, and measured as a function of crystal thickness at two neutron wavelengths by Freund *et al.* (1983), are shown in Fig. 3.2. These workers achieved a significant gain in the efficiency of utilization of these crystals by using the concept of vertical focusing; see Currat (1973). Such focusing is most advantageous in a variable-energy instrument when the curvature of the crystal assembly can be varied. The arrangement chosen by Freund *et al.* (1983) used five or eight rectangular crystals with dimensions $75 \times 15 \times 7\ \text{mm}^3$ mounted with their long axes horizontal on a flexible titanium backing plate, and this plate was curved vertically. The crystals were magnetized in a 0.12 T horizontal field supplied by $SmCo_5$ permanent magnets. Tests on the magnetized assembly showed that at a neutron wavelength $\lambda = 4$ Å the peak

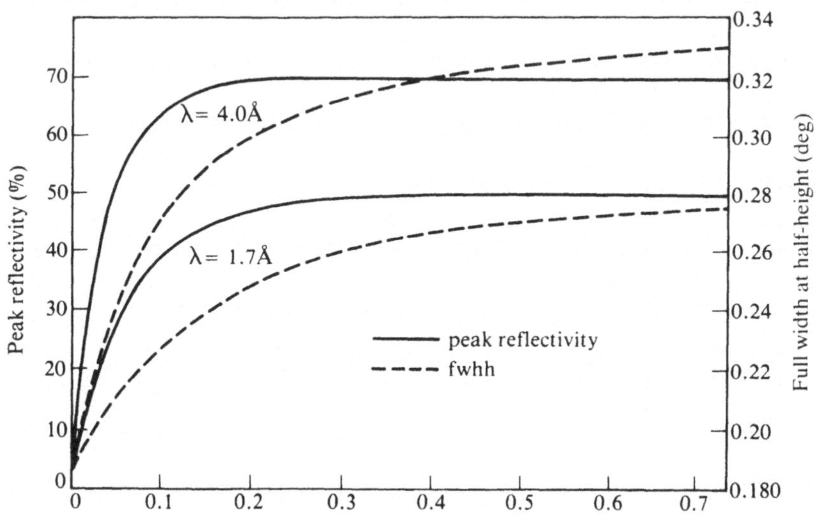

Fig. 3.2 Peak reflectivities (full curve) and profile widths (broken curve, f.w.h.h.) as a function of crystal thickness at two neutron wavelengths for a Heusler alloy polarizing monochromator with a 0.186° mosaic spread (from Freund *et al.* 1983).

TABLE 3.1 Properties of polarizing crystal monochromators

	$Co_{0.92}Fe_{0.08}$	Cu_2MnAl	Fe_3Si	$^{57}Fe:Fe$	$HoFe_2$
Matched reflection $\lvert F_N\rvert \sim \lvert F_M\rvert$	(200)	(111)	(111)	(110)	(620)
d-Spacing (Å)	1.76	3.43	3.27	2.03	1.16
Take-off angle $2\theta_B$ at 1 Å (deg)	33.1	16.7	17.6	28.6	50.9
Cut-off wavelength, λ_{max} (Å)	3.5	6.9	6.5	4.1	2.3

reflectivity was 73% for (−) spin neutrons, and with $|P| > 0.95$. The polarizing efficiency remained unchanged at $\lambda = 2.36$ Å, but the peak reflectivity reduced to ~50%.

One serious disadvantage with the Heusler alloy monochromator is its low take-off angle ($2\theta_B$) especially at neutron wavelengths ~1 Å and shorter. This is shown in Table 3.1, which is a compilation by Freund and Forsyth (1979) of some of the important parameters of five polarizing crystal monochromators. The *d*-spacings for the Cu_2MnAl and Fe_3Si (111) reflections are similar to those for the (002) reflection in pyrolytic graphite and the (111) reflection in Fe, so these crystals can be combined to provide an approximately focused geometry in a double-crystal arrangement.

Iron is an alternative to $Co_{0.92}Fe_{0.08}$, but it has much less absorption. Structure factor matching is best achieved by using Fe mixed with ^{57}Fe and 3% Si (see Reed *et al.* (1973)). The polarizing efficiency is high, but so also is the material cost, and this monochromator is unlikely to be used widely at neutron establishments.

There has also been some effort directed towards developing new rare-earth–transition metal alloys as 'large θ_B' crystal polarizers, but the results so far have been disappointing. The idea is to use rare-earth atoms with large moments (~10β) at relatively large κ or $(\sin\theta)/\lambda$ values, since their magnetic form factors fall off more slowly than those of 3d electrons. The suitability of $HoFe_2$ in this respect is discussed by Freund and Forsyth (1979).

3.1.2 Polarizing filters

Polarizing filters operate by selectively removing one of the neutron spin states from an incident beam, and allowing the other spin states to be transmitted with only moderate attenuation. The spin selection process can be either by preferential scattering or by preferential absorption. The polarizing action is in general neutron-energy-dependent, and several filters have now been developed which give excellent broad-band or white-beam polarizing characteristics.

The important parameters which determine the effectiveness of a neutron polarizing filter are:

(i) the polarizing efficiency P, i.e. the polarization attained by an unpolarized neutron beam on being transmitted through the filter,

$$P = (T_+ - T_-)/(T_+ + T_-); \qquad (3.1.3)$$

(ii) the total transmittance T, i.e. the (total number of transmitted neutrons)/(number of neutrons incident on filter) ratio,

$$T = (T_+ + T_-)/2; \qquad (3.1.4)$$

where T_+ and T_- are the transmittances for the 'up' and 'down' neutron spins. The filter designer has some control of these parameters, since increasing the thickness improves the spin selection or polarizing efficiency, whereas the transmittance decreases. The maximization of a quality factor $P\sqrt{T}$ is often used to optimize the filter thickness, though in general the optimum thickness should be evaluated by taking into account the nature of the experiment. In polarization analysis experiments where there is a small amount of spin-flip scattering, it is more beneficial to increase P at the expense of T, i.e. to use a thicker filter than that given by maximizing $P\sqrt{T}$ (Williams 1973).

We now evaluate these parameters for a generalized polarizing filter, for which the (+) and (−) spin total cross-sections are given by

$$\sigma_\pm = \sigma_0 \pm \sigma_p. \qquad (3.1.5)$$

Here σ_0 is a spin-independent cross-section and $\sigma_p = \frac{1}{2}(\sigma_+ - \sigma_-)$ is the polarization cross-section. Substitution into equns (3.1.3) and (3.1.4) gives a polarizing efficiency

$$P = -\tanh(\sigma_p N t) \qquad (3.1.6)$$

and a transmittance

$$T = \exp(-\sigma_0 N t)\cosh(\sigma_p N t), \qquad (3.1.7)$$

where N is the density of the atom or nucleus responsible for the spin selection, and t is the filter thickness. These two equations can be combined to give universal 'constant polarization' curves which characterize the behaviour of any polarizing filter in terms of the cross-section ratio (σ_p/σ_0); examples are shown in Fig. 3.3. It is assumed that the filter thickness, which varies as shown across the constant-P curves, can be increased or decreased at will. Obviously for positive σ_p/σ_0 values the polarizing efficiency P is negative, and vice versa. If we set the requirements that $|P| > 0.95$ and $T > 0.2$ in any practicable polarizing filter, then we see from Fig. 3.3 that this requires $(\sigma_p/\sigma_0) \gtrsim 0.65$, and this

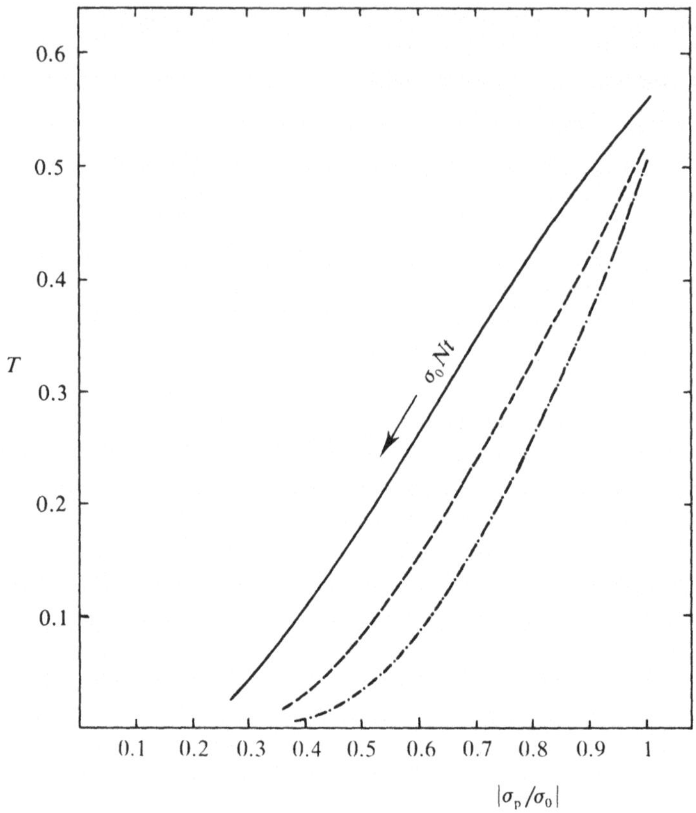

Fig. 3.3 Universal 'constant-P' curves which describe the properties of polarizing filters. Full curve, $|P| = 0.80$; broken curve, $|P| = 0.95$; chain curve, $|P| = 0.99$.

is a good starting point for assessing the potential of a polarizing filter material. We now describe three types of neutron polarizing filters: (a) the magnetized-iron filter, (b) filters based on spin-dependent absorption by polarized nuclear resonances, and (c) polarized-proton polarizing filters.

Magnetized iron filter

Our review of polarized neutron beam developments in Chapter 1 showed that the magnetized iron filter was the first neutron polarizer. The method depends on the interference between the nuclear and magnetic scattering in a magnetized polycrystalline block, and the polarization cross-section is given by

$$\sigma_p = N_c^2\lambda^2 \sum_{\substack{hkl \\ d \geq \lambda/2}} (2bp\boldsymbol{q} \cdot \boldsymbol{P})d_{hkl}, \tag{3.1.8}$$

where N_c is the number of unit cells per unit volume, d_{hkl} are interplanar spacings, and $\boldsymbol{q}$ is the magnetic interaction vector which equals $(\tilde{\boldsymbol{\kappa}}(\tilde{\boldsymbol{\kappa}} \cdot \tilde{\boldsymbol{\eta}}) - \tilde{\boldsymbol{\eta}})$. Here b and p are the nuclear and magnetic scattering amplitudes for iron respectively. This equation can be rewritten in terms of the b.c.c. unit cell dimension a_0 as

$$\sigma_p = (\lambda^2/a_0^3) \sum_{\substack{hkl \\ d \geq \lambda/2}} (2bp\boldsymbol{q} \cdot \boldsymbol{P})d_{hkl}. \tag{3.1.9}$$

By using the directional dependence of $\boldsymbol{q}$ it can be shown (see for example Bacon (1975)) that the average value of $\boldsymbol{q} \cdot \boldsymbol{P}$ in a saturated ferromagnetic powder is $\frac{1}{2}(1 + \sin^2\theta_B)$, where θ_B is the Bragg diffraction angle. The polarization cross-section then becomes

$$\sigma_p = (\lambda^2/a_0^3) \sum_{\substack{hkl \\ d \geq \lambda/2}} bp(1 + \sin^2\theta_B)d_{hkl}, \tag{3.1.10}$$

and this is easily evaluated as a function of λ if the $(\sin\theta_B)/\lambda$ dependence of the decrease in p due to the magnetic form factor is known. The calculations of Steinberger and Wick (1949) are shown in Fig. 1.9. We note that σ_p approaches 10 barns atom^{-1} near the Fe cut-off wavelength at 4 Å, and this should be compared with the combined scattering and absorption cross-section values near this wavelength of $\sigma_0 \simeq 16.8$ barns atom^{-1}; hence the maximum possible value of (σ_p/σ_0) in iron is ~0.59. Figure 3.3 shows that it should therefore be possible to achieve a polarizing efficiency $P \sim 0.80$ with a beam transmittance $T \sim 0.30$ in polycrystalline Fe at wavelengths close to the cut-off. This performance, however, is only a theoretical maximum, since in practice iron can never be completely saturated, and domain misalignment causes an appreciable depolarization of the neutron beam as it crosses the filter. Magnetized iron filters are used as the polarizer and spin analyser in the LONGPOL polarization analysis instrument first described by Campbell *et al.* (1974), which operates at a wavelength $\lambda \sim 3.6$ Å. The polarizing efficiency of a single filter is typically $P \sim 0.50$ with a transmittance $T \sim 0.25$. This polarizing filter method is useful in a continuous-source diffuse scattering instrument, but the wavelength over which it is effective is very restrictive and is limited to $\lambda \sim 3.5$–4 Å.

Filters using spin-dependent absorption by polarized nuclei

The second polarizing filter method uses the spin-dependence of the absorption cross-section of polarized nuclei at their nuclear resonance energies, and can provide good polarizing capabilities over a wide neutron energy range. The basic principle of the method has been described in Section 2.5, where the cross-sections for the two neutron spin states are given by equn (2.5.15):

$$\sigma_{A\pm} = \sigma_A(1 \pm \rho P_N).$$

The polarizing efficiencies and transmittances of this type of filter can be calculated using equns (3.1.6) and (3.1.7) with σ_p replaced by $\sigma_A \rho P_N$. Similarly, $|\sigma_p/\sigma_0|$ in Fig. 3.3 is equivalent to $|\rho P_N|$, thus the value of the product of the statistical spin factor and nuclear polarization can be used as a useful criterion for assessing the usefulness of these polarizing filters; following our earlier discussion we require $|\rho P_N| \gtrsim 0.65$. The need to produce a high nuclear polarization is the commonest limiting factor in determining the viability of this type of polarizing filter.

In the resonance absorption filters so far examined, the nuclear polarization has been created statically, i.e. under conditions in which the nuclear spins are cooled to the low temperatures of their lattice environments. The unequal populations of the magnetic substates results from the direct interaction of the nuclear magnetic moments with the magnetic field which acts at the nuclei. The nuclear polarization (P_N) produced depends on the ratio of the magnetic field B and the lattice temperature T, and is given by the Brillouin function

$$P_N = \frac{2i+1}{2i}\coth\left(\frac{2i+1}{2}\frac{\mu_N B}{k_B T}\right) - \frac{1}{2i}\coth\left(\frac{\mu_N B}{k_B T}\right) \qquad (3.1.11)$$

in which i and μ_N are the nuclear spin quantum number and the nuclear magnetic moment respectively; k_B is Boltzmann's constant. The large magnetic field can be provided in two ways. The first is to apply an external magnetic field; this is called 'brute force' polarization. The second is to utilize the nuclear hyperfine field due to the unpaired electrons in an atom; this is called hyperfine polarization. This second method frequently gives much higher nuclear polarizations than the first. Nuclear hyperfine fields can range up to ~700 T, as for example in ^{165}Ho, whereas laboratory-generated fields are limited to ~10 T. Fortunately there are several rare-earth nuclei which have large hyperfine fields and useful nuclear resonances.

The first practical filter using this method was proposed and constructed at the Rutherford Appleton Laboratory and contained polarized ^{149}Sm nuclei (see Williams (1973); Freeman and Williams (1978)). ^{149}Sm has a strong thermal neutron resonance centred at an energy $E_R =$

0.097 eV with a spin $(i+\frac{1}{2})$, where $i=\frac{7}{2}$. The polarizing filter material first tested was a deuterated cerous magnesium nitrate single crystal, $Ce_2Mg_3(NO_3)_{12}24D_2O$, in which about 6% of the Ce atoms were replaced by ^{149}Sm atoms (CSMN). This deuterated paramagnetic salt was used rather than the hydrated one so as to reduce the neutron-spin-independent attenuation. The ^{149}Sm nuclear polarization was produced statically by applying a magnetic field $B\sim 0.1$ T along the crystal c-axis and by cooling the crystal to temperatures $T<0.020$ K in the mixing chamber of a ^{3}He/^{4}He dilution refrigerator. The hyperfine interaction along the c-axis of CSMN is ~0.041 K in temperature units, and this corresponds to a hyperfine field at the nucleus ~300 T. Under the conditions of the experimental tests, ρP_N was estimated to be ~ -0.7. Extensive measurements of the performance of this filter were made at two wavelengths, and on the basis of the results the polarizing efficiency and transmittance variations over the whole thermal neutron range were estimated; these are shown in Fig. 3.4. Account was taken of the effect of the $(i-\frac{1}{2})$ capture in ^{149}Sm cross-section referred to in Section 2.5.

The main difficulty with CSMN is that it is an insulator, and it has long been recognized that a metallic samarium material would offer two main advantages: better thermal conductivity and shorter spin-lattice relaxation times (Williams 1974). Various types of metallic filter materials are potentially suitable such as saturated paramagnets or ferromagnets, and even permanent magnets. The latter, in the form of $SmCo_5$, is particularly attractive and its performance has been evaluated at the Rutherford Appleton Laboratory (see Mayers *et al.* (1986)). The best results obtained by these workers were with a filter of sintered $SmCo_5$ cooled to 0.015 K, and are shown as the broken lines in Fig. 3.5. Also shown, for comparison purposes, are the theoretical maximum P and T values for the atomic thickness of the filter investigated; the measured values are slightly lower due to the beam depolarization as it traverses the magnetic domains in the filter. The measured polarizing efficiency was between 98% and 67%, with corresponding total transmittances of 10% and 33% within the neutron energy range from 10 meV to 150 meV.

The use of polarized ^{149}Sm nuclei in neutron polarizing filters has also been pursued in the USSR. Zhuchenko *et al.* (1984) have tested a $Sm_{0.1}Y_{0.9}Co_5$ single crystal at a temperature $T\sim 0.169$ K, and concluded that the depolarization in this material was acceptably small at 2% in applied fields $\gtrsim$0.4 T.

The only other resonance filter that has been examined experimentally is polarized europium, of which the main isotope is ^{151}Eu. This has two broad absorption resonances at energies $E_R=0.32$ eV and $E_R=0.46$ eV, and offers the possibility of providing high polarizing efficiencies up to neutron energies ~0.6 eV. The best material seems to be the ferromagnetic semiconductor EuS which has been tested by Felcher (private

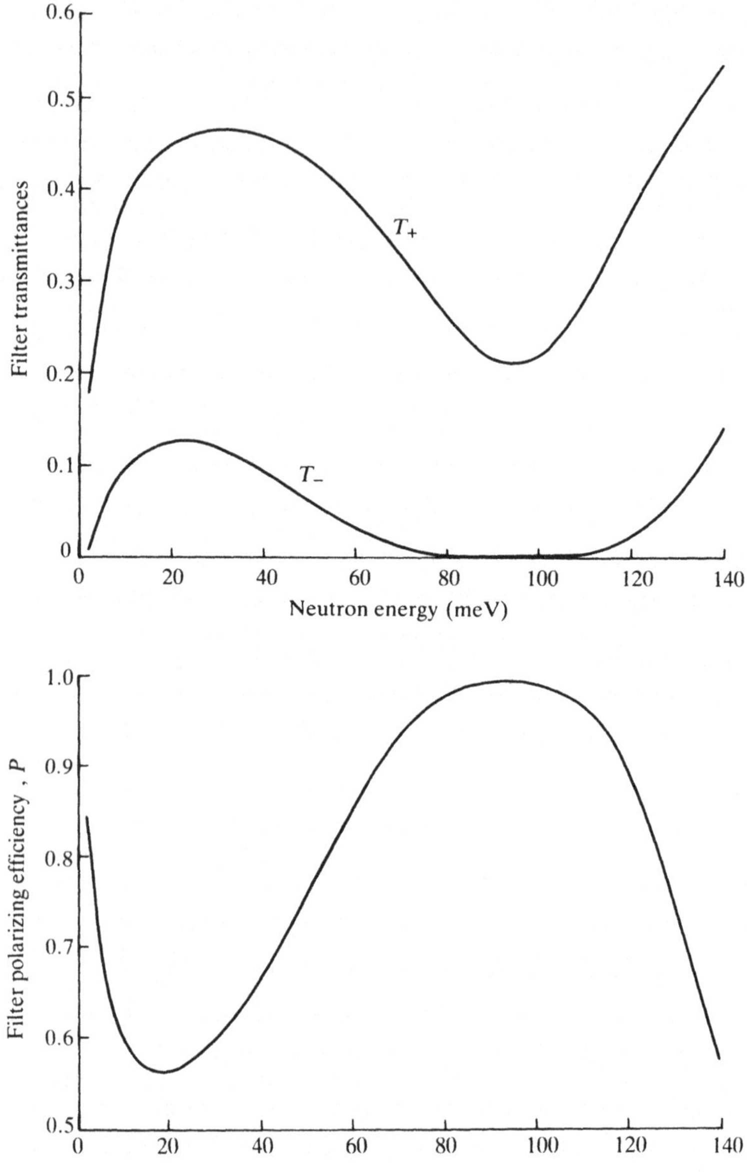

Fig. 3.4 Polarizing efficiencies and transmittances for a ^{149}Sm polarizing filter based on the measurements of Freeman and Williams (1978).

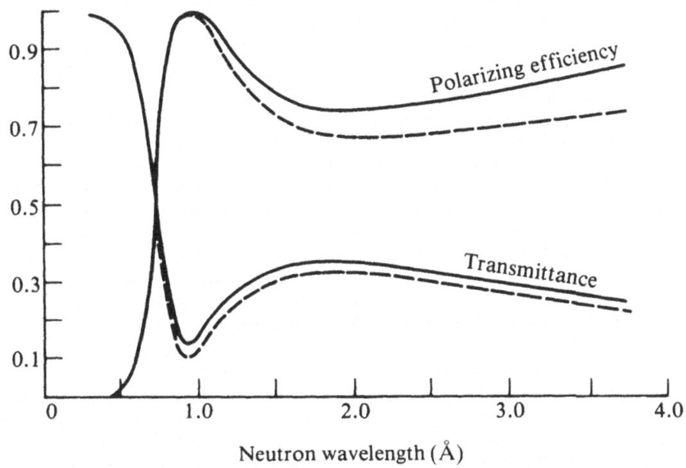

Fig. 3.5 Theoretical maximum polarizing efficiencies and transmittances (full curve) for a $SmCo_5$ polarizing filter as a function of the neutron wavelength compared with measured values (broken lines) (from Mayers *et al.* 1986).

communication, 1986), for which the ^{151}Eu hyperfine interaction is 4.003×10^{-2} K, i.e. almost identical to that for ^{149}Sm in CSMN. The measured transmittances and polarizing efficiencies were close to the theoretical predictions; see Fig. 3.6 for the calculated energy dependence of its polarizing efficiency.

Special mention should be made of the attractive features of polarized ^{3}He as a broad-band polarizing filter, though no experiments have been performed so far. It has excellent absorption characteristics up to a neutron energy ~1 eV and the capture proceeds via a $(i - \frac{1}{2})$ excited state, i.e. $\rho = -1$. The best prospect of achieving a high nuclear polarization in ^{3}He seems to be by using Pomeranchuk cooling in which a liquid sample below ~300 mK is compressed and solidified; temperatures ~2 mK can be attained starting from a liquid temperature ~50 mK. Heeringa (1985) calculated that if the sample cell is placed in a strong external magnetic field a ^{3}He nuclear polarization of 54.6% should be achievable at $T = 10$ mK with $B = 10$ T.

Cywinski *et al.* (1981) have estimated the expected performances of a number of polarizing filters based on isotopes that can be polarized by the hyperfine interaction method. The results for some of the more promising filters are summarized in Table 3.2, where the atomic thicknesses correspond to those which give a peak polarizing efficiency of 0.995. The energy dependences of the polarizing efficiencies in these filters are illustrated in Fig. 3.6. It was concluded that the method can provide a broad-band polarizing behaviour up to a neutron energy ~0.6 eV, but only 'monochromatic' polarization at higher energies.

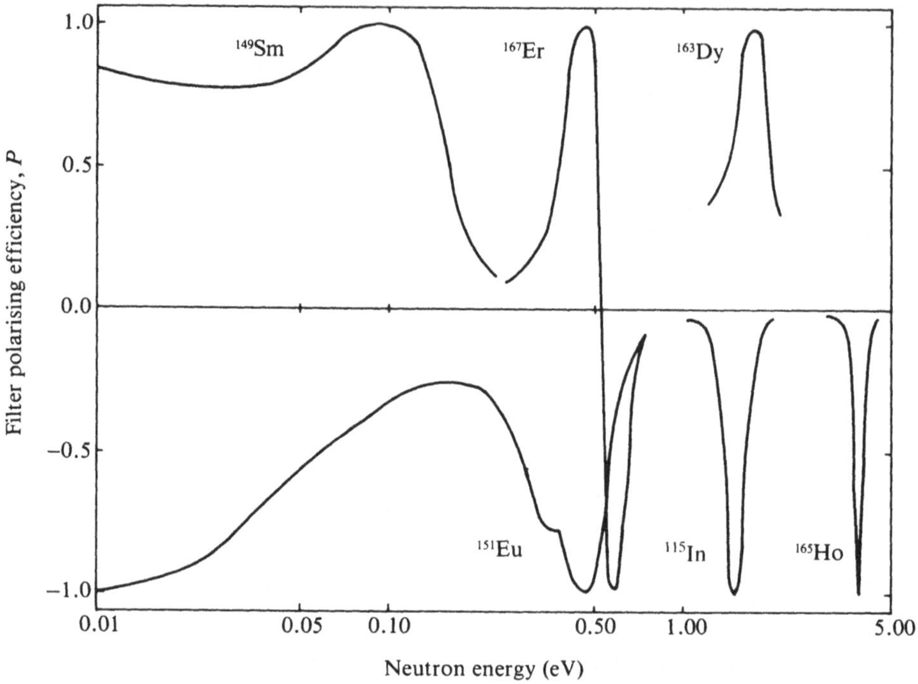

Fig. 3.6 Polarizing efficiencies as a function of the neutron energy for the polarizing filters specified in Table 3.2 (from Cywinski *et al.* 1981).

TABLE 3.2 Polarizing filter properties for several absorbing isotopes for which hyperfine nuclear polarization is feasible. The filter temperature is assumed to be 10 mK and the atomic thickness Nt, for the naturally-occurring element, is selected to give a neutron polarization $P = 0.995$ at the peak of the nuclear resonance (after Cywinski *et al.* 1981)

Isotope	Hyperfine field (tesla)	Nuclear polarization, P_N	Nt (at cm^{-2})	Transmittance at peak polarizing efficiency, $P = 0.995$
^{115}In	−30	0.922	1.3×10^{20}	0.188
^{149}Sm	−300	−0.959	2.5×10^{20}	0.181
^{151}Eu	−30	0.888	4.5×10^{20}	0.088
^{163}Dy	−720	1.000	3.0×10^{21}	0.500
^{165}Ho	−730	1.000	7.7×10^{20}	0.213
^{167}Er	−726	−0.996	1.7×10^{20}	0.097

Polarized-proton filters

The third type of polarizing filter is the polarized-proton filter, which utilizes the spin dependence of nuclear scattering, and is effective up to neutron energies ~1 keV. The polarization cross-section depends strongly on the neutron energy and its values have been determined by Lushchikov *et al.* (1970). At neutron energies which are large compared with the chemical binding energy of hydrogen ($E \gtrsim 1$ eV), the appropriate scattering process is that for a free particle and the spin-dependent cross-sections are given by equn (3.1.5) with $\sigma_p = -\sigma'_p P_H$, where P_H is the proton polarization. Lushchikov *et al.* (1970) give, for these cross-sections, $\sigma_0 = 20.35$ barn at.$^{-1}$ and $\sigma'_p = 16.7$ barn at.$^{-1}$. At low neutron energies ($E \lesssim 10$ meV) the cross-sections take the form (see equn (2.2.19) and also Schermer and Blume (1968)),

$$\sigma = \sigma_0[1 - (i/(i+1))P_H^2 - (1/(i+1))\boldsymbol{P}_H \cdot \boldsymbol{P}], \qquad (3.1.12)$$

or, using numerical values obtained by Lushchikov *et al.*,

$$\sigma_\pm = 79.7[1 - \tfrac{1}{3}P_H^2 \mp \tfrac{2}{3}P_H]. \qquad (3.1.13)$$

In the intermediate energy region between 10 meV and 1 eV all theoretical expressions for the cross-sections become very complex owing to interference and inelastic scattering effects. Lushchikov *et al.* (1970) fitted their data in this region to an equation of the same form as equn (3.1.13),

$$\sigma_\pm = \sigma_1 - \sigma_2 P_H^2 \mp \sigma_3 P_H, \qquad (3.1.14)$$

but with σ_1, σ_2, and σ_3 as empirical constants. Figure 3.7 shows polarizing efficiency and transmittance values for a polarized proton filter as a function of neutron energy, and calculated using the above expressions. The value used for the proton atomic thickness $(Nt)_H$ was 6×10^{22} at cm^{-2}.

The viability of the method again relies on achieving a high nuclear polarization P_H. A proton polarization $P_H = 0.7$ gives $|\sigma_p/\sigma_0| \sim 0.56$ in the cold neutron region, and provides a neutron polarizing filter with only the minimum acceptable performance. Proton polarizations ~80% are therefore needed to give a useful polarizing filter. There are three main methods for creating targets with high proton polarizations; two are dynamic and the third is static.

The first, and historically the most-tried technique, is to use magnetic resonance dynamic polarization as illustrated in Fig. 3.8. In this method some of the protons in the target are in dipolar coupling with the saturated electron spins of a paramagnetic dope, and by irradiating at well defined frequencies above and below that of the main ESR transition of the dope it is possible to create both positive and negative proton

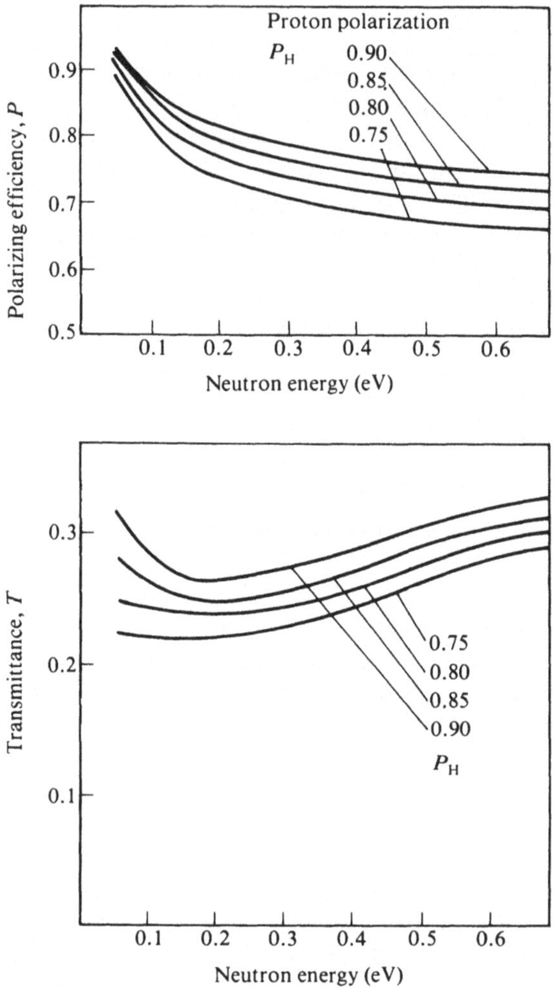

Fig. 3.7 Calculated polarizing efficiency and transmittances as a function of neutron energy and at different proton polarizations for a polarized proton filter with proton thickness $(Nt)_H \sim 6 \times 10^{22}$ at cm^{-2}.

polarizations. The overall result is that the high electron spin polarization produced in a magnetic field ~2.5 T and at a temperature ~1 K is transferred to the proton spin system. The first polarized proton target developed for high-energy physics experiments was $La_2Mg_3(NO_3)_{12}24H_2O$ (LMN) doped with Nd^{3+}; this was the material used by Lushchikov *et al.* (1970) to obtain their neutron cross-section data. The possibility now exists of using hydrogen-rich organic compounds doped with Cr^{5+} complexes as neutron polarizing filter materials. These are being used by Japanese workers (see Ishida *et al.* (1985)),

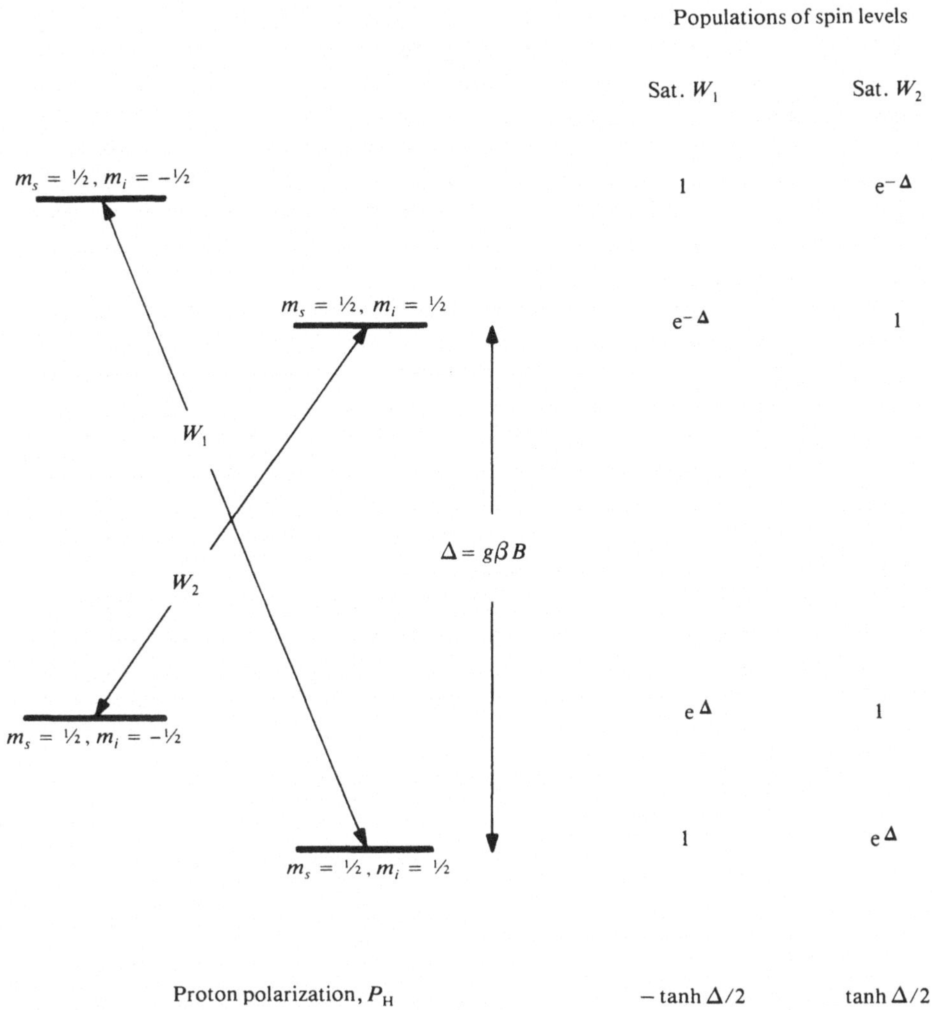

Fig. 3.8 Energy level scheme for dynamic proton polarization using magnetic resonance. Proton spin *i* and electron spin *s* in dipolar coupling, e.g. Nd^{3+} in LMN, Cr^{5+} in ethylene glycol. Saturation of transitions W_1 or W_2 enables a high electron spin polarization to be transferred to the proton spin system. The method is commonly known as the 'solid-state effect'.

though the performance so far achieved falls just short of the $P_H > 0.7$ criterion set above.

A simpler dynamic polarization method, which uses the spin refrigerator principle first introduced by Langley and Jefferies (1966) has been pursued by Felcher *et al.* (1984). In this method the proton spins are coupled to the electron spins of an anisotropic paramagnetic ytterbium dope in a yttrium ethyl sulphate (YES) single crystal. By rotating the crystal in a magnetic field, the large electron spin polarization that occurs for one crystal setting can be transferred into the proton spin system. A proton polarization $P_H = 0.41$ was obtained with the crystal cooled to ~1.2 K and rotated at a frequency of 150 Hz in a field of 1.46 T. The authors extrapolated their results to predict that a proton polarization $P_H \simeq 0.65$ was ultimately achievable using this method, and we have seen that even this produces a polarizing filter performance which is only marginally adequate.

Finally, it is worthwhile commenting on the possibility of using a statically polarized proton target as a neutron polarizing filter. There has been some remarkable progress in producing 'brute force' polarized-proton targets for nuclear physics applications, and Fig. 3.9 shows the results obtained by Herringa (1985) for TiH_2 in an applied field of 8.2 T and at a temperature around 12 mK. The predicted value of the proton polarization in a 10 T field at 10 mK is $P_H = 0.77$, and it would seem that this method is a promising one for the future, and should provide a viable white-beam neutron polarizing filter.

In conclusion, we have seen that the magnetic resonance dynamic polarization and 'brute force' static polarization methods offer the most promise for producing effective white-beam neutron polarizing filters with polarized protons. The operating conditions for both these filters are,

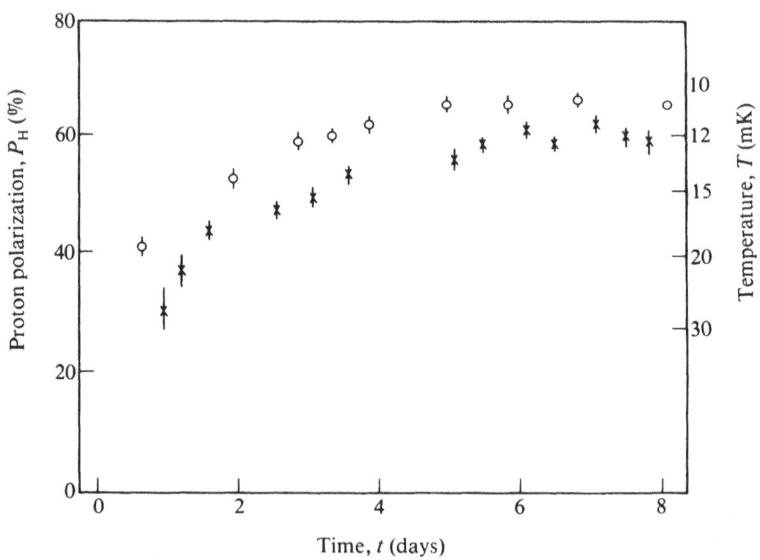

Fig. 3.9 Results of the build-up of proton polarization in TiH_2 by the brute-force method in a magnetic field $B = 8.2$ T. ○, ^{60}Co γ-anisotropy data; ×, neutron transmission data (from Heeringa 1985).

however, technically non-trivial to achieve, but of the two the static method is probably the easier.

3.1.3 Optical polarizers

The theory of neutron optical phenomena was discussed in Section 2.4, where it was stated that many of the effects described have been applied to the construction of neutron polarizers. In this section we describe some of these polarizers, emphasizing the improvements that have resulted from exploiting the advances made in the technology for producing thin metallic films.

The technique of polarizing cold neutron beams by mirror reflection has been widely used since Hughes and Burgy (1951) first produced a 100% polarized beam with a cobalt mirror. A perfect polarizing surface has $\bar{b} = |(B/B_s)p|$ in equn (2.4.10), and it follows that only one neutron spin state is reflected at all angles up to the critical reflection angle. An important limitation of the mirror reflection method is that the critical ratio $(\theta/\lambda)_c$ of glancing angle to wavelength for (+) spin neutrons is small, and only reaches ~1.8 mrad Å^{-1} in the best polarizing mirrors. Hence solid angle is often lost when using thermal and epithermal divergent beams. The first mirror polarizer with a useful beam area was a focusing system of cobalt mirrors built by Abrahams *et al.* (1966). This was followed by a curved polarizing neutron guide constructed from the polished magnetic alloy Permendur, $Co_{49}Fe_{49}V_2$ (Berndorfer 1971). The matching of the magnetic and nuclear scattering amplitudes is much closer in this alloy than in pure cobalt. Both these devices, with their associated magnets, are extremely large and heavy, and they have now been superseded by polarizers which use thin films of ferromagnetic alloys in place of the bulk material mirrors.

The most successful thin-film polarizing mirrors have been made from evaporated cobalt–iron alloys; these are easily saturated in magnetic fields ~0.1 T, and their assemblies therefore require only modest permanent magnet systems. The optimum composition for the alloy was investigated by Penfold *et al.* (1975). These authors calculated the critical (θ/λ) values for (+) and (−) spin neutrons for all alloy compositions using known scattering length, magnetization and density data. Their results for two values of the magnetic induction ratio (B/B_s) are shown in Fig. 3.10. An alloy composition $Co_{60}Fe_{40}$ was recommended as one that gives a low $(\theta/\lambda)_{c-}$ at a reasonably large $(\theta/\lambda)_{c+}$; this conclusion was also reached independently by Drabkin *et al.* (1975).

The (−) spin neutrons which are not reflected by the film must be removed, and this is achieved in one of two ways: by neutron absorption in a carefully prepared underlayer, or by incoherent scattering in the substrate material. A necessary optical requirement for the underlayer

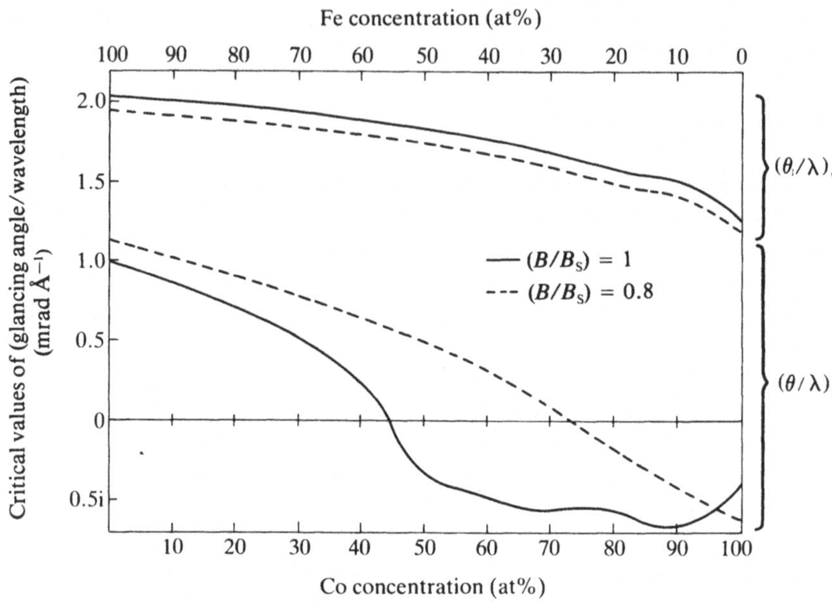

Fig. 3.10 Critical values of (glancing angle/wavelength) for (+) and (−) spin neutrons on magnetized Fe–Co mirrors as a function of the alloy composition (from Penfold *et al.* 1975).

and substrate materials must be that their refractive indices should match that for the (−) spin neutrons in the magnetized thin film. Drabkin *et al.* (1975) satisfied this requirement by using a $Gd_{15}Ti_{85}$ absorbing underlayer, whereas Hayter *et al.* (1978) used a hydrogenous plastic substrate to scatter out the unwanted neutrons. Later work by Childs *et al.* (1981) showed that an improved antireflecting sublayer performance was obtained by using a sublayer composed of a multilayer of the magnetic film sandwiched between thin Gd layers. The gadolinium layer thicknesses were gradually decreased as successive layers were evaporated, and the smooth refractive index change ensured good (−) spin neutron penetration and absorption in the sublayer with minimal reflectivity. Figure 3.11 shows polarization profiles measured by Childs *et al.* (1981) as a function of the ratio (θ/λ) for a series of $Fe_{0.3}Co_{0.7}$/Gd sublayers, where the total thickness of the Gd films in the sublayers was progressively increased. The polarization of the reflected beam is given by

$$P(\theta/\lambda) = (\mathcal{R}_+ - \mathcal{R}_-)/(\mathcal{R}_+ + \mathcal{R}_-), \tag{3.1.15}$$

where $\mathcal{R}_+$ and $\mathcal{R}_-$ are the reflectivities for the (+) and (−) spin neutrons. The improved performance observed in Fig. 3.11(d) is due to a more effective removal of the (−) spin neutrons. These polarization profiles

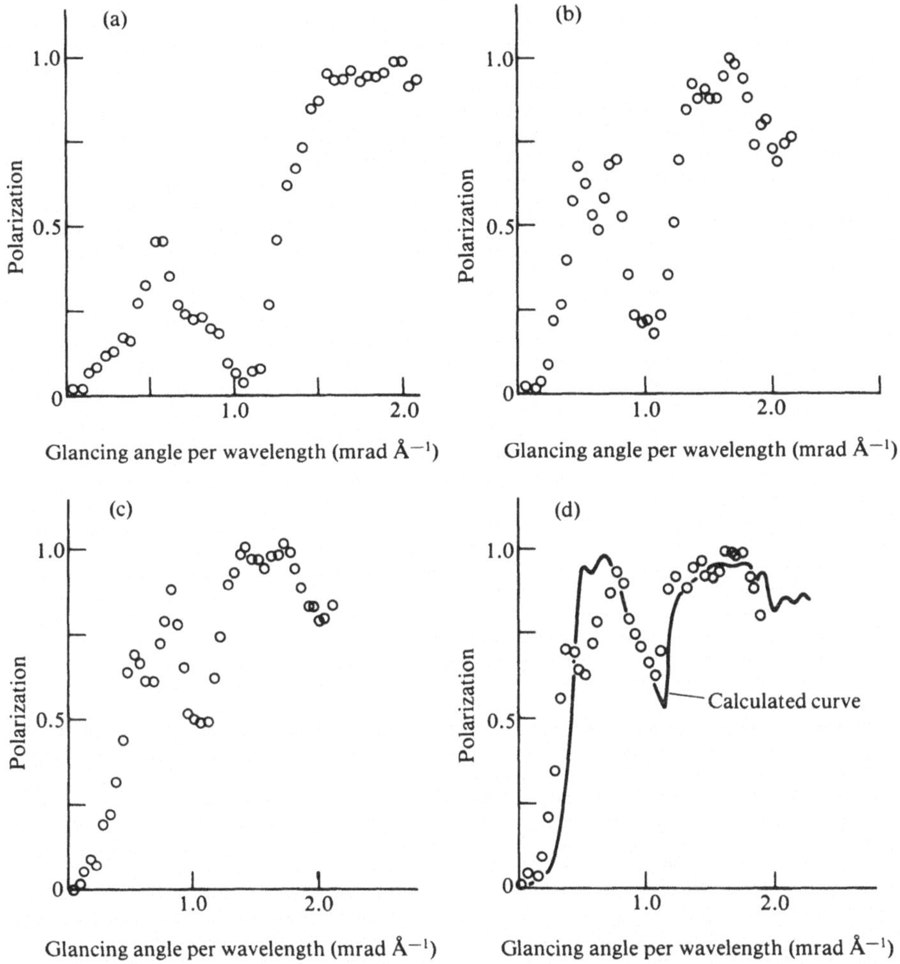

Fig. 3.11 Measured polarization profiles for $Fe_{0.3}Co_{0.7}$/Gd multilayer films with (a) 10 Å total Gd thickness, (b) 30 Å total Gd thickness, (c) 60 Å total Gd thickness; (d) shows the measured and calculated polarizations for a multilayer film with a 100 Å total Gd thickness (from Childs *et al.* 1981).

were also calculated using the method described in Section 2.4, and there was good agreement with experiment. Excellent antireflecting layer properties were also observed for similar sublayer structures when the magnetic film layers were replaced by manganese, which has a negative average coherent scattering amplitude $\bar{b}$.

Following the advances made by the Leningrad group in producing efficient thin-film cobalt–iron polarizing mirrors on glass, Drabkin *et al.* (1976) built a small-aperture polarizing neutron guide. Of greater significance, however, was the development of compact polarizing Soller

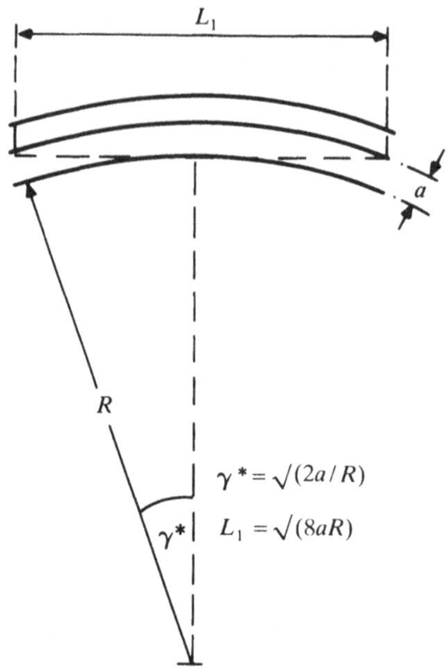

Fig. 3.12 Geometric parameters which define the characteristics of a curved polarizing Soller guide.

devices in which a large number of mirrors could be stacked together to form a series of polarizing and conducting channels. A curved Soller is a collection of one-dimensional curved neutron guides, the theory of which has been summarized by Jacrot (1970). A neutron guide is characterized by its radius of curvature R, its length L, and the channel width a; see Fig. 3.12. In order that all neutrons are reflected at least once within a channel, L must be greater than the 'length of direct view' L_1 given, in the approximation $a \ll R$, by

$$L_1 = (8aR)^{1/2}, \tag{3.1.16}$$

and this is related to a characteristic angle γ^* for the curved Soller, by

$$\gamma^* = (2a/R)^{1/2} = L_1/(2R). \tag{3.1.17}$$

Once the mirror material is specified (that is N, $\bar{b}$ and Bp/B_s are known), the behaviour of the guide is completely determined by the independent geometrical parameters a and R. The guide properties can then be usefully summarized by a single parameter λ_+^*, which is the wavelength at which θ_{c_+} in equn (2.4.10) is equal to the characteristic angle γ_+^*. This characteristic wavelength may be written as

$$\lambda_+^* = \{2\pi a/[RN(\bar{b} + Bp/B_s)]\}^{1/2}. \tag{3.1.18}$$

The characteristic angle γ_+^* is the largest reflecting angle at which (+)

spin neutrons can be guided by zig-zag reflections through the conducting channel, therefore (+) spin neutrons with wavelengths $\lambda < \lambda_+^*$ are transmitted only by single or multiple (garland) reflections on the concave surface of the channel, and those with wavelengths $\lambda > \lambda_+^*$ are transmitted by zig-zag (from both concave and convex surfaces) as well as single and garland reflections. The transmittance through the Soller guide decreases at short neutron wavelengths, since the smaller θ_{c_+} values mean that neutrons can only emerge from the guide at smaller solid angles near the concave surface. Polarizing guides have the property $\lambda_+^* \ll \lambda_-^*$, and maximum polarization will always be obtained at wavelengths $\lambda \sim \lambda_+^*$.

The first polarizing Soller guide was constructed by Hayter *et al.* (1978), and the stacked mirrors were thin films of $Fe_{40}Co_{60}$ alloy evaporated on to a plastic substrate material, poly(methylpentene), commercially known as TPX. The average $\bar{b}$ value in this substrate material is small and negative (-0.0277×10^{-12} cm), and this gives a neutron refractive index which is very similar to that for the (−) spin neutrons in the magnetized thin film. The main design parameters for the guide were: $a = 0.39$ mm; $R = 6.5$ m; length $L = 1.26L_1 = 180$ mm; number of channels = 85; beam area = 50 mm (height) × 40 mm (width): characteristic wavelength $\lambda_+^* = 6.52$ Å. The Soller guide was magnetized using a light-weight, yoked permanent magnet which produced a vertical applied field ~0.14 T in the direction of the film planes. The polarizing efficiency and transmittance values measured for two identical Sollers at a wavelength $\lambda = 6.5$ (±1.0) Å were 0.95 and 0.40 respectively, a significant improvement on the more bulky polarizing mirror systems described earlier.

A second type of polarizing Soller device, also using standard polarizing mirror surfaces, was developed by Rekveldt and Kraan (1983). In this ingenious system, thin Si crystal wafers with dimensions $(50 \times 25 \times 0.2)$ mm^3 are used both as substrates for iron–cobalt thin film mirrors, and as a conducting channel for the polarized neutrons. The geometrical parameters of this polarizer are shown in Fig. 3.13(a). The (−) spin refractive indices must be matched to the Si refractive index at one interface, and must be close to unity at the vacuum interface. This was arranged by using two magnetic alloy compositions separated by an absorbing antireflecting layer; details are shown in Fig. 3.13(b). The polarizing efficiency measured for a device with $2\beta = 16$ mrad was close to 100% at a neutron wavelength $\lambda = 1.08$ Å. Rekveldt and Kraan (1983) showed how the transmittance depends on θ_{c_+}, the bend angle 2β, and on the incident beam divergence α. The measured transmittance, however, was approximately one-fifth of the theoretical prediction; this was attributed to a much greater attenuation in the silicon crystal wafers then had been anticipated.

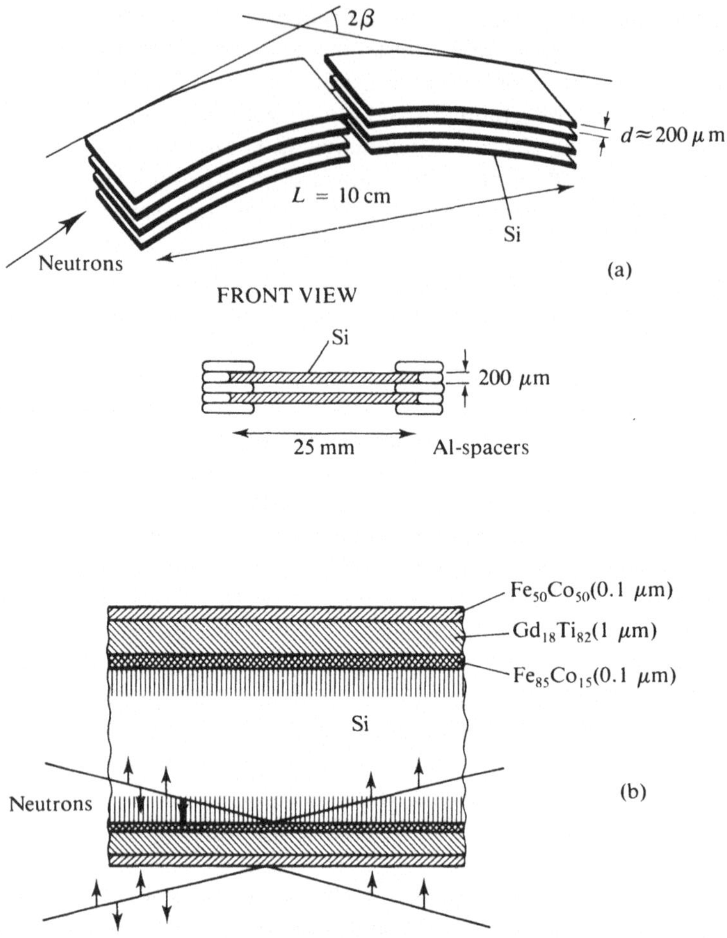

Fig. 3.13 Diagrams of (a) the general assembly, and (b) a cross-section through a Si wafer with its thin-film layers, for the curved mirror polarizing device described by Rekveldt and Kraan (1983).

We now consider the methods available for increasing the wave vector transfer or (θ/λ) range of mirror reflections by using thin-film interference in multilayers. The simplest of these multilayer structures is the multibilayer shown in Fig. 2.17(b); when one of the layers is magnetic this produces a polarized monochromatic beam. The method was first realized in practice by Lynn *et al.* (1976a) who evaporated about 30 alternating and equally thick layers of iron and germanium on to a glass substrate. For 80 Å layer thicknesses the polarized beam reflected from their device occurs at $(\theta/\lambda) \sim 3\,\text{mrad}\,\text{Å}^{-1}$, and their results are shown in Fig. 1.15. The resolutions of the monochromatic beams obtained from multilayer monochromators and mosaic crystal monochromators are determined in different ways. In the latter, the wavelength resolution is given by

$$\Delta\lambda/\lambda = \cot(\theta_B)\Delta\theta_B, \qquad (3.1.19)$$

where θ_B is the Bragg angle, and $\Delta\theta_B$ is its spread due to the incident beam divergence and the crystal mosaic. However, in the case of bilayers there are two terms,

$$\Delta\lambda/\lambda = \cot(\theta)\Delta\theta + \Delta d/d, \qquad (3.1.20)$$

with $\Delta\theta$ given mainly just by the beam divergence, and $\Delta d/d$ arising from the distribution of the bilayer spacings. From the point of view of reflected intensity it is important to equate these two contributions, and this is done by creating a distribution of bilayer thickness to match the known divergence characteristics of the neutron beam.

If we label the non-magnetic and magnetic components in the bilayer as A and B respectively, and consider the case in which a magnetizing field is applied perpendicularly to the scattering vector $\boldsymbol{\kappa}$, then the (+) and (−) spin reflectivities are proportional to

$$\mathcal{R}_{\pm} \propto [N_A b_A - N_B(b_B \pm p_B)]^2, \qquad (3.1.21)$$

where N_A and N_B are atomic densities, and b and p denote the nuclear and magnetic coherent scattering amplitudes. By a suitable choice of N and b it can be arranged that neutrons in one of the spin states experience a uniform refractive index, i.e. $\mathcal{R} = 0$, while the opposite-spin neutrons see a periodic modulation of the refractive index, and hence have a finite reflectivity when the Bragg condition is satisfied. Figure 3.14

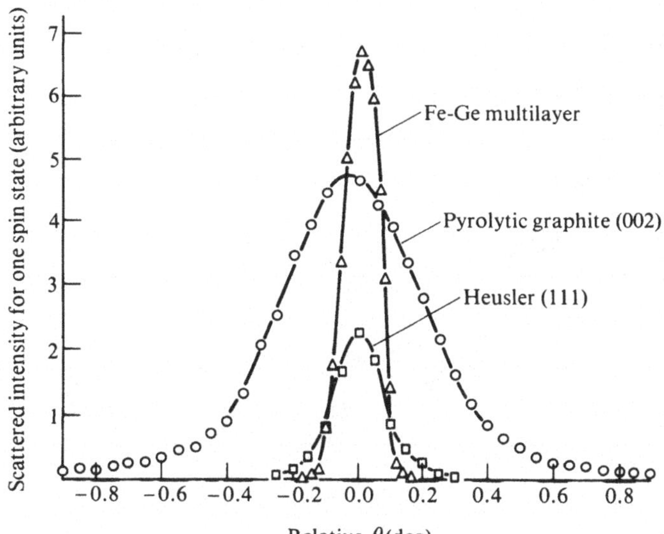

Fig. 3.14 Reflectivities and angular acceptances measured for an Fe–Ge multilayer, a pyrolytic graphite crystal (002) reflection, and a Heusler crystal (111) reflection at $k = 2.609$ Å^{-1} (from Majkrzak and Passell 1985).

compares the reflectivity observed by Majkrzak and Passell (1985) for a structure composed of 1500 FeGe bilayers of average thickness 55 Å, with those for the best non-polarizing mosaic crystal monochromator pyrolytic graphite, and also the polarizing monochromator Heusler alloy; the neutron wave vector for these measurements was $k = 2.609$ Å^{-1}. For the mosaic crystals, θ corresponds to the angle between the incident beam and reflection planes in a rocking curve measurement, whereas for the multibilayer it is the angle relative to the incident beam in a θ–2θ scan. It was concluded that the synthetic structure has a quarter of the integrated reflectivity of pyrolytic graphite, but is clearly superior to the Heusler alloy monochromator. Another important property of multibilayers is that the reflectivities of higher-order wavelengths are typically ~1% or less of that of the primary wavelength; this usually eliminates the need to use them with higher-order filters, and also means they can be conveniently combined, if required, with a pyrolytic graphite crystal to provide improved wavelength resolution and good intensity.

The (+) and (−) neutron spin intensities observed from a 10-bilayer Fe--Ge structure with $d = 108$ Å are shown in Fig. 3.15. The flipping ratio $R = I_+/I_-$ determined at $\kappa = 0.057$ Å^{-1} was 237, and corresponds to a polarizing efficiency in excess of 99.5%. It should be noted that the (+)

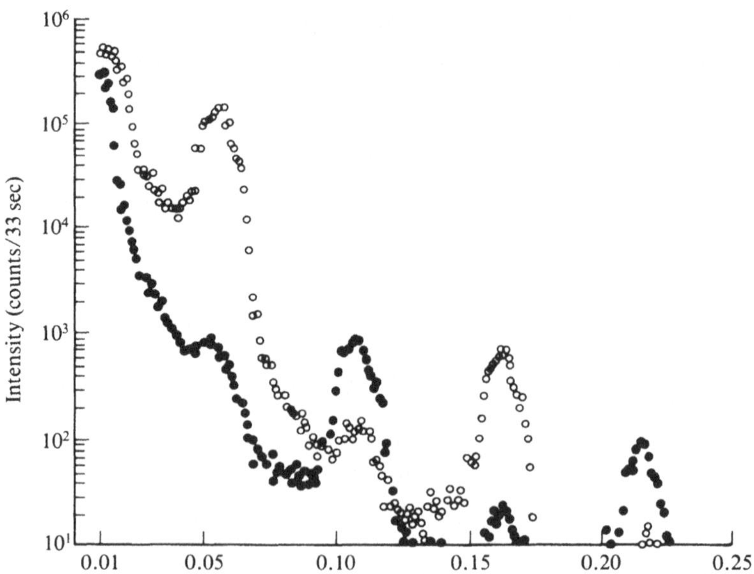

Fig. 3.15 (+) and (−) spin intensities as a function of the scattering vector κ for a Fe–Ge multibilayer with ten bilayers and an average bilayer thickness of 108 Å; ○, (+) spinstate; ●, (−) spin state (from Majkrzak and Passell 1985).

spin reflectivities become less than those for the (−) spins at the second- and fourth-order positions, and this is most easily explained by invoking the assumption that some interdiffusion takes place between the Fe and Ge layers. Interdiffusion is also important in determining the ultimate (θ/λ) values obtainable in multibilayers. The smallest d-spacings so far achieved with Fe–Ge are $d = 40$ Å (Majkrzak and Shirane 1982); with a multilayer which is 50 cm long this would reflect a 1 cm wide beam at a wavelength of $\lambda = 1.5$ Å.

An alternative multibilayer polarizer which uses the isotopic alloy mixture $^{63}Ni_{0.66}{}^{54}Fe_{0.34}$ as the magnetic layer B, and vanadium, with its low coherent scattering length $\bar{b} = -0.05 \times 10^{-12}$ cm as layer A, was proposed and tested by Gukasov *et al.* (1979). A multibilayer mirror of this composition with a d-spacing of 200 Å gave a 'spin' reflectivity ~75% and a polarizing efficiency $P \sim 90$–95% at the peak wavelength $\lambda \sim 3.5$ Å when $\theta = 0.5°$.

A different type of multilayer system in which the d-spacings between the layers are changed in a controllable way provides, through constructive interference, high reflectivities at scattering vectors up to two or three times the critical (θ/λ) values of normal mirrors. These devices, now called 'supermirrors', were proposed by Turchin (1967) and Mezei (1976a) and, again by using alternate layers of ferromagnetic and non-magnetic materials, can also act as polarizers. The optical principles by which they operate are described in Section 2.4.2. The first polarizing supermirror was constructed by Mezei and Dagleish (1977) who used 37 bilayers of iron and silver evaporated on to a glass substrate. This gave a reflectivity profile similar to that shown in Fig. 3.16(b); note the more extensive (θ/λ) range compared with Fig. 3.16(a). A serious drawback of the Fe–Ag supermirror is its relatively high $(\theta/\lambda)_{c_-}$ value, which gives $\mathcal{R}_+ = \mathcal{R}_- = 1$ and $P = 0$ at all (θ/λ) values up to approximately one-third of $(\theta/\lambda)_{c_+}$. This therefore only acts as a polarizer over the range $1.2 < \theta/\lambda$ (mrad Å^{-1}) < 3.6. The best polarizing supermirrors now use alternate layers of cobalt and titanium deposited on to glass, and were developed by Schärpf (1982). The (θ/λ) range of the polarizing reflection extends from essentially zero up to nearly 3.3 mrad Å^{-1} (see Fig. 3.16(c)). An important feature of this supermirror is the need to provide a special Ti–Gd antireflecting layer which is itself a 'concentration gradient' multilayer, in which thick Gd layers are first evaporated on to the substrate and are subsequently gradually made thinner. This antireflecting layer therefore operates using the same principle as used by Childs *et al.* (1981) for conventional iron–cobalt polarizing mirrors.

The supermirrors produced by Schärpf are deposited on to a glass substrate which is thin and flexible, and this is an ideal blade material for the curved Soller devices which were described earlier. Schärpf (1982) described a compact supermirror Soller in which the radius of curvature

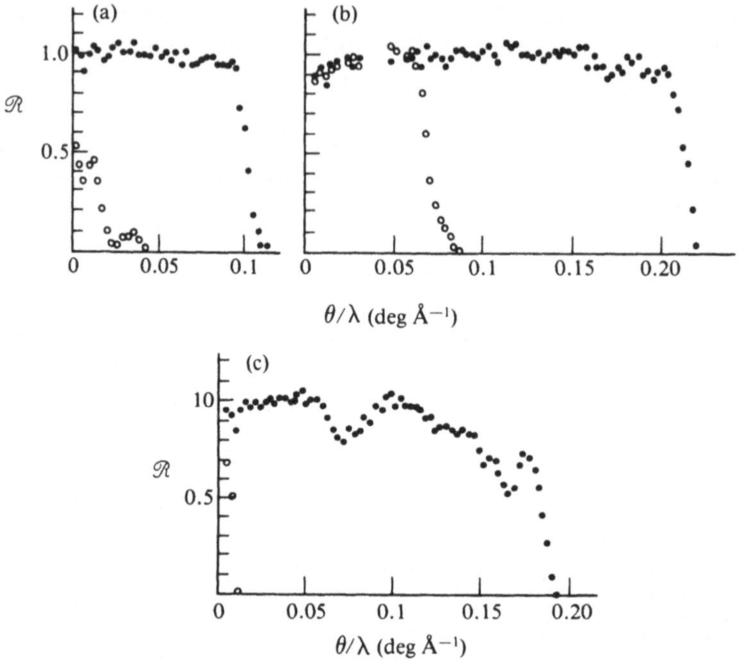

Fig. 3.16 Spin up (●) and spin down (○) reflectivities for (a) an Fe–Co alloy mirror, (b) a 80-layer Fe–Ag supermirror, and (c) a 80-layer Co–Ti supermirror. These measurements were made by Schärpf using 7 Å neutrons and are reproduced from Pynn (1984).

is adjustable, and which has excellent polarization and transmittance properties down to neutron wavelengths ~2 Å.

In concluding this section on optical neutron polarizers it should be remarked that the most versatile device so far produced is the 'superbender' of Schärpf, and that this is the product of a combination of several of the developments discussed earlier. It has been argued, see Majkrzak and Passell (1985), that multilayer monochromators operate at larger (θ/λ) values than supermirrors. The maximum attainable (θ/λ) is, however, simply a function of the lowest d-spacing incorporated into the multilayer structure, and this can in principle be built into a supermirror as well as a multilayer monochromator. The progress which has been achieved in developing optical neutron polarizers is impressive, and it is widely accepted that they are the best broad-band polarizers at cold and thermal neutron energies. Calculations show that it should also be possible to build supermirror benders and guides with lengths between 2 m and 100 m, which would form effective polarizers up to neutron energies ~0.5 eV. Such polarizers are ideal for long incident beam spectrometers such as exist at pulsed neutron sources, but they are too unwieldy to be used as spin analysers.

3.1.4 Field gradient devices

These polarizers use the Stern–Gerlach effect to spatially separate the two neutron spin states in a beam. The magnetic potential acting on the ($\pm$) spin neutrons in a magnetic field is $\pm |\mu_n| B$, where $|\mu_n|$ is the magnitude of the neutron magnetic moment. In the presence of a field gradient these neutrons experience opposing forces given by

$$F_{\pm} = \mp |\mu_n| (\mathrm{d}B/\mathrm{d}r), \tag{3.1.22}$$

where the direction of r is perpendicular to the beam direction.

The special advantage of the Stern–Gerlach polarimeter (see Sherwood *et al.* (1954), Jones and Williams (1980)) is that it can be used to measure directly the polarization of cold neutron beams. Similarly, it can also determine the efficiencies of other instrument components such as spin-flippers and depolarizing shims which act on the neutron spins. Its main disadvantage is that it can only be used to spin-analyse relatively small beam areas, though this property can sometimes be exploited, e.g. when it is required to make a detailed examination of the polarization profile across the beam from a polarizer.

In the Stern–Gerlach polarimeter, neutron beams travelling in, for example, the z-direction are subjected to a transverse field gradient which is designed to be predominantly along a suitable direction for analysis, say y, with $|\partial B/\partial y| \gg |\partial B/\partial x|$, and the opposing forces spatially separate the spin polarization states in this direction. The magnitude of the separation δ is determined by the field gradient integral along the neutron trajectory in the polarimeter magnet and by the time t the neutron spends in the magnet. The separation is proportional to the following quantities:

$$\delta \propto t^2 \propto l^2 v^{-2} \propto l^2 \lambda^2, \tag{3.1.23}$$

where l is the magnet length, v is the neutron velocity and λ the neutron wavelength. The above considerations apply for parallel beams of infinitesimal width. In practice, although the two spin-state beams are mutually divergent as they emerge from the magnet, they have finite widths determined by collimating apertures and neutron velocity spreads, and in a typical practical system remain unresolved at this point. The two 'spin' beams are resolved spatially by providing a drift space behind the magnet.

The 1 m long magnet system built by Jones and Williams (1980) contained pole tips made from a high-saturation iron–cobalt alloy which were shaped to form a 90° apex angle. A tapering pole gap geometry was used to prevent collisions with the pole pieces; the pole gaps were 5.5 mm and 9.0 mm at the entrance and exits of the magnet. The maximum resultant field gradients were ~0.15–0.10 T mm^{-1}. A complete separation of the spin states in cold beams required a drift space ~1–2 m beyond the magnet exit. The deflection direction (y) was vertical and typical beam apertures had dimensions ~0.6 mm high and 4 mm wide. Figure 3.17 shows results obtained using this polarimeter to measure the beam polarization at a neutron wavelength $\lambda = 6.7$ Å from a polarizing Soller constructed from Co–Fe thin films deposited on to TPX plastic foils. Also illustrated is the use of the polarimeter to analyse the depolarization of the beam caused by a 250 μm thick mild steel shim. Integration of the peak intensities gave a polarizing efficiency $P = 0.946 \pm 0.003$, and by defining the depolarizing shim efficiency S as

$$P' = P(1 - S), \tag{3.1.24}$$

where P' is the polarization which emerges after the shim, a depolarizing efficiency $S = 0.985$ was measured. The beam separation at $\lambda = 6.7$ Å was 9.3 mm.

Apart from the beam area limitation of the Stern–Gerlach magnet, the other drawback in polarizing neutron beams by this method is that the divergences of the separated beams are asymmetric. This problem can be overcome by using a multipolar magnet geometry, for example a hexapole magnet acts as a converging lens for one neutron spin and a divergent lens for the other; the result is a focused polarized beam. The equations of motion for the two neutron spins in a hexapole have been given by Hayter (1978a). This author concluded that for neutrons which are incident parallel to the axis of a hexapole of length L, the (+) spin neutrons diverge at the exit, while the (−) spin neutrons focus on to an axial point at a distance $|L \operatorname{cotan}(\omega L v)/(\omega L v)|$ from the magnet exit; this result was found to be independent of their entry point. The parameter ω is given by

$$\omega^2 = 2\mu_n B_0/(mR^2), \tag{3.1.25}$$

where R is the inner radius of the magnet conductor, and B_0 is the field at R. Just as for the Stern–Gerlach magnet, hexapoles only become practical devices for cold neutrons. A 2 m long hexapole with $B_0 = 4$ T and a 30 mm aperture produces, at a wavelength $\lambda = 8$ Å, a 100% polarized beam at 0.3 m from the exit, and with a 60% transmittance loss for the correct spins. A 1 m long prototype superconducting hexapole has been built, see Coupland and Stovold (1977), but its neutronic performance was never evaluated.

3.2 CONTROL OF POLARIZATION DIRECTION

The equation of motion for the expectation value $\langle \hat{\boldsymbol{\sigma}} \rangle$ of the Pauli neutron spin operator in a magnetic field $\boldsymbol{B}$ may be expressed, following

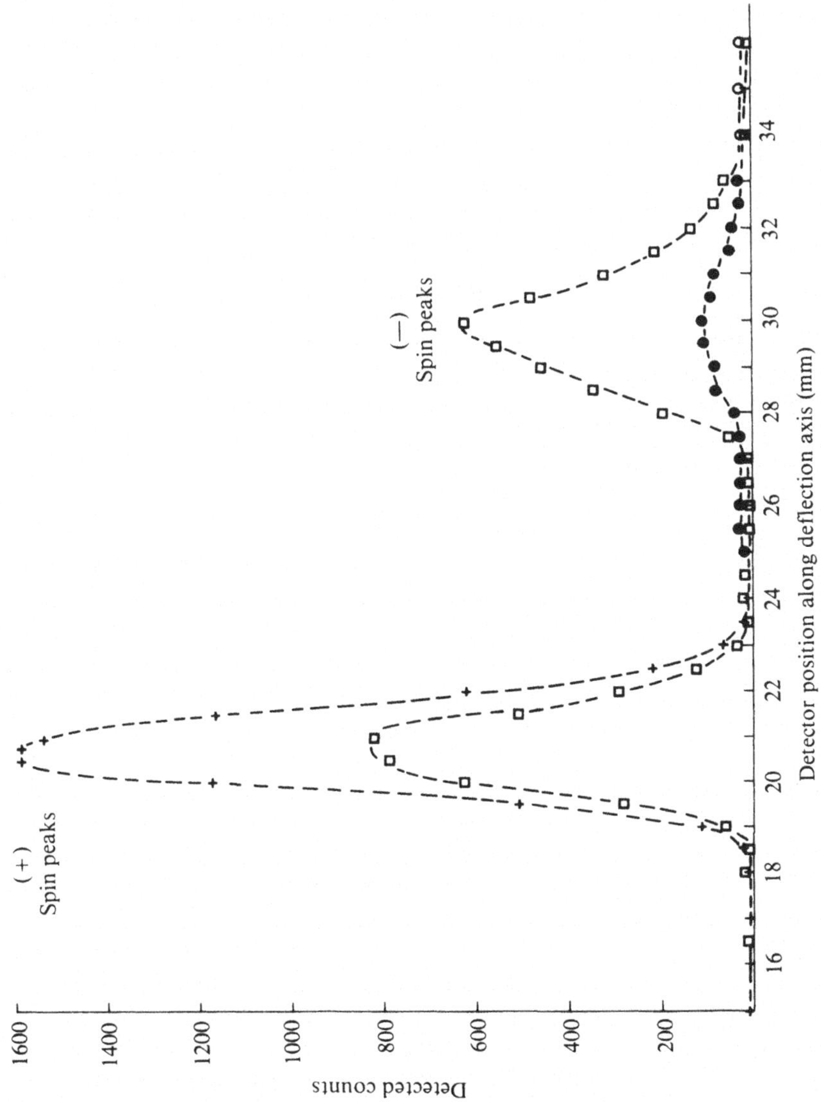

Fig. 3.17 Spin analysis using a Stern–Gerlach polarimeter of a 6.7 Å polarized beam from a polarizing Soller and of its depolarization using a 250 μm thick mild steel shim; +, polarized beam, 2 min per point; ⊕, polarized beam, 5 min per point; □, depolarized beam, 2 min per point (from Jones and Williams 1980).

equation (1.4.13) with $\langle\hat{\boldsymbol{\sigma}}\rangle \equiv \langle\hat{\boldsymbol{s}}'\rangle/\frac{1}{2}$, by

$$\mathrm{d}\langle\hat{\boldsymbol{\sigma}}\rangle/\mathrm{d}t = \langle\hat{\boldsymbol{\sigma}}\rangle \times |\gamma|\,\boldsymbol{B} = \langle\hat{\boldsymbol{\sigma}}\rangle \times (2\,|\mu_n \boldsymbol{B}|/\hbar), \tag{3.2.1}$$

where μ_n is the neutron magnetic moment and $\gamma = -1.833 \times 10^8\,\mathrm{rad}^{-1}\,\mathrm{T}^{-1}$ is its gyromagnetic ratio. If the magnetic field is homogeneous, this relation is identical to the classical equation of motion for a free magnetic dipole and, since $\langle\hat{\boldsymbol{\sigma}}\rangle$ is equivalent to the polarization vector $\boldsymbol{P}$, it also describes the precession of $\boldsymbol{P}$ around the field direction at the Larmor frequency $\omega_L = |\gamma|\,B$.

If the field direction as seen by the neutron's rest frame varies along the beam trajectory, then changes in the direction of $\boldsymbol{P}$ also become possible. The parameter which determines how $\boldsymbol{P}$ evolves is then the frequency Ω with which the direction of $\boldsymbol{B}$ changes,

$$\Omega = \mathrm{d}(\boldsymbol{B}/|\boldsymbol{B}|)/\mathrm{d}t, \tag{3.2.2}$$

and there are two limiting cases which are of practical importance in polarized neutron beam instruments.

The first is when $\Omega \ll \omega_L$ and the polarization vector follows the field rotation adiabatically. The polarization component parallel to the field, P_Z, is then conserved, that is

$$P_Z = \boldsymbol{P} \cdot \boldsymbol{B}/B. \tag{3.2.3}$$

This slow change in the local field direction experienced by the neutron spin is an important criterion which must be satisfied when designing polarized-beam equipment, since instrumental depolarization effects should ideally be zero. In practice the beam polarization generated by a polarizer is maintained by using magnetic guide fields produced either by permanent magnets or current-carrying coils (see Abrahams *et al.* (1962)).

The second limiting case is when $\Omega \gg \omega_L$, and this occurs when the magnetic field direction changes so rapidly that the polarization vector $\boldsymbol{P}$ remains spatially fixed; the phenomenon is known as non-adiabatic fast passage. We shall see that there are several types of white-beam neutron spin-flippers which use this principle to reverse the neutron polarization direction.

The spin-flipper is in fact the only device, other than the polarizer itself, which is a common feature of all polarized-beam instruments. Several different high-efficiency spin-flipping methods are available, and all will be described shortly. Another, though less common, spin-turn device is the $\pi/2$ spin-turner, which is used in spin echo spectrometers and in three-dimensional polarization analysis instruments. We deal first with spin-flippers, and define the parameter conventionally used to denote their efficiency.

Let us consider a one-dimensional polarization experiment in which the beam polarization can be treated as a scalar $P = \phi_+ - \phi_-$, and ϕ_+ and ϕ_- are the fractions of (+) and (−) spin neutrons in the beam i.e. $\phi_+ + \phi_- = 1$. The flipper efficiency f is defined as the fraction of both ϕ_+ and ϕ_- which undergo a spin reversal, i.e.

$$\phi_+ \longrightarrow \phi_- f + \phi_+(1-f)$$

$$\phi_- \longrightarrow \phi_-(1-f) + \phi_+ f.$$

The final polarization P^* is

$$P^* = \phi_+(1-2f) - \phi_-(1-2f) = P(1-2f). \qquad (3.2.4)$$

Thus the action of a flipper with efficiency f is to change the beam polarization by a multiplicative factor $(1-2f)$.

There are five basic types of neutron spin-flippers which, when properly tuned, can give flipping efficiencies close to 100%. These are (a) non-adiabatic coil flippers with either coaxial or coplanar coil pairs, (b) the Dabbs foil or superconducting sheet flippers, (c) the r.f. resonance flipper, (d) the magnetic wiggler flipper, and (e) precession coil or Mezei flippers. Flippers (a) and (b) both contain fast field-reversal regions which satisfy the non-adiabatic spin rotation condition, and their efficiencies are nearly independent of the neutron wavelength. They are therefore particularly suitable for white-beam applications. In all the other flippers the neutron spins undergo a well-defined number of Larmor precessions over a fixed distance or time, and their efficiencies are wavelength dependent. This sensitivity to the neutron wavelength is at its most extreme in the wiggler flipper, in which the neutrons traverse a magnetic field which oscillates in space, and the method is frequently used to provide monochromatization.

The main factors which influence the choice of a spin-flipper for a particular application are its wavelength characteristics, its active area, and whether the device has to be pulsed. We now describe the various methods.

3.2.1 Non-adiabatic coil flippers

A non-adiabatic neutron spin-flip is achieved by reversing the direction of the magnetic guide field as seen by the neutron in a short time compared with its Larmor period, thus they are often termed 'field flippers'. The condition for a non-adiabatic transition may be expressed by

$$\left(\frac{\mathrm{d}\theta_\mathrm{B}}{\mathrm{d}Z}\right)\left(\frac{\mathrm{d}Z}{\mathrm{d}t}\right) \gg |\gamma|\, B, \qquad (3.2.5)$$

where θ_B represents the angle between the direction of the resultant guide field B and the neutron beam direction Z, and the other parameters are as previously defined. This relationship shows that 'field flipping' is most easily achieved for fast neutrons in low magnetic fields.

The advantages of non-adiabatic spin flippers are (i) that their performances are relatively insensitive to the neutron velocity so that they can be used to spin-flip white beams, (ii) they are relatively simple to construct, (iii) they have no special magnetic field homogeneity and stability requirements, and (iv) they are efficient over large beam areas ($\sim 10\,\text{cm}^2$).

The concept of using two d.c. coils to spin-flip neutrons non-adiabatically was first discussed by Hughes and Burgy (1951) and later used to spin-flip a 5 Å neutron beam by Drabkin *et al.* (1969). The two coils were arranged coaxially about the neutron beam so as to produce magnetic fields of opposite polarity, and the flipping region was roughly midway between the coils where these fields cancel. It is important to minimize any stray magnetic fields in the critical flipping region, since these increase the neutron's Larmor precession frequency ω_L, thus reducing the non-adiabatic transition probability. The entire region between the coils is therefore surrounded by magnetic shielding. This flipper offers a special advantage in that it does not require any material to be placed in the neutron beam path.

The two-coil flipper design adopted by Jones and Williams (1978) is shown in Fig. 3.18, together with its axial magnetic field profile under typical operating conditions. Extensive measurements on the performance of the flipper were reported by these authors, and the effects of instrumental depolarization and imperfect spin flipping were separated in a definitive manner by using positive and negative efficiency polarization analysers, see Section 3.3.6. The instrumental depolarization was found to increase, and the flipping efficiency to decrease, as the beam distance away from the centre axis was increased. These effects were attributed to inefficient adiabatic spin guiding at the ends of the flipper, and the presence of a radial magnetic field component at the non-axial points in the flipping region. By integrating the measurements over various circular areas it was shown that a carefully designed two-coil spin-flipper can have an efficiency comparable with r.f. and Mezei coil spin-flippers ($\langle f \rangle > 0.995$), at least for beam diameters less than 30 mm.

A second type of two-coil non-adiabatic spin flipper, in which two large rectangular coils are placed in a coplanar 'wing' arrangement transverse to the polarized beam direction, was proposed by Korneev (1980). The high-efficiency flipping region is then a thin vertical strip, which is planar with and centrally disposed with respect to the two coils. This is preferable to the circular geometry of the axial two-coil flipper in applications with polarized beams produced by polarizing mirrors and guides.

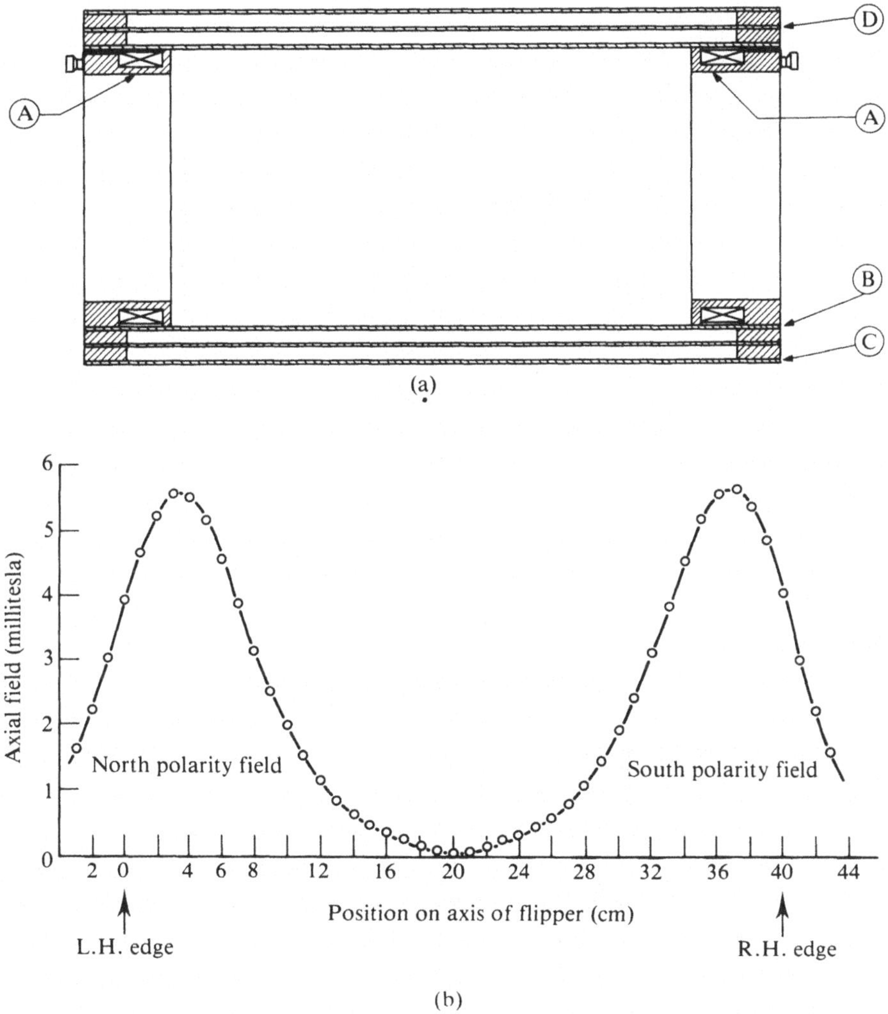

Fig. 3.18 (a) Longitudinal section of a two-coil spin flipper showing A, the copper coils in their insulating formers; B and C, inner and outer high-saturation annealed iron magnetic shields; D, a central high-permeability mu-metal screen. (b) The measured axial magnetic fields in the flipper when the coil currents are 2 amps. (From Jones and Williams 1978).

3.2.2 The Dabbs foil and superconducting sheet flippers

The method of using an aluminium foil carrying a current ~400 amps to create an abrupt field-reversal region in a neutron beam path, so causing the neutron spin to field-flip, was first suggested by Dabbs *et al.* (1955); it is often called a Dabbs foil. Abrahams *et al.* (1962) later used the current

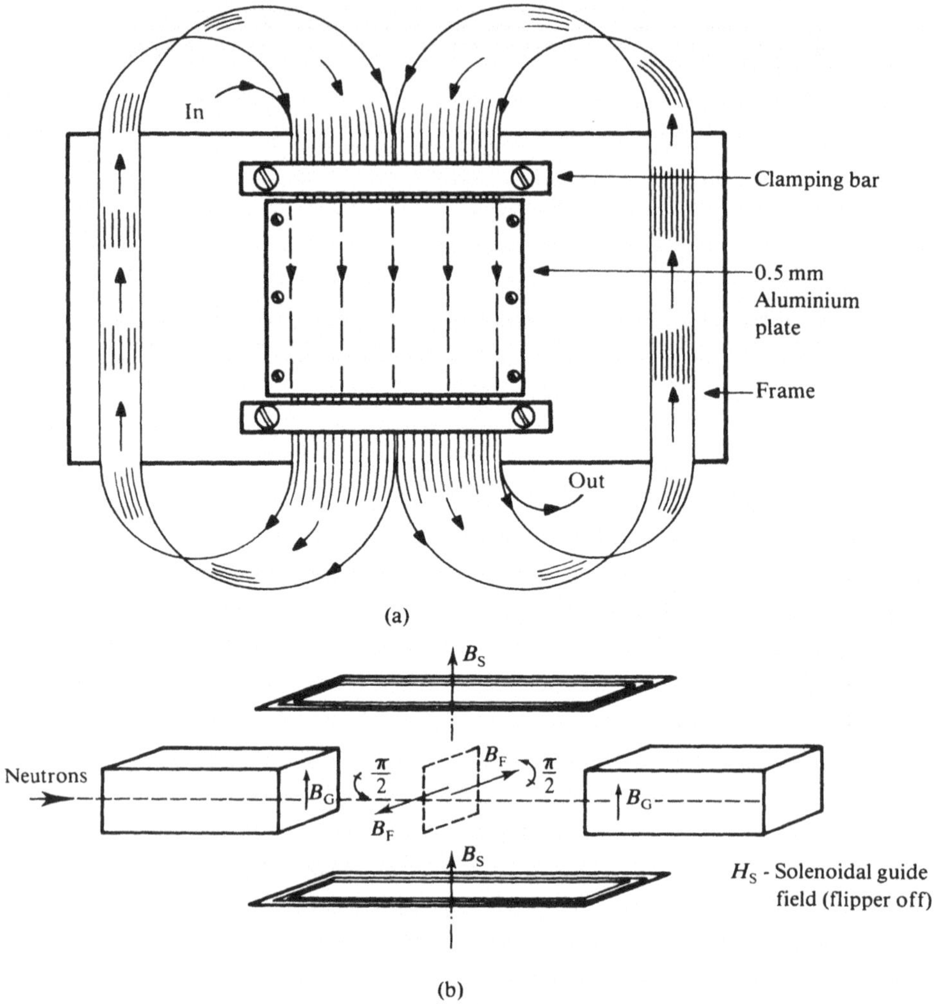

Fig. 3.19 (a) Multiwire current sheet spin flipper. (b) Suggested guide field (B_G) and flipper field (B_F) configurations for this flipper (from Jones and Williams 1978).

sheet idea in the form of a layer of closely wound 0.3 mm diameter enamelled copper wire, and successfully spin-flipped neutrons with a current ~0.5 amps. Jones and Williams (1978) later constructed a multiwire current sheet flipper and measured its flipping efficiency for different arrangements of the flipper and magnetic guide fields. The spin flipper was constructed in the 'figure eight' design described by Abrahams *et al.* (1962) and had an active area ~100 mm × 100 mm; it is shown in Fig. 3.19(a).

The reversing fields of a current sheet flipper must of necessity be directed perpendicular to the neutron beam direction, and this causes an incompatibility in the requirements for good spin guide conditions using transverse guide fields in the 'flipper off' state, and good 'field-flip' conditions (with a minimum perturbation from the guide fields) in the 'flipper on' state. Various guide field arrangements were examined experimentally by Jones and Williams (1978) and a configuration was found which gave a flipping ratio $f \sim 0.995$. These authors suggested, however, that a preferable solution to the problem associated with using permanent magnetic transverse guide fields with a current sheet is to use a solenoid guide field in the flipping region which is only powered in the 'flipper off' state; the guide field perturbation to the flipper fields in the 'flipper on' state would then be minimal. A magnetic field arrangement of this type is shown in Fig. 3.19(b).

A similar non-adiabatic sheet flipper formed from a superconducting niobium foil has been successfully operated at the D3 polarized beam diffractometer at the Institut Laue–Langevin. The principle of using the zero-field region in a superconductor for field-flipping was suggested and developed by Tasset, and is described by Forsyth (1979). The operation of the flipper depends critically on how the electromagnetic windings of two short guide field coils, placed before and after the cryostat which contains in superconducting foil, are activated. In the flipper 'off' condition the fields produced by these coils are arranged to be parallel with the other permanent magnet guide fields in the diffractometer. In the flipper 'on' condition the fields from these coils are antiparallel. The separation of the coils is made sufficiently large to permit an adiabatic π-rotation in the gap containing the superconducting foil. These two field arrangements therefore provide alternately both excellent guide and flipping field conditions.

3.2.3 Radiofrequency spin-flippers

These flippers consist of a coil which generates a radiofrequency (r.f.) field $B_1 \cos \omega t$ perpendicular to a high, uniform magnetic field B_0. The usual arrangement has the coil axis along the neutron beam path, and the steady field B_0 is applied transverse to the beam. Typical values for the applied magnetic field are $B_0 \simeq 0.01$ T, where the corresponding r.f. is $\omega \simeq 300$ kHz. By optimizing the strength of the r.f. field B_1, a complete spin reversal can be effected in a well-defined flipper distance, or, for a monochromatic beam, transit time t. Magnetic resonance flipping is an adaptation of an early experiment by Alvarez and Bloch (1940) to determine the neutron magnetic moment μ_n. It has since been used extensively in polarized neutron diffractometers, see for example that described by Nathans *et al.* (1959). The spin-flip probability of the flipper is given by

$$f = \sin^2\{(\mu_n B_1 t/2h)[1 + (2\delta B_0/B_1)^2]^{1/2}\}/[1 + (2\delta B_0/B_1)^2], \qquad (3.2.6)$$

with $\delta B_0 = (B_0 - B_0^*)$, where $B_0^* = h\omega/(4\pi\mu_n)$ is the resonance field at which the neutron spin's Larmor frequency equals the radiofrequency field. The conditions for complete spin reversal are therefore, first that $B = B^*$, but also that the amplitude term $\mu_n B_1 t/2h$ must equal an odd multiple of $\pi/2$, e.g. $B_1 = h\pi/\mu_n t$. The first condition imposes the requirement of high homogeneity for the applied field B_0, and the second condition is responsible for the neutron-wavelength-dependence of the flipping efficiency.

The flipping efficiency f is reduced if either there are variations in B_0 across the coil, or alternatively the degree of monochromaticity of the beam is worsened. These effects were examined by Jones and Williams (1977), and one of the main conclusions was that in polarized neutron diffractometers, where the wavelength resolution $\delta\lambda/\lambda$ of the beams are often less than ~2%, the main cause of any deviation in f from unity is due to inhomogeneities in B_0. The variation of $f_{(B_0 \pm \delta B_0)}$, calculated as a function of the fractional deviation in the magnitude of the optimum field $(\delta B_0/B_0)$ for the r.f. flipper, is shown in Fig. 3.20. In practice the most extreme magnetic field variation that can be tolerated in an r.f. spin

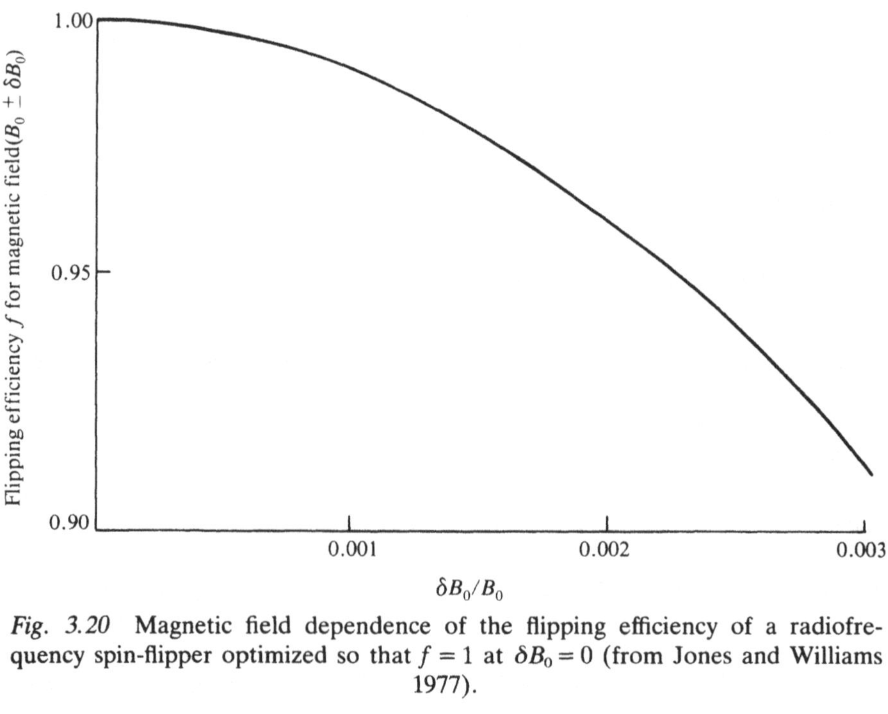

Fig. 3.20 Magnetic field dependence of the flipping efficiency of a radiofrequency spin-flipper optimized so that $f = 1$ at $\delta B_0 = 0$ (from Jones and Williams 1977).

flipper is $\delta B_0 \sim \pm 0.003 B_0$, and this gives average flipping efficiencies of the order $\langle f \rangle \simeq 0.996$.

The most fundamental difference between the r.f. flipper and all the other neutron spin flippers is that there is a photon exchange in the excitation process, which means that the neutron energy is not conserved. Alefeld *et al.* (1981) showed that the energy difference between the two sub-beams which leave a magnetic resonance coil is $4\mu_n B_0$, and demonstrated this experimentally using a high-resolution back-scattering crystal energy analyser. This energy difference amounts to 0.48 μeV in a field $B_0 = 2\ T$.

Radiofrequency spin-flippers give high efficiencies for monochromatic beams and at one time were the most commonly used flippers in polarized-beam diffractometers. They are, however, somewhat more difficult to construct and operate than precession coil flippers and are being superseded by the latter, especially in new instruments.

3.2.4 The magnetic wiggler flipper

This flipper has similarities with the r.f. flipper since it is also a resonance device, though it uses a magnetic field which oscillates in space rather than in time as is the case for the r.f. spin flipper. The principle of the method, which was first proposed by Drabkin (1963), is shown in Fig. 3.21. The alternating magnetic field with spatial period $2a$ and amplitude $\boldsymbol{B}_1$ is produced by a current-carrying aluminium foil in the form of a sinusoid, and is oriented perpendicular to a homogeneous field $\boldsymbol{B}_0$. The field $\boldsymbol{B}_0$ is parallel to the neutron polarization vector $\boldsymbol{P}$ and determines the Larmor precessional frequency through $\omega_0 = |\gamma|\ B_0$. The $\boldsymbol{B}_1$ field

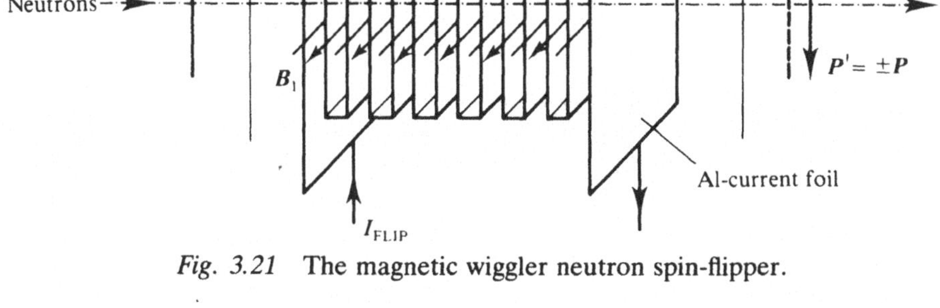

Fig. 3.21 The magnetic wiggler neutron spin-flipper.

appears in the neutron's local reference frame as a resonating field with angular frequency

$$\omega = \pi v / a \tag{3.2.7}$$

and one result of this is that the flipping properties of the device are very sensitive to the neutron velocity v. The resonant velocity of the neutron then becomes

$$v_0 = \gamma B_0 a / \pi \tag{3.2.8}$$

and the spin-flip probability in this system of magnetic fields has been given by Drabkin *et al.* (1968) as

$$f = \frac{(\gamma B_1)^2}{(\omega - \omega_0)^2 + (\gamma B_1)^2} \sin^2\left(\frac{[(\omega - \omega_0)^2 + (\gamma B_1)^2]^{1/2}}{2} \frac{l}{v}\right), \tag{3.2.9}$$

where l is the length of the resonant system. As well as satisfying the frequency matching condition in equn (3.2.9), optimal flipping also requires the amplitude of $\boldsymbol{B}_1$ to be chosen so that the $\sin^2$ term in equn (3.2.9) is unity, i.e.

$$\gamma B_1 l / (2 v_0) = \pi / 2,$$

or

$$B_1 = \pi v_0 / (\gamma l). \tag{3.2.10}$$

Drabkin *et al.* (1968) showed that the flipping efficiency expression given by equn (3.2.9) yields a response as a function of wavelength with a width $\delta\lambda/\lambda_0 \sim 1.59 a/l$; hence the wavelength resolution of the wiggler flipper is easily changeable. The main use of this flipper is in fact as a monochromator when it is placed with a broad-band spin flipper between two polarizers. Its advantages are, first, that the wavelength λ_0 can be quickly and easily set by changing B_0 and B_1 according to

$$a B_0 = l B_1, \tag{3.2.11}$$

and, secondly, that the resolution $\delta\lambda/\lambda_0$ can be varied between large limits by the choice of a and l. Its main disadvantage is the relatively high intensity of its satellite spin-flip peaks. Agamalian *et al.* (1979) showed how this defect can be overcome by using two magnetic monochromator systems in series.

3.2.5 Precession coil spin-flippers

Mezei (1972) was the first to propose using a thin rectangular coil to rotate the neutron polarization direction into any general direction with respect to a guide field $\boldsymbol{B}_0$. Devices in which the angular change is π radians are now commonly known as Mezei spin-flippers. The action of

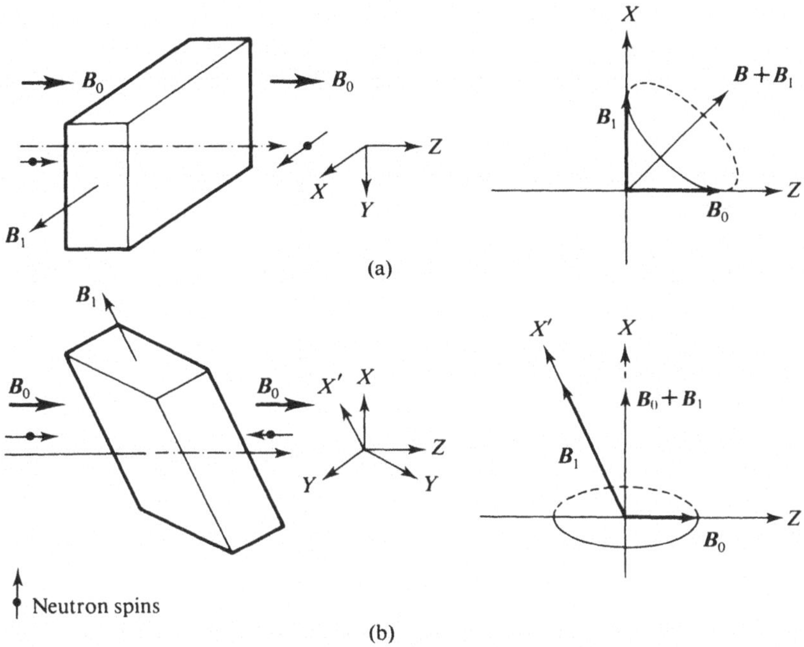

Fig. 3.22 Mezei precession coils showing (a) a spin-turn through 90°, and (b) a 180° spin-flip (from Hayter 1978a).

these coils can be most easily understood by reference to Fig. 3.22. A direct current is passed through the coil to produce an internal field $\boldsymbol{B}_1$ which adds to $\boldsymbol{B}_0$ within the coil, but leaves the $\boldsymbol{B}_0$ field outside the coil essentially unchanged. The neutron spin enters and exits the coil non-adiabatically, i.e. without spatially changing its direction, whereas inside the coil the polarization vector components precess around the resultant field $(\boldsymbol{B}_0 + \boldsymbol{B}_1)$ according to the relationships given in equn (1.4.11).

Figure 3.22(a) shows how these principles can be applied to rotate the neutron polarization vector by $\pi/2$ radians with respect to the guide field B_0. In this diagram we assume $\boldsymbol{B}_0 \cdot \boldsymbol{B}_1 = 0$ and $|B_0| = |B_1|$; thus precession takes place within the coil around an axis which makes an angle $\pi/4$ with $\boldsymbol{B}_0$. If the coil thickness a is chosen so that the transit time across the coil corresponds to half a Larmor period, i.e.

$$a = \pi v/\omega_L = \pi v/[\gamma(B_0^2 + B_1^2)^{1/2}], \tag{3.2.12}$$

then the neutron polarization vector in the exit beam is rotated by exactly $\pi/2$ with respect to $\boldsymbol{B}_0$. A rotation by π is easily achieved by using two identical coils of this type in series, and with their $\boldsymbol{B}_1$ fields oppositely directed. The original Mezei spin-flipper (Mezei 1972), was designed in this way. The arrangement was shown experimentally (Williams and Penfold 1973), to give flipping efficiencies ~99.5%, comparable with that attainable using r.f. spin-flippers.

Most precession coil flippers constructed today use a single coil and a magnetic field arrangement similar to that shown in Fig. 3.22(b). Here the relative magnitudes and directions of the $\boldsymbol{B}_0$ and $\boldsymbol{B}_1$ fields are selected to give a resultant field $(\boldsymbol{B}_0 + \boldsymbol{B}_1)$ which is perpendicular to $\boldsymbol{B}_0$. The path length in the coil is now chosen to correspond to the distance travelled by the neutron of velocity v during half a Larmor period, and the result is a polarization reversal in the $\boldsymbol{B}_0$ or Z direction.

A general description of the action of neutron spin-turn coils using a rotation matrix formalism has been developed by Hayter (1978b) in the context of their use in neutron spin echo spectrometers. A detailed calculation of the effect of polychromaticity on the performance of these coils was also given, and the result for the flipping efficiency of a π-turn coil is

$$f = 0{\cdot}5 + (1 - \cos \pi\delta)/(\pi^2\delta^2), \qquad (3.2.13)$$

where $\delta = \delta\lambda/\lambda_0$. A comparison of the results of this calculation with

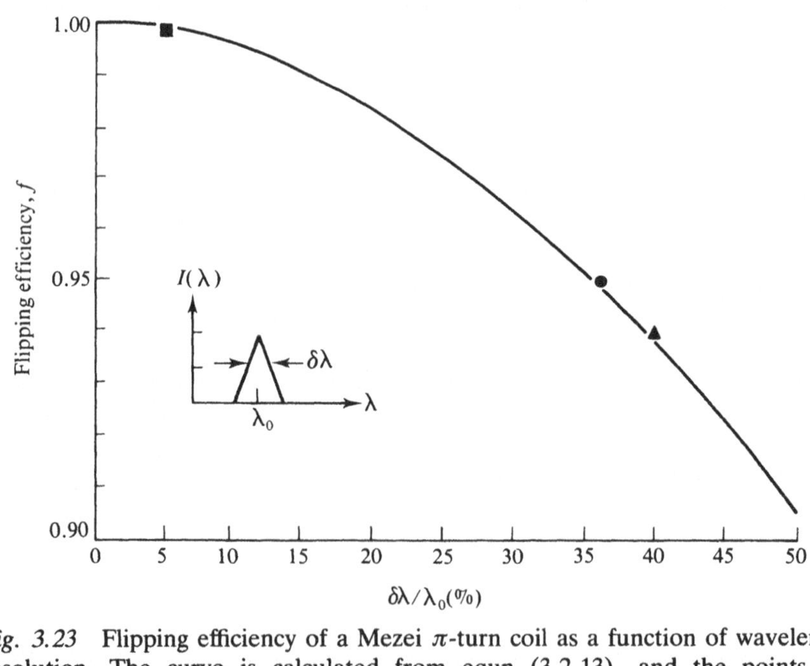

Fig. 3.23 Flipping efficiency of a Mezei π-turn coil as a function of wavelength resolution. The curve is calculated from equn (3.2.13), and the points are measured values using monochromatic beams from a graphite crystal monochromator (■) and a helical velocity selector (●, ▲) (from Hayter 1978b).

experimentally measured flipping efficiencies at different beam monochromaticities is shown in Fig. 3.23, where the wavelength distributions are taken to have a triangular form; the agreement is seen to be excellent. A similar result for the flipping efficiency dependence on $\delta\lambda/\lambda_0$ was obtained by Jones and Williams (1977) for a Gaussian distribution of wavelengths, and these authors also showed that the same wavelength-dependent behaviour is expected for r.f. flippers.

3.2.6 Pulsed spin-flippers as choppers

Electronic chopping by a spin-flipper which is placed between two neutron polarizers can provide an elegant and easily controllable means of modulating the intensity of a neutron beam. Pulse widths of a few microseconds are attainable, and can be used for either conventional or statistical time-of-flight experiments. Spin-dependent scatterers may replace the second polarizer, and time-of-flight scattering experiments then become possible by placing the pulsed flipper close to the scatterer. Similarly, the correlation time-of-flight method can be applied in polarization analysis instruments when the chopped flipper is placed in front of the sample.

Badurek (1981) has discussed the suitability of all the available types of spin-flippers for operation in a dynamic mode. The rise and fall times of the neutron bursts depend on the rates at which the flipping condition can be established and destroyed, and the neutron time-of-flight through the active flipping region. It was concluded that only the r.f. and precession coil flippers are suited to dynamic operation. In the non-adiabatic flippers, for example, although the flipping distance is short in the static mode of operation, it becomes much longer ($\sim$10 cm) in the dynamic mode because the change between non-adiabatic and adiabatic guide field conditions requires distance to establish itself.

There are basically two different types of chopped r.f. spin-flippers. In the first the r.f. field B_1 is periodically switched on and off over a short coil distance. Friesleben and Rauch (1972) described such a spin-flip chopper which gave a peak-to-background ratio of $\sim$25:1 for repetition frequencies $\sim$20 kHz, a 50% duty cycle, and pulse lengths $\sim 50\ \mu s$. In the second type of r.f. spin-flip chopper a pulsed d.c. field δB_0 is applied in the opposite direction to the main B_0 d.c. field so that the magnetic resonance condition (with B_1 optimized for $\delta B_0 = 0$) is violated. The chopped spin-flipper constructed by Kendrick (1970) used a large oscillating magnetic field amplitude $B_1 \sim 0.005$ T over a resonance coil length $\sim$2.8 cm, and the peak-to-background ratios measured with a $Co_{0.92}Fe_{0.08}$ crystal polarizer and spin analyser were also $\sim$25:1 with pulse times $\sim 40\ \mu s$.

Precession coil flippers also have suitable characteristics for rapid pulsing, since they can be designed to give rapid rise and fall times, and they also have a sufficiently short active flipping distance. The first spin-flip chopper based on Mezei's original double-coil flipper was built by Badurek and Westphal (1975); their device flipped the neutron spin in a distance ~6 mm and gave pulses with widths ~6–8 μs. An improved version which used two single-coil $\pi/2$ turners in series was later developed by Badurek (1981). By carefully phasing the activation of the two $\pi/2$ coils, the neutron pulse widths were significantly shortened and were also more symmetrically shaped. The switching ratios achieved with two $Fe_{0.08}Co_{0.92}$ polarizing crystal monochromators were ~25 with pulse widths ~5 μs at repetition frequencies up to 50 kHz. The fact that the total polarization precession angle in the new device was $(2 \times \pi/2)$ radians, compared with $(2 \times \pi)$ radians in the original design, meant that for identical flipping distances the coil fields could be reduced by one-half in the former. The $(2 \times \pi/2)$-coil chopper built by Badurek was installed and used as a statistical chopper with supermirror polarizers at the diffuse scattering spectrometer D7 at the Institut Laue–Langevin at Grenoble.

Badurek (1981) also examined the importance of the pulsed flipper efficiency f_p on the switching ratio R observed in the dynamic mode. This ratio depends strongly on the polarizing efficiencies of the two polarizing elements, and if, for simplicity, we assume that these are equal and given by P, then ignoring backgrounds,

$$R = \frac{I_{OFF}}{I_{ON}} = \frac{1 + P^2}{1 + (1 - 2f_p)P^2}. \tag{3.2.14}$$

The switching ratios calculated by Badurek for representative values of P and f_p are shown as the solid lines in Fig. 3.24, and are seen to be very sensitive to the pulsed flipper efficiency. For a dynamically driven spin-flipper it is not, unlike in the static mode, straightforward to achieve $f_p \to 1$ without incurring long switching times, and Badurek suggested combining the pulsed flipper with a static flipper for which $f \to 1$. The switching ratio (now inverted) becomes

$$R' = \frac{I_{ON}}{I_{OFF}} = \frac{1 + (1 - 2f)(1 - 2f_p)P^2}{1 + (1 - 2f)P^2} \tag{3.2.15}$$

and typical values for R' are shown as the dotted lines in Fig. 3.24 where $f = 0.999$. The modulation depth R' is much less sensitive than R to the dynamic flipper efficiency f_p, and in a two-flipper arrangement a large-area pulsed flipper with a relatively low efficiency can give a tolerably good switching ratio.

The values of R and R' depend largely on the smallness of the denominators in equns (3.2.14) and (3.2.15), and only the former contains f_p. This property of the flipping ratio equations is also exten-

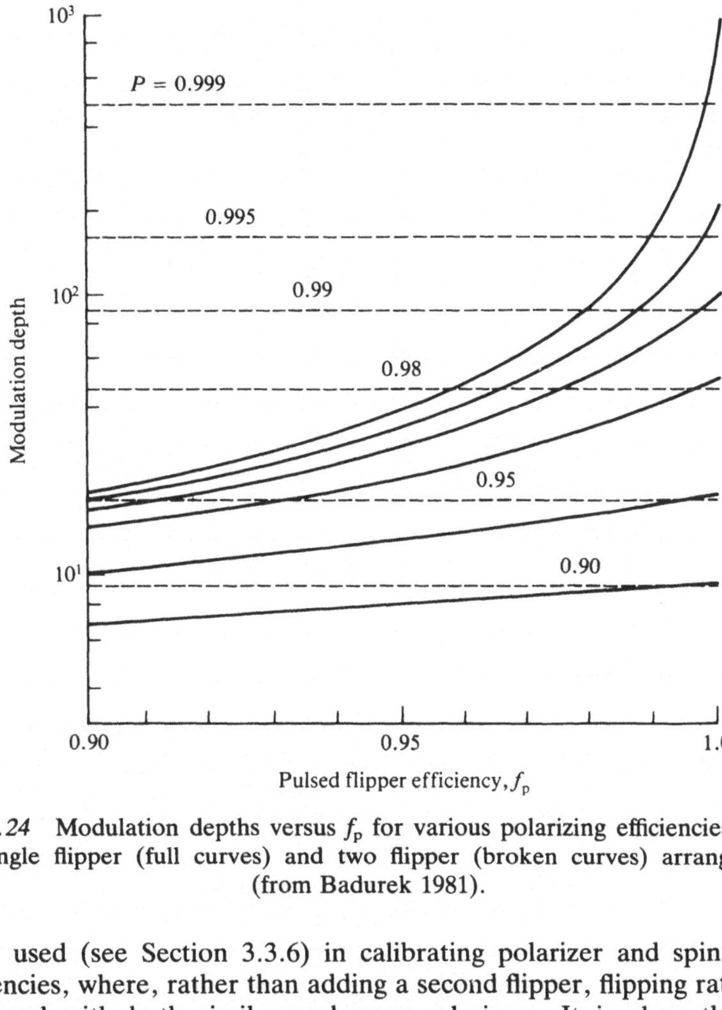

Fig. 3.24 Modulation depths versus f_p for various polarizing efficiencies P and the single flipper (full curves) and two flipper (broken curves) arrangements (from Badurek 1981).

sively used (see Section 3.3.6) in calibrating polarizer and spin-flipper efficiencies, where, rather than adding a second flipper, flipping ratios are measured with both similar and cross-polarizers. It is clear that two cross-polarizers and a single relatively inefficient pulsed flipper can also provide high switching ratios.

3.2.7 Spin turners

The method of rotating the polarization vector by $\pi/2$ radians by using a precession coil is discussed in Section 3.2.5 and illustrated in Fig. 3.22(a). We shall see that such coils, which must be tuned for a particular monochromatic wavelength, form an integral part of all neutron spin echo spectrometers. The first device using this principle was built by Rekveldt (1971), who used three mutually perpendicular coils to rotate the polarization vector to one of three orthogonal directions. This type of spin turner has since been successfully used in many three-dimensional depolarization studies of magnetic domain properties.

A different type of spin turner, which uses three mutually orthogonal Helmholtz coil pairs, has been used by Schärpf (1982) for three-dimensional polarization analysis scattering experiments. In this case the spins are rotated adiabatically into one of the (X, Y, Z) coordinate directions, i.e. there is a spatial rotation of the polarization. In the Rekveldt and Mezei coil devices, the $\pi/2$ rotations take place with respect to the guide field directions outside the coils.

3.3 POLARIZED-BEAM INSTRUMENTS

The polarized-neutron instruments described in this section are all concerned with linear polarization, i.e. the beam may be considered as being composed of different numbers of spin 'up' (+) or spin 'down' (−) neutrons, and the experiments all rely on the use of a spin-flipper to switch the polarization between the two states. We defer discussing another class of polarized-beam instruments, which use the precession of the neutron polarization, until Chapter 6.

3.3.1 Two-axis diffractometers

The general form of the classical polarized-beam diffractometer, first described by Nathans *et al.* (1959) and shown in Fig. 2.6, has been retained in all subsequent instruments. Significant advances, however, have been made in the methods for spin-flipping, in the provision of suitable high magnetic fields at the specimen, and in the experiment control and data-handling. Measurements are generally needed up to κ values of the order $10\,\text{Å}^{-1}$, and thus the neutron wavelengths used are $\lambda \sim 1\,\text{Å}$ or less. Extinction is a major source of error in single-crystal experiments, and since this effect decreases with the neutron wavelength, the ability to perform variable-wavelength experiments below 1 Å is important.

The simultaneous polarization and monochromatization provided by polarizing crystals means that these are the ideal polarizing elements in steady-state-source diffractometers. The majority of the earlier instruments used the (200) polarizing reflection in f.c.c. $Co_{0.92}Fe_{0.08}$ with the crystal oriented in the symmetric transmission geometry. The high absorption cross-section of Co is a disadvantage in this monochromator, and crystal thicknesses are restricted to 2–3 mm, which means that the (+) spin reflectivity $\mathcal{R}_+$ rarely exceeds 20%, even in the best mosaic crystals. The more modern diffractometers use the Heusler alloy Cu_2MnAl as the polarizing monochromator. The (111) reflection of Heusler alloy gives a good reflectivity $\mathcal{R}_-$; however, it suffers from having a low take-off angle (16.7° at 1 Å) and also an appreciable $\lambda/2$ contamination (see Delapalme *et al.* (1971)). This contamination can be reduced by using resonance foil filters which have sharp nuclear resonances in the 0.2–0.6 Å range to suppress preferentially the $\lambda/2$ component in the polarized beam. A list of suitable nuclear resonances for this purpose has been compiled by Forsyth (1979). In analysing data, it is always desirable to correct for the $\lambda/2$ contamination. This can be particularly large at hot-source instruments where there are copious numbers of short-wavelength neutrons.

Figure 3.25 shows a diagram of the D3 polarized neutron diffractometer at the Institut Laue–Langevin, Grenoble. It is currently positioned at a thermal source and a series of wedges allows a wavelength range between 0.6 Å and 1.5 Å to be utilized using a $Co_{0.92}Fe_{0.08}$ polarizing monochromator. The instrument is constructed in the conven-

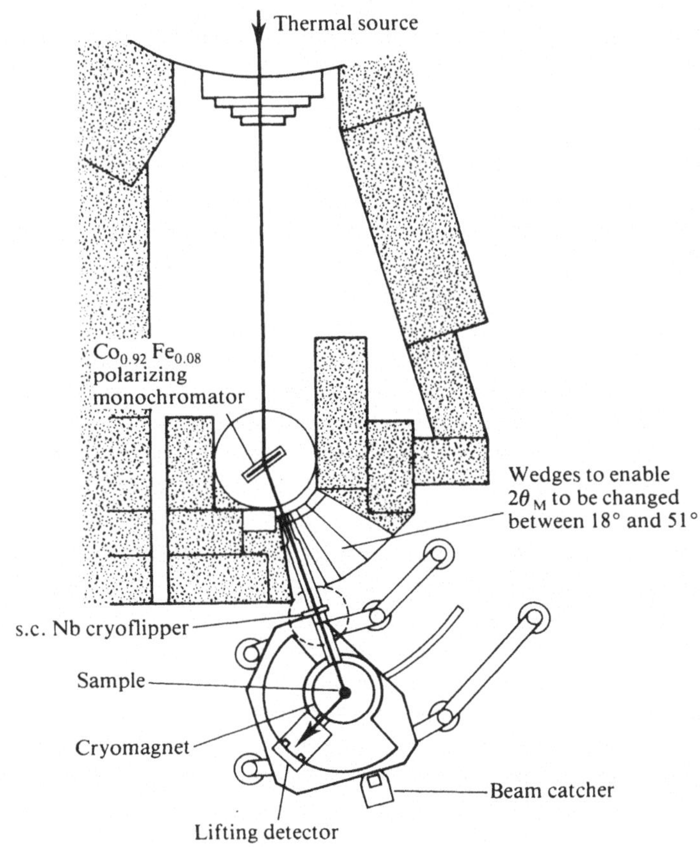

Fig. 3.25 The D3 polarized neutron diffractometer at the Institut Laue–Langevin, Grenoble (from Maier 1983).

tional geometry, with the incident beam polarization direction either parallel or antiparallel to the vertical rotation axis, which is also the magnetizing field direction at the specimen. The detector can be raised out of the horizontal plane to allow non-equatorial reflections to be examined. Spin flipping is achieved using a superconducting niobium sheet cryoflipper.

The majority of experiments require low sample temperatures and the application of high magnetic fields. These conditions are most easily satisfied by using superconducting Helmholtz coils housed in a helium cryostat. In polarized-beam instruments with vertical magnetic fields, it is essential to design the Helmholtz pair to be asymmetric, otherwise the null-field region in the central plane between two symmetrical coils would cause depolarization. Detailed aspects of the design of such magnets have been discussed by Gilbert *et al.* (1973). The D3 diffractometer at Grenoble is equipped with such an Nb–Ti superconducting coil pair, which provides a magnetic field up to 4.6 T.

The availability of superconducting magnets which can produce fields in the range 5–10 T has enabled very weak magnetic effects, e.g. induced moments in paramagnets, to be studied. In many of these experiments the flipping ratio R has to be measured to accuracies of the order 1 part in 10^4. Moon *et al.* (1975) showed that at this level of accuracy it is necessary to consider the magnetic force which acts on the neutron due to magnetic field gradients in the sample magnet. This force, which is exploited in the Stern–Gerlach polarimeter, is proportional to the field gradient, and is oppositely directed for the two neutron spin states. It can therefore influence both the velocity and positional distributions of the (+) and (−) spin neutrons at the scattering sample. Moon *et al.* (1975) calculated representative values of the velocity changes, and hence Bragg peak position changes, in a high-field superconducting split-coil solenoid. The typical shift in the Bragg angle is 10^{-4} degrees, and though this is too small to be seen in normal rocking curve measurements, its effect on the flipping ratio is easily measurable. The magnitudes of the peak shifts will in general be anisotropic in the scattering plane, and also depend on the detailed magnetic field profile of the magnet. It was concluded that the effect can be minimized by carefully setting the crystal on the flat region of its rocking curve to an angular accuracy of better than 0.02°; this is, however, a non-trivial task when a crystal is mounted at the centre of a superconducting magnet. Moon *et al.* (1976) have also calculated the vertical deviation expected for the two spin components as they traverse the vertical field gradient in a split-coil magnet. The vertical angular deviation in a 6 T magnet is about 6×10^{-4} degrees, and this causes no difficulty unless the crystal is slightly misaligned so that the reciprocal lattice vector is out of the horizontal plane. An experimental demonstration on the (002) reflection from Ti showed differences in the residual flipping ratio $(R-1)$ of some 20 parts in 10^{-4} due to this effect.

Single-crystal experiments involve measuring the flipping ratio R_{hkl} for a series of Bragg peaks, where

$$R_{hkl} = (N_+ - N_+^B)/(N_- - N_-^B) \tag{3.3.1}$$

and N_+, N_- are the peak count rates for spin 'up' and spin 'down' neutrons respectively; the superscript B denotes the equivalent background count rates below the Bragg peaks. The time taken to measure R_{hkl} depends on the accuracy to which the ratio is required, on the intensity of the reflection, and on the background. In instruments like D3 at the Institut Laue-Langevin, in which the appropriate diffractometer shafts are under computer control, it is straightforward to arrange for the counting times for the various conditions of the flipper, sample orientations and detector angles to be optimized so as to maximize the statistical accuracy of the measured R_{hkl} values.

A two-axis polarized-beam instrument at the Oak Ridge National Laboratory, which uses a ^{57}Fe polarizing monochromator, has been adapted for inelastic scattering measurements by Mook, see for example Mook and Tsuei (1977), Mook and Lynn (1984). The diffractometer is suitable for measuring excitations in magnetized scatterers; the original instrument used a rapidly pulsed r.f. spin-flipper which was activated by a pseudorandom pulsing sequence. The neutrons were detected as a function of the time-of-flight originating from the 10 μs long flipper pulses, and the excitation spectra were obtained by cross-correlating the detected counts with the spin-flipper sequence for a large number of sequences.

The (110) reflection from the ^{57}Fe polarizing monochromator produces (−) spin neutrons, and in the usual experimental arrangement these are scattered from a magnetized sample with the horizontal magnetizing field applied along the direction of the nominal (elastic) scattering vector. The spin-flipper is interfaced to a computer which outputs a pseudorandom sequence of 1s and 0s. A 1 corresponds to the flipper 'on' condition or a (+) incident spin state, and a 0 gives correspondingly a (−) spin incident neutron. The neutron beam is utilized continuously during the experiment, only the neutron spin polarization state is modulated, and the technique is therefore much more economical in the use of neutrons than conventional time-of-flight methods which use mechanical choppers. We now illustrate its use in measuring magnon scattering. The (+) spin neutron can only cause magnon creation in the sample, which it does by neutron energy loss, and we denote the scattering function for this process by $S^{+-}(\boldsymbol{\kappa}, \omega)$. Similarly the (−) spin neutron causes magnon annihilation by gaining energy, with a scattering function $S^{-+}(\boldsymbol{\kappa}, \omega)$. The cross-correlation technique (see Mook and Lynn (1984) and also Section 3.3.4) yields a function which is related to the difference between these two quantities, i.e.

$$I(\boldsymbol{\kappa}, \omega) = c(k'/k)[S^{+-}(\boldsymbol{\kappa}, \omega) - S^{-+}(\boldsymbol{\kappa}, \omega)] + b, \tag{3.3.2}$$

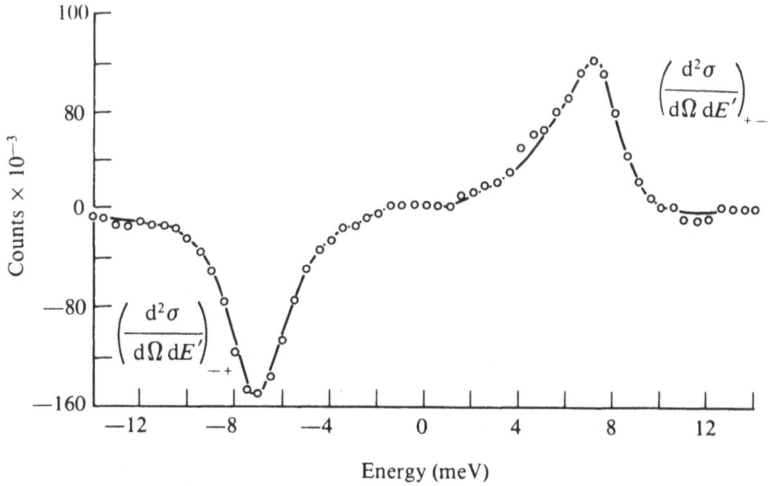

Fig. 3.26 The observation of inelastic magnetic scattering in a $Ho_{0.88}Tb_{0.12}Fe_2$ single crystal using the cross-correlation polarized beam time-of-flight method (from Mook and Lynn 1984).

where c and b are constants which are independent of $\boldsymbol{\kappa}$ and ω. We therefore see that for neutron energy gain $S^{-+}(\boldsymbol{\kappa}, \omega)$ provides the non-zero cross-section in this equation, and the cross-correlated data is then negative. For neutron energy loss, $S^{+-}(\boldsymbol{\kappa}, \omega)$ is finite and the cross-correlation function is positive.

One of the demonstration experiments showing the feasibility of the technique, described by Mook and Lynn (1984), was the measurement of spin waves in a single crystal of $Ho_{0.88}Tb_{0.12}Fe_2$. This is a ferrimagnet which is easily saturated magnetically at room temperature, and the data obtained at Oak Ridge in a 12-hour measuring time at an incident energy of 37 meV are shown in Fig. 3.26. The negative energy scale corresponds to neutron energy gain, and as expected gives negative cross-correlated counts. The technique has been widely used in the study of excitations in amorphous ferromagnetic solids.

3.3.2 Triple-axis polarization analysis spectrometers

The first triple-axis polarization analysis spectrometer was built at the Oak Ridge National Laboratory (ORNL) and is described by Moon *et al.* (1969a). The basic layout is shown in Fig. 2.9 and the principles by which it operates are given in Section 2.3. The polarizer and analyser were originally $Co_{0.92}Fe_{0.08}$ polarizing monochromators mounted in permanent magnets, though these were later replaced by ^{57}Fe polarizing crystals. The scattering sample is placed between the poles of an electromagnet which has a horizontal rotation axis; this facilitates a rapid interchange between the two commonly used $\tilde{\boldsymbol{P}} \cdot \tilde{\boldsymbol{\kappa}} = 1$ and $\tilde{\boldsymbol{P}} \cdot \tilde{\boldsymbol{\kappa}} = 0$ geometries. The permanent magnet guide fields 2 and 3 in Fig. 2.6 are rotatable, so that the neutron spins can enter the sample magnetic field in either a vertical or horizontal orientation. The spin-flippers were radiofrequency coils with a vertical guide field. The overall efficiency of the instrument can be assessed by rotating the analysing crystal into the direct beam, and measuring the flipping ratio. This varied between 60 and 120 depending on the wavelength and beam area, and indicated that the beam polarization product PP_A and flipping efficiency were in the range 0.98–1. The performances in the horizontal and vertical field geometries were nearly identical, and this showed that excellent adiabatic guide field conditions were achieved in the horizontal guide field arrangement. The reflectivities of the polarizing crystals used in this instrument are only of the order of 10%, and it is therefore evident that polarized triple-axis spectrometers require a high-flux reactor really to be effective.

A similar spectrometer, designated D5, was built in the early 1970s to view the hot source at the Institut Laue–Langevin high-flux reactor at Grenoble and is shown in Fig. 3.27. The main differences between this and the original ORNL spectrometer were that it was equipped with Heusler alloy polarizing crystals as well as $Co_{0.92}Fe_{0.08}$, and also the available wavelength range, $\lambda = 0.4$–1.1 Å, was more extensive.

Both these spectrometers were generally operated at neutron wavelengths ~1 Å and less. A further development has been the adaption of cold neutron triple-axis spectrometers for polarization analysis applications. The methods for doing this have been discussed by Stirling (1982),

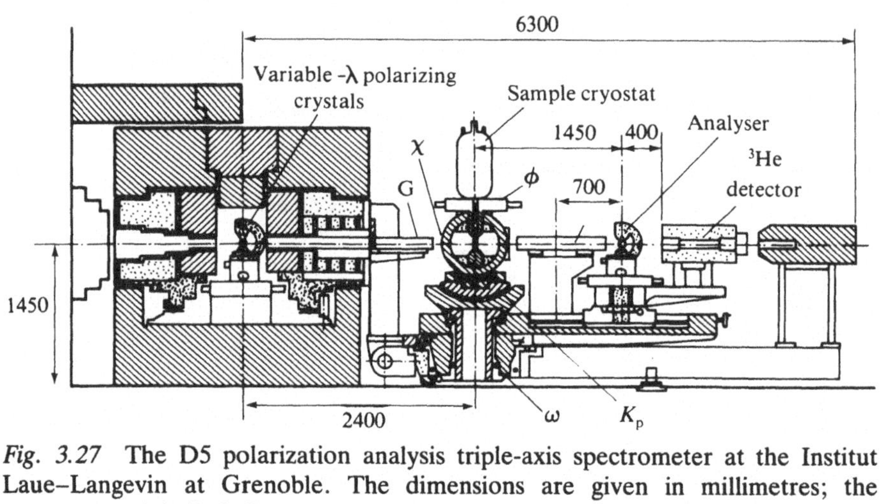

Fig. 3.27 The D5 polarization analysis triple-axis spectrometer at the Institut Laue–Langevin at Grenoble. The dimensions are given in millimetres; the components labelled G are magnetic guide field tubes which use permanent magnets; the symbols ϕ, χ, K_p and ω indicate the various motions available on this instrument (from Brown and Forsyth 1977).

with particular emphasis on how they relate to the Institut Laue–Langevin cold neutron triple-axis spectrometer IN12. The most straightforward and favoured modification is that with a pyrolytic graphite monochromator and Heusler alloy analyser, since this only involves changing the secondary spectrometer. If the instrument is placed on a doubly-curved cold neutron guide, as occurs for IN12, then there is very little $\lambda/2$ contamination for primary wavelengths less than 4 Å, and for operation at longer wavelengths a cooled Be filter can be placed in either the incident or scattered beam. Unfortunately, this simple modification only provides a means of performing partial polarization analysis experiments, in which the samples themselves must have a preferred orientation axis. For full polarization analysis experiments on a cold neutron triple-axis spectrometer, the following polarizer/monochromator systems may be envisaged for either the polarizer or the analyser: (a) Heusler alloy crystals, (b) a pyrolytic grahite and optical polarizer combination, and (c) optical polarizers combined with a magnetic wiggler spin flipper for the energy selection. These alternatives provide progressively better resolutions. The limitation in the resolution with method (c) is determined by the flipper, and this solution is favoured by the Brookhaven National Laboratory group (see for example Majkrzak and Shirane (1982)). The polarizer–monochromator scheme proposed by these authors using two polarizing multilayers is illustrated in Fig. 3.28.

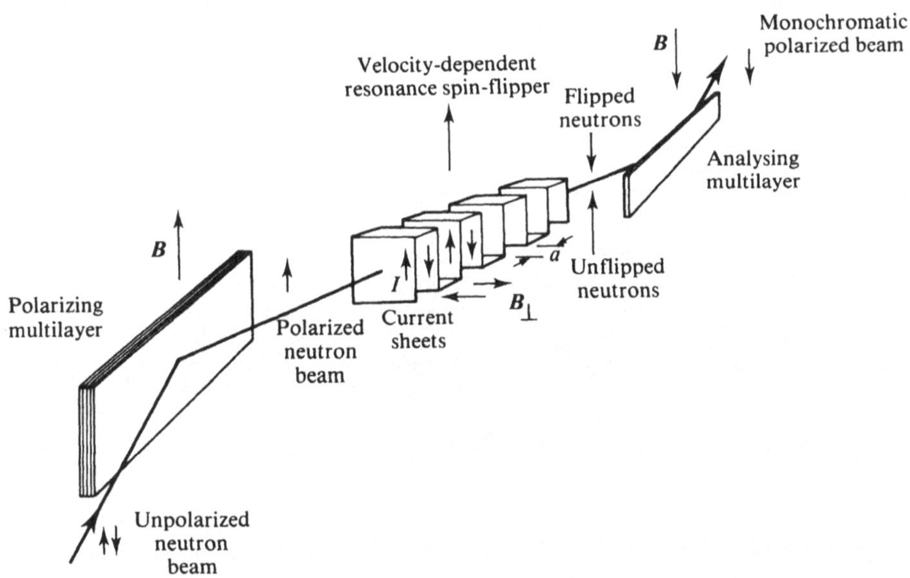

Fig. 3.28 Polarizing-monochromator unit for a cold neutron beam triple-axis polarization analysis spectrometer proposed by Majkrzak and Shirane (1982).

3.3.3 Spectrometers with optical polarizers

We have seen that optical polarizers are effective at all wavelengths longer than about 2 Å, and are therefore suitable for broad-band or low-resolution cold neutron instruments. The pulsed-source polarization analysis spectrometer TOP built at KENS uses curved conventional polarizing mirror Sollers, and the time-of-flight polarization analysis instrument D7 at the Institut Laue–Langevin uses similar devices, but with supermirror polarizing coatings.

The TOP spectrometer is described by Endoh *et al.* (1985), and is shown in Fig. 3.29. The magnetic alloy used for the polarizing mirrors in the Soller polarizer and analysers was $Fe_{0.35}Co_{0.65}$ deposited on to a 50 μm thick oriented polypropylene substrate. The Sollers, which have a cross-sectional area $20 \times 50\,\text{mm}^2$, were magnetized in 0.15 T fields and gave polarizing efficiencies ~95%. The spin flipper is of the Drabkin two-coil axial type, and in normal operation the current in one of the coils is switched every 2.5 s. For pulsed operation this flipper can be replaced by a Mezei coil π flipper in which the current is phased with respect to the source pulse, and has a time-dependence which gives a resultant magnetic field B_R that satisfies the π-turn criterion over a wide neutron velocity range, i.e.

$$B_R(t) = (\pi/a\gamma)v = (\pi L/a\gamma)t^{-1} \propto t^{-1}. \qquad (3.3.3)$$

L is the distance of the coil from the source, and the other terms are as defined in Section 3.2.5. The majority of the early experiments carried

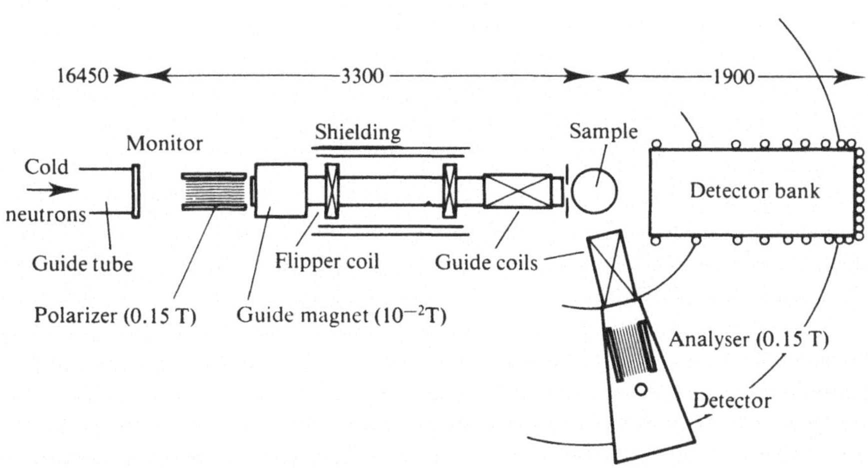

Fig. 3.29 The cold neutron polarization analysis spectrometer TOP at the KENS pulsed neutron source (dimensions in millimetres) (from Endoh *et al.* 1985).

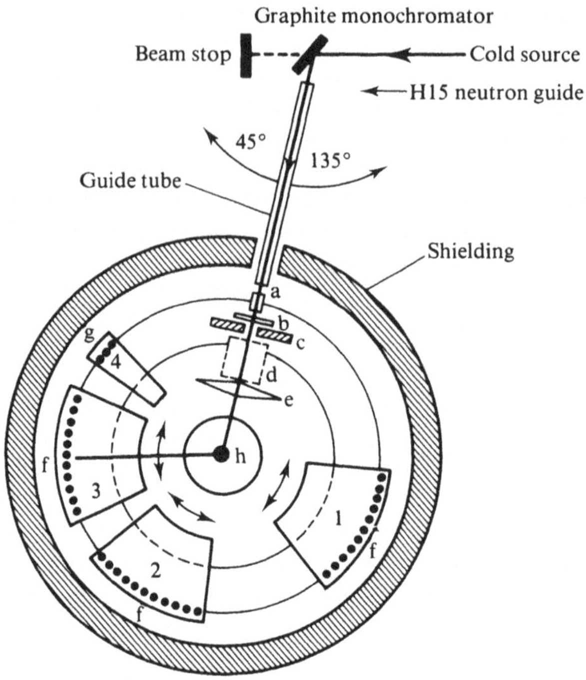

Fig. 3.30 Schematic diagram of the diffuse scattering instrument D7 for polarized and unpolarized neutrons at the Institut Laue–Langevin, Grenoble a, Be filter or polarizer; b, main monitor; c, slit system; d, flipper and second polarizer for pseudostatistical chopper; e, disc chopper or flipper; f, detector banks 1–3 with supermirror polarizing Sollers; g, vertical detector bank 4; h, sample. (From Maier 1983).

out on this instrument used the analyser and detector in the straight-through position to measure the depolarization of polarized beams transmitted through magnetic materials.

The Institut Laue–Langevin instrument D7 shown in Fig. 3.30 is a time-of-flight polarized and unpolarized beam spectrometer for elastic and quasielastic diffuse scattering measurements, and is described by Schärpf (1985). This instrument can be operated in several modes and those of primary interest here are: (a) its multidetector polarization analysis capability, (b) its pulsed operation by activating the spin-flipper *d* pseudostatistically, and (c) the fact that it contains three large and orthogonal Helmholtz coil pairs outside the sample cryostat which allows experiments to be performed with three orthogonal incident polarization directions.

Three incident wavelengths $\lambda = 3.1$ Å, 4.8 Å and 5.7 Å are made available by rotating the main part of the spectrometer on air cushions around the graphite monochromator. There are four motor-driven detector banks which scan the whole angular range on both sides of the incoming beam. All the polarizing elements are the cobalt–titanium supermirror Sollers described in Section 3.1.3; each has an area $50 \times 100\,\text{mm}^2$ and on the analyser side is associated with its own detector. The spin-flipper, whose activation can be programmed pseudostatistically, is a Mezei π coil which produces pulsed beams with widths down to $5\,\mu\text{s}$ duration, and allows elastic and inelastic scattering processes to be distinguished. Details of the cross-correlation technique which makes this possible are discussed, for example, by Von Jan and Scherm (1970). An equivalent method which utilizes a modulation of the polarization rather than the intensity of the neutron beam will be described in the following section. The D7 instrument may therefore be used in a static or dynamic mode, and either with or without three-dimensional polarization analysis. This makes it the most versatile of all the polarized-beam scattering instruments so far constructed, although it only operates in the cold neutron energy region.

3.3.4 Instruments with polarizing filters

Polarizing filters have similar functions in neutron scattering instruments to those of optical polarizers, though they can be effective over much larger energy ranges. Polarized-beam instruments have been constructed which use all the polarizing filter methods described in Section 3.1.2, though only the simplest, the magnetized iron filter, has so far been used extensively in scattering experiments.

The polarization analysis instrument LONGPOL, which uses magnetized iron filters as the polarizer and analyser, has been in operation at the HIFAR Reactor, Lucas Heights, Australia, since the early 1970s; its original design is described by Campbell *et al.* (1974). It has been used primarily to investigate the atomic ordering and magnetic moment distributions in spin-glass alloys and disordered antiferromagnets. These diffuse scattering experiments can be performed perfectly adequately with very relaxed wavelength resolutions ($\delta\lambda/\lambda \sim 0.10$), and the mean wavelength chosen for LONGPOL was $\lambda = 3.6\,\text{Å}$, where the polarization cross-section of the Fe filter, (see Fig. 1.9), is close to its maximum. The polycrystalline iron filters were 6 mm thick and when magnetically saturated gave a beam polarization, as measured by the double transmission effect, of 42%. The original instrument used a mechanical velocity selector to define the monochromatic beam; this has now been replaced by a pyrolytic graphite double-crystal monochromator; the current design of LONGPOL is shown in Fig. 3.31. Higher-order wavelength contamination at λ/n in the incident beam is reduced first by a set of 5° mosaic graphite crystals tuned to attenuate at $n = 3, 4, \ldots$, etc, which

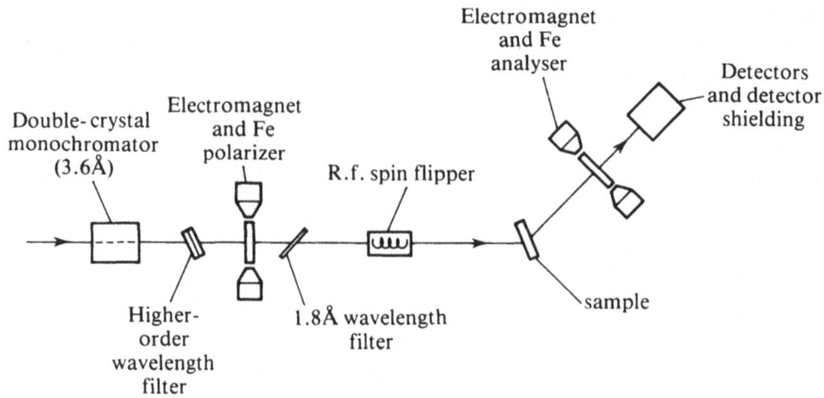

Fig. 3.31 Schematic diagram of the long-wavelength polarization analysis instrument LONGPOL at the HIFAR reactor, Lucas Heights (from Davis *et al.* 1982).

are placed before the first filter. The $\lambda/2$ component is eliminated by an additional 3.5° mosaic pyrolytic graphite crystal placed behind the flipper. The spin-flipper is a 50 mm long 500 kHz r.f. coil with a flipping efficiency $f = 0.98 \pm 0.03$. For most experiments the interest is in measuring the magnetic spin-flip elastic scattering, and this is observed with the applied field at the sample directed along the scattering vector.

An important extension to the use of LONGPOL has been its adaption to operate in a time-of-flight mode, since this has introduced the possibility of examining dynamic scattering processes. Elastic and inelastic scattering can be separated by using a cross-correlation procedure in which the net polarization, i.e. that resulting from the combined action of the polarizer, scatterer and analyser, is modulated according to a pseudorandom sequence. This modulation is accomplished by switching the spin flipper 'on' and 'off' in a well-characterized sequence; the method is described by Davis *et al.* (1982). The flipping sequence uses a finite and repeating string s_L of L bits, i.e. $s_{i+L} = s_i$. Of the L bits in the string K are 1 and $(L-K)$ are 0, and the duty cycle c of the string is defined by

$$c = (K-1)/(L-1). \tag{3.3.4}$$

If a measured count rate $Z(\tau)$ in a particular time-of-flight channel τ is cross-correlated with the pseudorandom binary string sequence s_i, and the sequence is repeated r times during the course of the experiment, then this produces a cross-correlation function

$$C_\tau = \sum_{j=1}^{rL} s_j Z(\tau + t_j), \tag{3.3.5}$$

which, as shown by Davis *et al.* (1982), can be written in the condensed form

$$C_\tau = rK\left[(1-c)D_\tau^* + cp^{-1}\sum_{\tau=1}^{N} S_\tau^* + (1-c)p^{-1}\sum_{\tau=1}^{N} N_\tau^* + b\right], \qquad (3.3.6)$$

where p is an integer ratio of the modulating pulse width to the time width of channel τ, N is the total number of τ channels, and b is the uncorrelated time-independent background per channel. S_τ^* and N_τ^* are spin-flip (SF) and non-spin-flip (NSF) sample transfer functions which have been convoluted with the pulse shape of the spin-flipper modulator, and D_τ^* is the difference function

$$D_\tau^* = S_\tau^* - N_\tau^*. \qquad (3.3.7)$$

The last three terms in equn (3.3.6) yield a flat 'background of ignorance' which should be completely independent of the channel number. Thus we see that this cross-correlation procedure produces a term which is proportional to the difference between the SF and NSF scattering cross-sections of the sample. In practice, rather than storing the scattered neutron counts in a multichannel analyser and cross-correlating these with the pseudorandom sequence at the end of the experiment, the cross-correlation is performed 'on-line' for each detected neutron, and D_τ^* is constructed dynamically.

Figure 3.32 shows the cross-correlation difference function D_τ^*

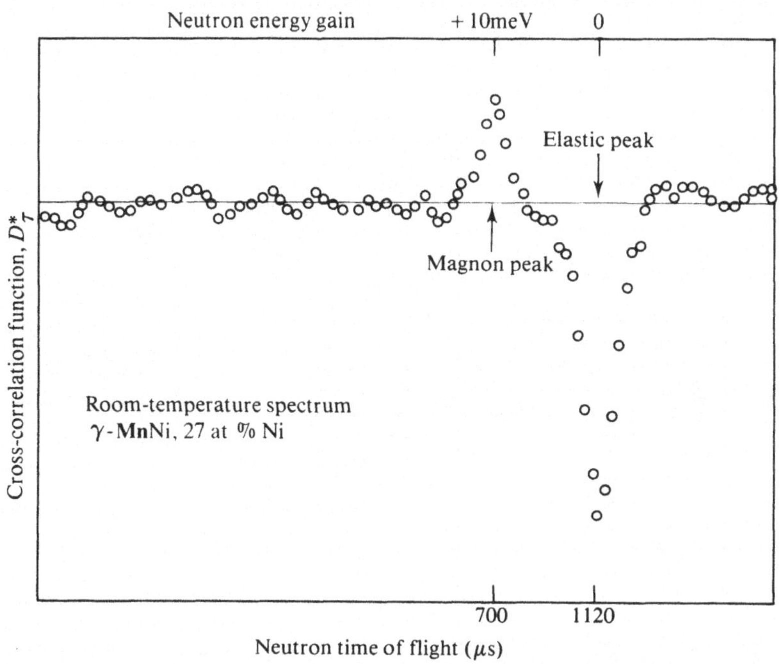

Fig. 3.32 Cross-correlation function for γ-MnLi observed on LONGPOL with channel widths of 25 μs, $L = 4095$, $c = 0.5$, $p = 2$ (from Davis *et al.* 1982).

measured on LONGPOL for an antiferromagnetic γ-MnNi single crystal; the 'background of ignorance' has been subtracted for clarity of presentation. At small energy transfers, and with the neutron polarization direction parallel to the elastic scattering vector, S^*_τ represents mainly magnetic scattering, whereas N^*_τ is predominantly a result of nuclear scattering. The elastic scattering signal in Fig. 3.32 appears as a negative quantity since the nuclear elastic scattering cross-section is much larger than its magnetic counterpart, i.e. $N^*_{\tau(\mathrm{E1})} > S^*_{\tau(\mathrm{E1})}$. The reverse is true for the inelastic peak, and this can therefore be identified unambiguously as being due to spin-flip scattering following magnon annihilation.

The second polarized-beam spectrometer which uses polarizing filters is the POLARIS instrument at the Rutherford Appleton Laboratory spallation neutron source ISIS. The first version of this instrument is a broad-band time-of-flight polarized-beam diffractometer, and is shown in Fig. 3.33. The incident beam is polarized by the spin-dependent absorption of neutrons by nuclei which are themselves polarized by the hyperfine interaction method at millikelvin temperatures. Initial experiments have used polarized ^{149}Sm nuclei in $SmCo_5$, and this is effective at all neutron energies up to about 200 meV. The spin-flipper can be either a Dabbs foil current sheet, or, for rapid pulsing applications, a Mezei precession coil in which the current is carefully phased with respect to the source pulse, and is decreased in inverse proportionality to the neutron time-of-flight. A full description of the functional design of the first phase of POLARIS has been given by Cywinski (1983).

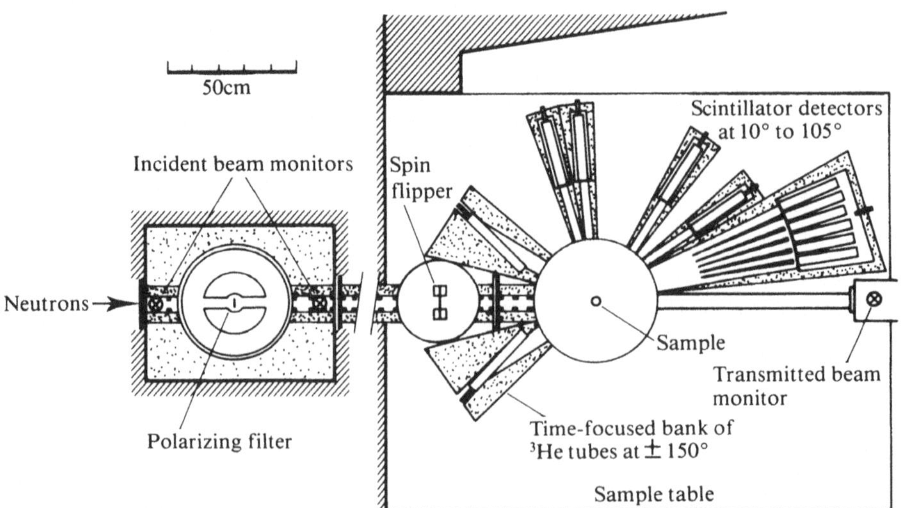

Fig. 3.33 The polarization spectrometer POLARIS at the Rutherford Appleton Laboratory pulsed neutron source ISIS.

The two major developments intended for POLARIS are its extension to a two-filter polarization analysis spectrometer as described by Mayers and Williams (1981), and also the implementation of a method for performing inelastic polarization analysis experiments. Three methods are available for the latter: (a) the use of polarizing crystal monochromators in a crystal analyser spectrometer, (b) a direct geometry chopper spectrometer operating at incident (resonance) energies at which the first filter's polarizing efficiency is high, and (c) a white-beam polarization-modulated cross-correlation method.

The most versatile method for inelastic polarization analysis measurements on POLARIS is (c), and is described by Cywinski and Williams (1985). The instrument requires two broad-band polarizers, such as samarium filters, in the incident and scattered beams, and a pseudorandomly pulsed fast spin-flipper placed in front of the scatterer. It is the pulsed-source analogue of the technique devised by Davis *et al.* (1982) for LONGPOL. In the white-pulsed-beam method the cross-correlation of the measured intensity at a given total time-of-flight τ gives a similar cross-correlation function to that given in equn (3.3.6), and when this procedure is repeated for all total times of flight this produces a two-dimensional cross-correlation function

$$C_{\tau,\tau_2} = rK\left[(1-c)D^*_{\tau,\tau_2} + c\sum_{\tau_2=1}^{L} S^*_{\tau,\tau_2} + (1-c)\sum_{\tau_2=1}^{L} N^*_{\tau,\tau_2} + b_{\tau,\tau_2}\right], \qquad (3.3.8)$$

where τ_2 is now the scattered neutron time-of-flight over a secondary flight path L_2.

Figure 3.34(a) shows a distance–time diagram of three typical scattering events which may be detected at time τ with this experimental arrangement, and Fig. 3.34(b) shows the cross-correlated count rates C_{τ,τ_2} expected for these events. The difference function obtained by subtracting the 'background of ignorance' was shown by Cywinski and Williams (1985) to be given by

$$D^*_{\tau,\tau_2} \propto f(\tau_1)P_1(\tau_1)P_2(\tau_2)T_1(\tau_1)T_2(\tau_2)\left[\left(\frac{d\sigma}{d\Omega}\right)^{SF}_{\tau,\tau_2} - \left(\frac{d\sigma}{d\Omega}\right)^{NSF}_{\tau,\tau_2}\right] \qquad (3.3.9a)$$

$$\propto G(\tau_1, \tau_2)\left[\left(\frac{d\sigma}{d\Omega}\right)^{SF}_{\tau,\tau_2} - \left(\frac{d\sigma}{d\Omega}\right)^{NSF}_{\tau,\tau_2}\right], \qquad (3.3.9b)$$

where f, P and T are instrument component terms as previously defined: see equns (3.2.4), (3.1.3), and (3.1.4). The subscripts 1 and 2 denote these quantities in the incident and scattered beams and $\tau_1 = \tau - \tau_2$ is the incident neutron flight time over the distance L_1. The maximization of $G(\tau_1, \tau_2)$ for a two-filter spectrometer involves optimizing the filter thicknesses. This method of inelastic polarization analysis is very economical in neutron usage, and is ideally suited to pulsed neutron sources.

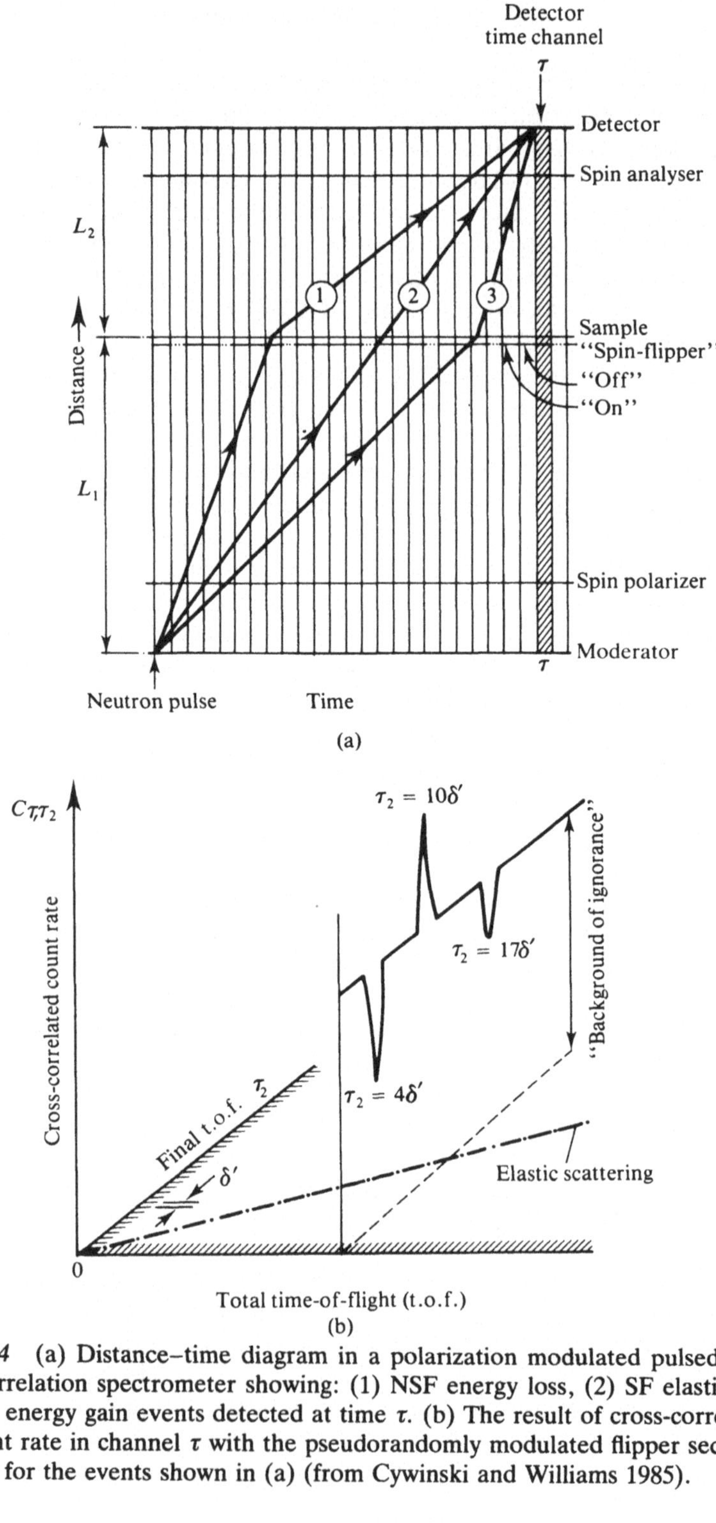

Fig. 3.34 (a) Distance–time diagram in a polarization modulated pulsed-beam cross-correlation spectrometer showing: (1) NSF energy loss, (2) SF elastic, and (3) NSF energy gain events detected at time τ. (b) The result of cross-correlating the count rate in channel τ with the pseudorandomly modulated flipper sequence for the events shown in (a) (from Cywinski and Williams 1985).

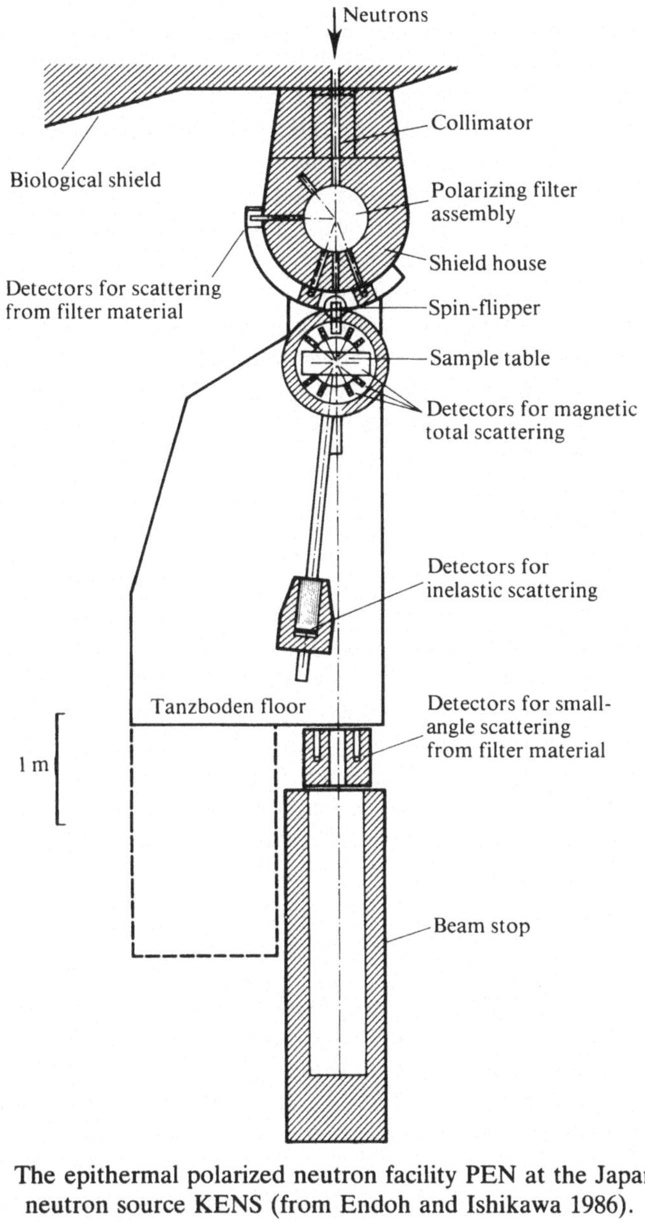

Fig. 3.35 The epithermal polarized neutron facility PEN at the Japanese pulsed neutron source KENS (from Endoh and Ishikawa 1986).

An epithermal polarized-beam spectrometer PEN, which utilizes the third polarizing filter method, i.e. the polarized-proton filter, has been built at the Japanese pulsed neutron source KENS; Fig. 3.35 shows a plan view of the facility. The polarizer is a dynamically polarized ethylene glycol target doped with a Cr(V) complex (see Ishida *et al.* (1985)) and gives approximately 60% proton polarization. It is used for a variety of scientific applications, for example the study of the process of proton dynamical polarization, magnetic structure determinations in amorphous magnets, high-energy magnetic excitation studies in metallic ferromagnets, and the study of the spin parity of nuclear states. For inelastic neutron scattering experiments the filter is combined with an incident beam chopper which is phased with the source pulse to provide a variable-energy monochromatic pulsed polarized beam at the scatterer. There are two types of spin-flipper: a non-adiabatic superconducting niobium plate in a helium cryostat for white-beam experiments, and a phased Mezei precession coil for inelastic scattering measurements with the chopper. The detector bank and sample table are mounted on air pads so that the incident and scattered beam flight-paths and scattering angles can be varied readily.

3.3.5 Polarized neutron reflectometers

The method of measuring the critical reflection of polarized neutrons to obtain information on the depth profile of surface magnetism was pioneered at the Argonne National Laboratory by Felcher (1981). An evaluation of the neutron reflectivity of a surface at angles slightly greater than the critical angle θ_c provides a very sensitive probe to the way that the refractive index n changes as a function of its distance z from the surface. For polarized neutrons and magnetic materials, we saw in Section 2.4 that n takes on two values, and the complete expression (cf. equation (2.5.6b) for their depth-dependences) has the form (see Parkin *et al.* (1986))

$$n_{\pm}(z) = 1 - \frac{\lambda^2}{2\pi}\left(\frac{\bar{b}(z)}{v} \pm cB(z)\right), \tag{3.3.10}$$

where the $\pm$ signs denote neutrons polarized parallel (+) and antiparallel (−) to the direction of $B(z)$, the magnetic induction in the material; $\bar{b}$ is the average nuclear scattering amplitude in the atomic volume v, and c is a constant equal to 2.3×10^{-6} Å^{-2} T^{-1}. It is assumed that the geometry is such that the material is magnetized in a field which is applied parallel to its surface.

The first evaluation of the potential of polarized neutron critical reflection for investigating surface magnetism was made by Felcher (1981). If we start by considering a semi-infinite slab of the magnetic material in which the magnetization is constant, then the spin reflectivities are given (see equn (2.4.11)) by

$$\mathcal{R}_{\pm} = \left|\frac{\sin\theta - (n_{\pm}^2 - \cos^2\theta)^{1/2}}{\sin\theta + (n_{\pm}^2 - \cos^2\theta)^{1/2}}\right|^2 \tag{3.3.11}$$

These reflectivities diminish rapidly above the critical glancing angles $\theta_{c_\pm}$, but they remain finite and detectable at least over the range $\theta_{c_\pm} < \theta < 2\theta_{c_\pm}$, where θ is the reflection angle. This critical reflection range is of particular interest because of its sensitivity to the magnetization of the material. The calculated dependence of the spin reflectivity ratio $\mathcal{R}_+/\mathcal{R}_-$ as a function of the reflection angle θ for ferromagnetic nickel with a constant magnetic moment of 0.6 β/atom, and at a wavelength of $\lambda = 5$ Å, is shown in Fig. 3.36. The reflectivities of both neutron spins are 100% at angles below θ_{c_-}, hence the unit reflectivity ratio. The ratio $\mathcal{R}_+/\mathcal{R}_-$ ratio increases dramatically between θ_{c_-} and θ_{c_+}, since $\mathcal{R}_+$ remains at 100% in this region, whereas $\mathcal{R}_-$ decreases. At $\theta > \theta_{c_+}$ the ratio decreases towards an asymptotic value since, although both $\mathcal{R}_+$ and $\mathcal{R}_-$ diminish in magnitude, $\mathcal{R}_+ > \mathcal{R}_-$ over the entire range.

Of particular interest are those reflecting surfaces for which the magnetization depth profile is non-uniform. The spin reflectivity ratio patterns expected for such surfaces can be evaluated by invoking a model to describe the depth dependence of the magnetic induction, i.e. $B(z)$ in

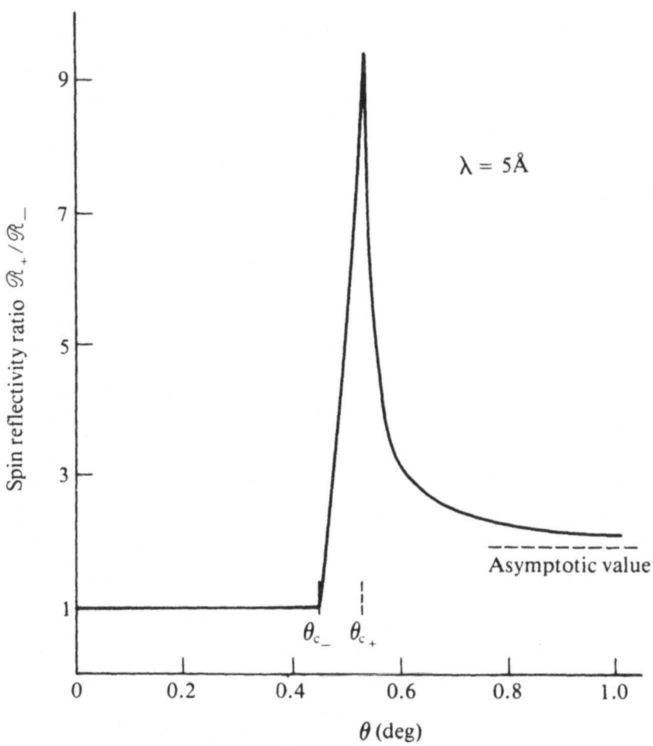

Fig. 3.36 Spin reflectivity ratio for specular reflection from ferromagnetic nickel with a uniform magnetic moment of 0.6 β at^{-1} (from Felcher 1981).

equn (3.3.10), and then subdividing the mirror material into a series of thin slices, with thicknesses down to atomic dimensions, each characterized by its own refractive index. The standard thin-film optical techniques described in Section 2.5.2 can then be applied to calculate $\mathcal{R}_+$ and $\mathcal{R}_-$, and hence their ratios. Felcher (1981) used this method to determine the spin reflectivity ratio deviations expected from the constant-magnetization result of Fig. 3.36 for a case in which the surface layer magnetizations were smaller than the bulk value. He showed that the effect of a reduced surface magnetization is to cause progressively larger decreases in the ratio as the (θ/λ) value increases. Examples of the experimental results obtained using the polarized neutron critical reflection technique are described in Section 4.6.

A polarized neutron reflectometer measures the flipping ratio as a function of the wave vector transfer $\boldsymbol{\kappa}$ perpendicular to the reflecting surface, and this ratio can then be corrected to give $\mathcal{R}_+/\mathcal{R}_-$. The specular reflection occurs by elastic scatttering with $|\boldsymbol{\kappa}| = 4\pi \sin(\theta)/\lambda$. Data over a wide range of $\boldsymbol{\kappa}$ is needed, and this is achieved either by using monochromatic beams and scanning a large number of angles (this has been done with unpolarized beams at reactor sources), or by using a broad-band time-of-flight method to determine λ at fixed θ. The latter technique has been used at both reactor sources (Farnoux 1985), and pulsed neutron sources (Felcher *et al.* 1985); the method is particularly well suited to pulsed sources.

Two polarized beam reflectometers have been built, both at pulsed neutron sources. The first was constructed at the Argonne National Laboratory and is described by Felcher *et al.* (1985); it is shown schematically in Fig. 3.37. The instrument views a cold moderator with effective temperature 57 K and the beryllium-filtered beam used for the measurements has a wavelength band of 3.8–8 Å. The polarizing mirror was originally a bulk Co mirror with polarizing efficiency $P \sim 0.75$; this was later replaced by an Fe–Ag polarizing supermirror with $P \sim 0.90$. The polarizing and sample mirrors are placed with their planes vertical and with a reflection angle at the sample mirror which is typically 0.35°. The beam divergence is of the order of 0.02°. The spin flipper used originally in the reflectometer was an axial two-coil non-adiabatic flipper, though this has been superseded by a magnetic wiggler flipper whose operation is phased with respect to the neutron source pulse. The spin intensities J_+ and J_- are measured as a function of time-of-flight or neutron wavelength at fixed reflection angles with the spin-flipper alternately 'off' and 'on'. These polarized beam intensities must then be corrected for the polarizing efficiency P of the polarizer and the spin-flipper efficiency to provide the spin reflectivity ratio $\mathcal{R}_+/\mathcal{R}_-$ as a function of the wave vector transfer. Many experiments were performed with a single ^{3}He detector containing a narrow vertical slit opening,

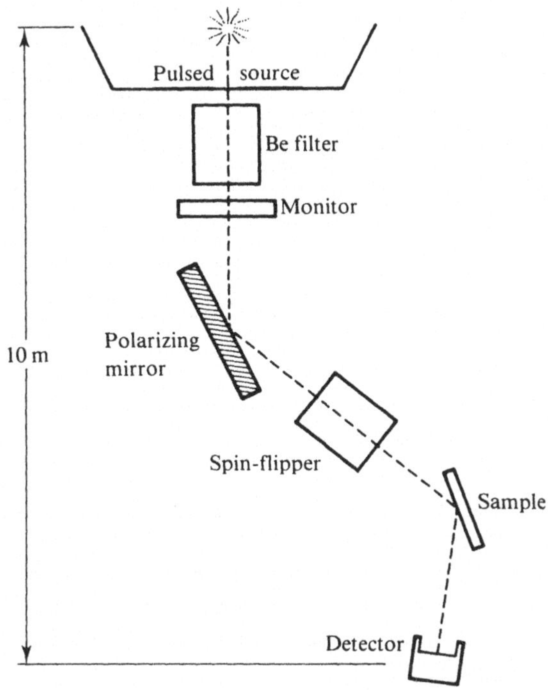

Fig. 3.37 The cold neutron time-of-flight polarized-beam reflectometer at the Argonne National Laboratory pulsed neutron source IPNS (from Felcher *et al.* 1985).

though the ease of operation of the instrument was later substantially improved by introducing a lithium-glass scintillator area detector.

A second polarized beam reflectometer, CRISP, was subsequently designed for the Rutherford Appleton Laboratory pulsed neutron source ISIS. The design concept is described by Penfold and Williams (1985), and a sketch of the instrument is shown in Fig. 3.38. The reflectometer is also required for unpolarized beam measurements from liquid surfaces, and this is facilitated by using a horizontal plane mirror geometry with an incident beam which is inclined at 1.5° to the horizontal. Fixed-reflection-angle measurements are taken over a range of wavelengths ($\Delta\lambda \sim 7$ Å with $\lambda_{min} \sim 2$ Å) defined by an in-shield disc chopper. Component A is a series of thin-film Ni mirrors set at a reflection angle of 3°, which reflect very-long-wavelength 'frame overlap' neutrons ($\lambda > 26$ Å) out of the main reflectometer beam. The polarizer (B) is a cobalt–titanium supermirror, and the spin-flipper can be of the two-coil non-adiabatic type (C) or a phased Mezei precession coil device. The reflectometer has been designed to utilize a very broad wavelength band, from 2.0 Å to 26 Å, from the ISIS 26 K hydrogen moderator, and contains a removable beryllium

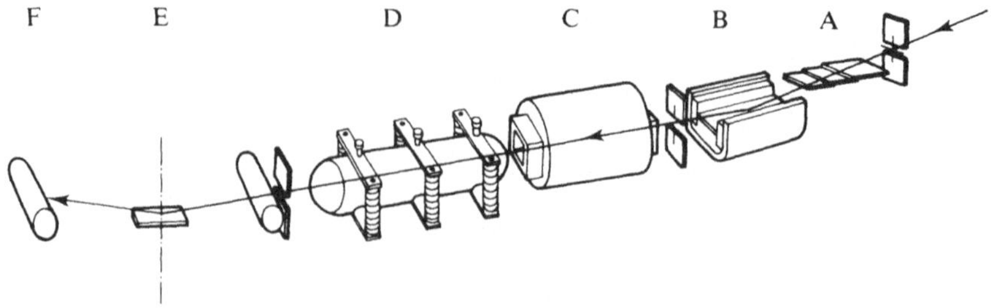

Fig. 3.38 Schematic layout of the CRISP polarized-beam reflectometer at the ISIS pulsed neutron source, Rutherford Appleton Laboratory. The principal components are: A, the frame-overlap Ni mirrors set at $\theta = 3°$; B, a polarizing supermirror; C, a two-coil non-adiabatic spin-flipper; D, permanent magnet guide fields; E, the test magnetized mirror; F, the neutron detector.

filter for operation at the shortest wavelengths. The $\boldsymbol{\kappa}$ range available for measurements is from $0.012\,\text{Å}^{-1}$ to $0.6\,\text{Å}^{-1}$, and the resolution in $\boldsymbol{\kappa}$ is determined essentially by the collimation angles defined by the horizontal collimating slits. This can readily be varied over the range $\delta\kappa/\kappa \sim 0.02$–$0.10$ or greater. The instrument is capable of measuring reflectivities down to values in the region 10^{-5}–10^{-6}.

3.3.6 Calibration of polarized-beam instruments

The efficiencies of polarized-beam instrument components like the polarizer and spin-flipper are in practice always less than 100%, and it is important to correct experimental data to allow for these imperfections. Before describing the methods available for calibrating these parameters, we must first establish how the combined action of two polarizing elements, for example with efficiencies P and P_A, affects the observed intensity. It is convenient to introduce the factors $\phi_\pm$ which represent the fractions of (+) and (−) spin neutrons in a beam after it has been polarized. For a polarizer with efficiency P we therefore have

$$P = \phi_+ - \phi_-, \tag{3.3.12a}$$

$$1 = \phi_+ + \phi_-, \tag{3.3.12b}$$

i.e. $\phi_+ = \frac{1}{2}(1+P)$ and $\phi_- = \frac{1}{2}(1-P)$. A polarizer can thus be considered as a device whose transmission for (+) spin neutrons is proportional to $(1+P)$, and for (−) spin neutrons is proportional to $(1-P)$. The (+) spin neutron intensity following the action of two (+) spin polarizers in series is therefore

$$I_+ = I'(1+P)(1+P_A), \tag{3.3.13}$$

where the intensity factor I' depends on the reflectivities or transmittance properties of the polarizers. The corresponding (−) spin neutron intensity from the same two polarizers is

$$I_- = I'(1-P)(1-P_A). \quad (3.3.14)$$

The total intensity of the beam is therefore

$$I = 2I'(1+PP_A) \quad (3.3.15)$$

and the resultant polarization P_R is

$$P_R = (P+P_A)/(1+PP_A). \quad (3.3.16)$$

In the simplest polarized-beam experiment, P represents the polarizer and P_A may be regarded as the specimen whose spin-dependent response $\rho_\pm$ relates to P_A as follows:

$$P_A = (\rho_+ - \rho_-)/(\rho_+ + \rho_-). \quad (3.3.17)$$

The object of the experiment is to determine P_A, and by way of illustration we refer to two examples. The first is polarized neutron single-crystal diffraction for which (see Section (2.2.2))

$$\rho_\pm = [F_N(\boldsymbol{\kappa}) \pm F_M(\boldsymbol{\kappa})]^2 \quad (3.3.18)$$

and $F_N(\boldsymbol{\kappa})$ and $F_M(\boldsymbol{\kappa})$ are the nuclear and magnetic structure factors for a particular Bragg reflection where the reciprocal lattice vector $\boldsymbol{\tau} = \boldsymbol{\kappa}$. The second is polarized neutron critical reflection, for which

$$\rho_\pm = \mathcal{R}_\pm \quad (3.3.19)$$

and here ρ equals the 'spin' reflectivities of a magnetized reflecting surface, as given for example by equn (3.3.11). The most accurate method for determining P_A is to determine the ratio

$$(I/I^*) = (1+PP_A)/(1+P^*P_A), \quad (3.3.20)$$

where I is the observed intensity with the polarizer and specimen as given by equn (3.3.15), and I^* is measured under identical conditions other than that P is changed in a known and controllable way to P^*. The most sensitive of these ratio techniques is that which uses a spin-flipper to reverse the polarization P, and where, (see equn (3.2.4))

$$P^* = P(1-2f), \quad (3.3.21)$$

where f is the spin flipper efficiency. An alternative, though seldom used, method is to use a depolarizing iron shim (see for example Nathans *et al.* (1959)) to act on the incident polarized beam. The gives a shim ratio R_S:

$$R_S = (1+PP_A)/(1+(1-S)PP_A), \quad (3.3.22)$$

where S is the depolarizing shim efficiency.

Precision measurements of P_A by the flipping ratio method therefore require accurate calibrations of P and f, and this is the subject of this section. The procedure by which data is corrected for inefficiencies in P and f in diffraction experiments is described in Chapter 4. We shall see that the best methods for determining these quantities are to use flipping ratio measurements with standard scattering or reflecting specimens for which the P_A values are well known.

Polarized-beam diffractometers

One of the most precise procedures for determining the beam polarization and flipping efficiency in polarized neutron diffractometers was proposed and implemented by Brown and Forsyth (1964). Their method relies on the fact that thin polarizing crystal analysers are available for which, to a very good approximation, $P_A = +1$ and $P_A = -1$. The former condition pertains for the (200) reflection in a single crystal of $Co_{0.92}Fe_{0.08}$, and the latter for the (111) reflection of the Heusler alloy Cu_2MnAl; we shall use the subscripts + and − to denote the flipping ratios R measured with these spin analysers. If the polarizer produces a positively polarized beam, then

$$R_+ = (1+P)/[1+(1-2f)P], \tag{3.3.23}$$

from which it is apparent that, if P and f approach unity, the ratio provides a sensitive measure of deviations of the product $(1-2f)P$ from unity. The Heusler analyser flipping ratio

$$R_- = (1-P)/[1+(2f-1)P] \tag{3.3.24}$$

is very insensitive to small differences in f from unity; however, it has excellent sensitivity to the value of P. A combination of R_+ and R_- measurements therefore enables both P and f to be evaluated to a high degree of accuracy. Typical values, with errors, are $P \sim 0.99 \pm 0.01$ and $f = 0.995 \pm 0.001$ for a $Co_{0.92}Fe_{0.08}$ polarizing monochromator and a radiofrequency spin-flipper. An extension of this treatment to allow for the $\lambda/2$ component in the monochromatic beam has been developed by Boucherle and is described by Forsyth (1979).

Calibration of non-adiabatic spin-flipper efficiencies

The above principle was also used by Jones and Williams (1978) to evaluate the flipping efficiencies and depolarization effects in non-adiabatic spin-flippers. Instrumental depolarization, which we express by the parameter D, affects the polarization in the same way as S of equn (3.3.22), and the full expression for the flipping ratios measured with $P_A = +1$ and $P_A = -1$ spin analysers is

$$R_\pm = [1 \pm P(1-D)]/[1 \pm P(1-D)(1-2f)]. \tag{3.3.25}$$

After rearrangement we obtain

$$P(1-D)=(R_+ + R_- - 2R_+R_-)/(R_+ - R_-) \qquad (3.3.26)$$

and

$$f=(R_+ - 1)(1-R_-)/(R_+ + R_- - 2R_+R_-). \qquad (3.3.27)$$

Thus, by separately measuring R_+ and R_- with thin $Co_{0.92}Fe_{0.08}$ and Heusler alloy analysing crystals, Jones and Williams (1978) were able to make an unambiguous distinction between inefficient spin flipping and instrumental depolarization due to imperfect guide fields. In an investigation of the properties of the axial two-coil non-adiabatic spin-flipper, these authors observed decreases in both the $(1-D)$ and f terms for non-axial beam trajectories through the coils.

Triple-axis polarization analysis spectrometers

In classical polarization analysis experiments the instrument parameters which require calibration are $PP_A(1-D)$ and the flipper efficiency f. Triple-axis polarization analysis spectrometers have large crystal analysers, for which in general $P_A \neq 1$. The best method for determining the polarization term is to measure the flipping ratio R' with an analysing crystal whose polarizing efficiency is of opposite sign to that of the polarizer, and which is placed in the direct beam from the polarizer. The general expression for the flipping ratio is then

$$R'=[1+PP_A(1-D)]/[1+PP_A(1-D)(1-2f)], \qquad (3.3.28)$$

which, after rearrangement, gives

$$PP_A(1-D)=(1-R')/[R'(1-2f)-1]. \qquad (3.3.29)$$

If $PP_A(1-D)$ is negative, we can use a similar argument to that used with equn (3.2.24) to show that R' is insensitive to f, thus this polarization term can be evaluated to a high degree of accuracy. This method was used by Freeman *et al.* (1978) in their calibration of the D5 polarization analysis triple-axis spectrometer at the Institut Laue–Langevin, Grenoble. The determination of the flipping efficiency is less straightforward, though flipping ratio measurements with $Co_{0.92}Fe_{0.08}(R'_c)$ and Heusler alloy (R'_H) crystal analysers still provide a similar relationship to equn (3.3.27), i.e.

$$f=(R'_c - 1)(1-R'_H)/(R'_c + R'_H - 2R'_cR'_H), \qquad (3.3.30)$$

provided that $|P_A|$ $(Co_{0.92}Fe_{0.08}) = |P_A|$ (Heusler alloy).

Instruments with optical polarizers

Optical polarizers all have positive polarizing efficiencies; thus the procedures described so far cannot be used for their calibration.

However, their polarizing efficiencies are still best determined using flipping ratio measurements, and the method used by Childs *et al.* (1981) for investigating the (θ/λ)-dependent polarizing properties of various single mirrors typifies the procedures available. A 'standard' mirror is set to reflect a monochromatic beam at a high (θ/λ) value at which its polarizing efficiency P_1 approaches unity. The reflected beam then passes through a flipper with efficiency f, and is reflected by the test mirror which has a polarizing efficiency P_2 at a reflection angle θ_2; flipping ratios R are measured as a function of θ_2.

The polarization profiles for the test mirrors investigated by Childs *et al.* (1981) were evaluated by using these R values together with the maximum flipping ratio R_0 which was observed at high test mirror reflection angles θ_2 where $P_2 = 1.0$. R_0 was used essentially to calibrate the performance of the polarizer/flipper combination and, since this involves two unknowns, two limiting approximations may be considered. In the first it is assumed that the spin-flipper is 100% efficient and by using the standard flipping ratio equation, e.g. equn (3.3.28), we obtain

$$P_1(1-D) = (R_0 - 1)/(R_0 + 1). \qquad (3.3.31)$$

The test mirror polarization is then given by

$$P_2' = (R_0 + 1)(R - 1)/[(R_0 - 1)(R + 1)]. \qquad (3.3.32)$$

The second approximation is to assume that the polarizer is 100% efficient, i.e. $P_1 = 1$, that there is no beam depolarization, and that the flipper efficiency $f < 1$. The flipper efficiency then becomes

$$f = (R_0 - 1)/R_0 \qquad (3.3.33)$$

with a test mirror polarization

$$P_2'' = R_0(R - 1)/(R_0 + RR_0 - 2R). \qquad (3.3.34)$$

The real test mirror polarization P_2 must clearly be intermediate between P_2' and P_2'', and P_2 can be arbitrarily defined as

$$\begin{aligned} P_2 &= (P_2' + P_2'')/2 \\ &= \frac{(R-1)[R_0^2(1+R) - R(R_0+1)]}{(R_0-1)(R+1)[R_0 + RR_0 - 2R]}, \end{aligned} \qquad (3.3.35)$$

together with a systematic error

$$\begin{aligned} \Delta P_2 &= (P_2' - P_2'')/2 \\ &= \frac{(R-1)(R_0-R)}{(R_0-1)(R+1)[R_0 + RR_0 - 2R]}. \end{aligned} \qquad (3.3.36)$$

The maximum systematic fractional error in P_2 using these assumptions is

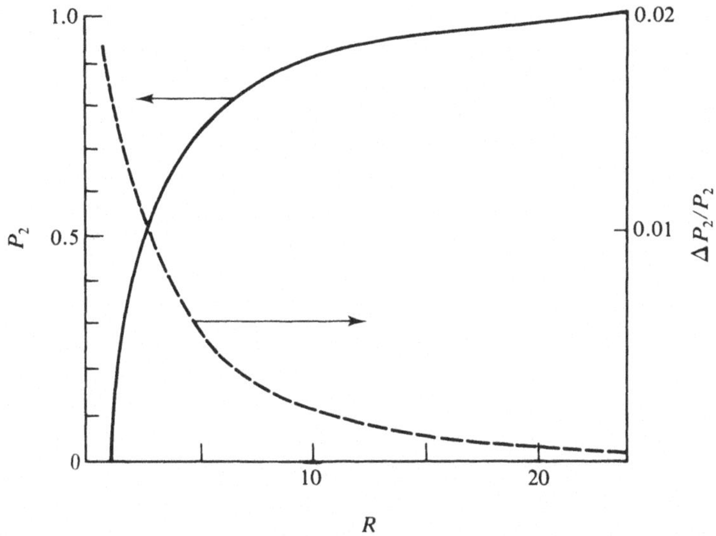

Fig. 3.39 Calibration of test mirror polarization P_2 (full curve) with the systematic error $\Delta P_2/P_2$ (broken curve) using the procedure described by Childs *et al.* (1981).

therefore

$$(\Delta P_2/P_2) = (R_0 - R)/[R_0^2(1 + R) - R(R_0 - 1)]. \tag{3.3.37}$$

Figure 3.39 shows values of P_2 and $\Delta P_2/P_2$ as a function of R evaluated for a maximum flipping ratio $R_0 = 24$, which was typical in the measurements described by Childs *et al.* (1981). The curves show that the maximum systematic error obtained using this method is ~2%, though over most of the range pertinent to actual measurements, where $R \gtrsim 2$ and $P_2 \gtrsim 0.3$, this error is less than 1%. The value of the flipper efficiency f is determined between the bounds of the minimum limit set by equn (3.3.33), and unity.

The use of standard scatterers

In polarization analysis instruments it is often possible to use a standard sample, for which the spin-dependence of the scattering is well characterized, to determine the effective polarizing efficiency product $PP_A(1 - D)$. It is convenient to express the flipping ratio observed in polarization analysis experiments in terms of the fraction of spin flip scattering r which occurs in the scattering process. The effect of r on the beam polarization is precisely the same as that of the flipping efficiency f, i.e. $P \rightarrow P(1 - 2r)$, therefore we obtain a flipping ratio

$$R = \frac{1 + PP_A(1 - D)(1 - 2r)}{1 + PP_A(1 - D)(1 - 2r)(1 - 2f)}. \tag{3.3.38}$$

One obvious advantage of using a scattering standard is that it can provide the effective polarization term $PP_A(1-D)$ over the same incident and scattered beam paths as are used in actual experiments.

Freeman *et al.* (1977) examined the possibility of using a number of Bragg peaks from a pyrolytic graphite single crystal to check the calibration of the D5 polarization analysis instrument at Grenoble. The scattering at nuclear Bragg peaks is purely non-spin-flip, i.e. $r=0$, and in a cross-polarizer arrangement the determination of $R_{r=0}$ provides an excellent means of calibrating the effective polarization product. Another useful scattering standard is vanadium, for which the r factor for single scattering is accurately known to be 0.663. Multiple scattering, however, always reduces the r value, though a calculation of the multiple scattering does allow an effective r to be determined, and calibrations using vanadium are convenient and reliable. Ahmed *et al.* (1974) used vanadium scattering to calibrate the polarization term in the LONGPOL instrument, and showed that, for low flipping-ratio conditions, it is a sufficiently good approximation to assume $f=1$ in the data analysis procedure. Vanadium has also been proposed as a calibrant for the pulsed-source polarization analysis instrument POLARIS (Mayers and Williams 1981).

Calibration of polarizing filters

The efficiencies of polarizing filters can be evaluated by transmission measurements. The double transmission effect τ_D, as defined in equn (1.5.5), for two filters with different thicknesses, is related to the polarization product P_1P_2 as follows,

$$\tau_D = 2\tanh(nt_1\sigma_p)\tanh(Nt_2\sigma_p) = 2P_1P_2, \tag{3.3.39}$$

and this effect was used by Ahmed *et al.* (1974) to assess the optimum thicknesses for the iron filters in the LONGPOL polarization analysis instrument.

Mayers *et al.* (1984) used the equivalent of the single transmission effect, τ_S, to measure the polarizing efficiencies of polarized samarium resonance absorption filters under various conditions. In resonance absorption filters containing polarized nuclei, the ratio R of the total transmittance T_C when the filter is cold to the transmittance T_H at a higher temperature when the nuclear polarization P_N is zero, is given by

$$R = T_C/T_H = \cosh(\rho P_N\sigma_A Nt) \equiv 1+\tau_S, \tag{3.3.40}$$

from which it is readily shown that the polarizing efficiency of the filter is

$$P = \pm[1-(T_C/T_H)^2]^{1/2}. \tag{3.3.41}$$

The provision of thin neutron monitors before and after the filter cryostat provides an easy method of detecting any changes in the polarizing efficiency of this type of polarizing filter when it is installed in a scattering instrument; this is done at the ISIS pulsed-source polarization analysis instrument POLARIS. Transmission measurements should not, however, be considered to be a substitute for the calibration procedures described earlier using standards since they cannot be used to reveal instrumental depolarization effects, and they cannot supply the sign of the polarizing efficiency.

Direct calibration of cold polarized neutron beams

Finally, before completing this section, reference should be made again to the Stern–Gerlach polarimeter, which spatially separates the two spin components in polarized beams, and can be used to measure directly the polarizing, spin flipping, or depolarizing efficiencies of instrument components. The method is only really practicable for small areas of cold neutron beams, though, since the spatial separations increase with wavelength, it may sometimes reveal wavelength-dependent effects in these efficiencies. Figure 3.17 shows the results of the use of a Stern–Gerlach polarimeter to measure the polarization of a beam from a polarizing Soller, and also the efficiency of a depolarizing mild-steel shim which was used to depolarize the beam.

4

Polarized neutron scattering

The polarized neutron diffraction method, introduced by Nathans *et al.* (1959), is still the most commonly performed polarized neutron scattering experiment. The specimens are generally single crystals, and the end-product of the many investigations so far carried out is a detailed understanding of the distribution of magnetism at the atomic level. The method has been applied to study metals, compounds, and alloys in all the transition metal, lanthanide and actinide series. There have also been important polarized-beam experiments on disordered magnetic solids, for example structural investigations of amorphous materials and magnetic-defect diffuse scattering in alloys. These will all be discussed in this chapter. We also include two more recent innovations: the study of nuclear magnetism which has resulted from the advances made in the methods for polarizing nuclei, and the study of surface magnetic phenomena by polarized neutron reflection. The latter technique is, experimentally, very similar to the diffraction method. Typical scientific results will be presented of all these applications, though space limitations have dictated that it has been necessary to exercise stringency with the number of examples selected.

In most of the experiments to be described, the observable quantities are the 'flipping ratios', which are the intensity ratios detected following scattering with 'unflipped' and 'flipped' incident polarized beams from scattering samples which have a preferred axis. In some of the experiments it is sufficient to measure the difference scattering $(d\sigma/d\Omega)_+ - (d\sigma/d\Omega)_-$. The raw data from both the ratio and difference methods must be corrected for instrumental deficiences in the polarizer and spin-flipper, and if the polarizer is a crystal monochromator account must also be taken of higher-order wavelength contamination. There are also two important sample-dependent effects which influence the measurements: extinction, which occurs in most single-crystal scatterers, and spontaneous depolarization, which is frequently present in multidomain magnetic specimens. We start by describing some of the procedures available for making these corrections; more complete accounts are given by Forsyth (1979), and by Moon *et al.* (1972).

4.1 FLIPPING RATIO CORRECTIONS

4.1.1 Instrumental factors

Let us consider a typical polarized-beam flipping ratio measurement on a magnetized single crystal for which the plane of scattering is horizontal, and the applied magnetic field on the sample is vertical. During the experiment the flipping ratio is determined for a large number of reflections, i.e. over as large a range of $(\sin\theta_B)/\lambda$ or $\boldsymbol{\kappa}$ as possible. In the ideal case of perfect beam polarization, perfect polarization reversal, zero $\lambda/2$ and sample-dependent corrections, and for a centrosymmetric crystal structure, the flipping ratio for a particular reflection is given by

$$R = \frac{F_N^2 + 2F_N F_M + F_M^2}{F_N^2 - 2F_N F_M + F_M^2} = \left(\frac{1+\gamma}{1-\gamma}\right)^2, \qquad (4.1.1)$$

where $\gamma = F_M/F_N$. The magnetic and nuclear structure factors F_M and F_N are defined by equns (2.2.23) and (2.2.21), and both depend on the $\boldsymbol{\kappa}$ or $\boldsymbol{\tau}$ value appropriate for the Bragg reflection being measured.

We should note at this point one of the reasons why the flipping ratio technique has been so successful. This follows because the accurate evaluation of a ratio is much easier than determining an integrated intensity; furthermore, it is not necessary to record counts over a complete diffraction peak. It is sufficient simply to detect the two spin-state count rates at the reflection maximum, and then to offset the crystal by a few degrees to obtain the backgrounds. The absorption is, other than in very exceptional circumstances when the sample contains a polarized absorbing nucleus, neutron-spin-independent; thus it has no effect on R.

The diffracted intensity from a beam with polarization P is proportional to

$$\begin{aligned} I_{OFF} &\propto [\phi_+(F_N^2 + 2F_N F_M + F_M^2) + \phi_-(F_N^2 - 2F_N F_M + F_M^2)] \\ &\propto (F_N^2 + 2F_N F_M P + F_M^2), \end{aligned} \qquad (4.1.2)$$

where the spin fraction parameters $\phi_\pm$ are as defined in Section 3.3.6. With a spin flipper of efficiency f 'on' this intensity becomes

$$I_{ON} \propto [F_N^2 + 2F_N F_M P(1-2f) + F_M^2] \qquad (4.1.3)$$

and the flipping ratio is

$$R = \frac{1 + 2P\gamma + \gamma^2}{1 + 2P(1-2f)\gamma + \gamma^2}, \qquad (4.1.4)$$

which has roots

$$\gamma = \frac{P[1 - R(1-2f)] \pm \{P^2[1 - R(1-2f)]^2 - (R-1)^2\}^{1/2}}{R-1}. \qquad (4.1.5)$$

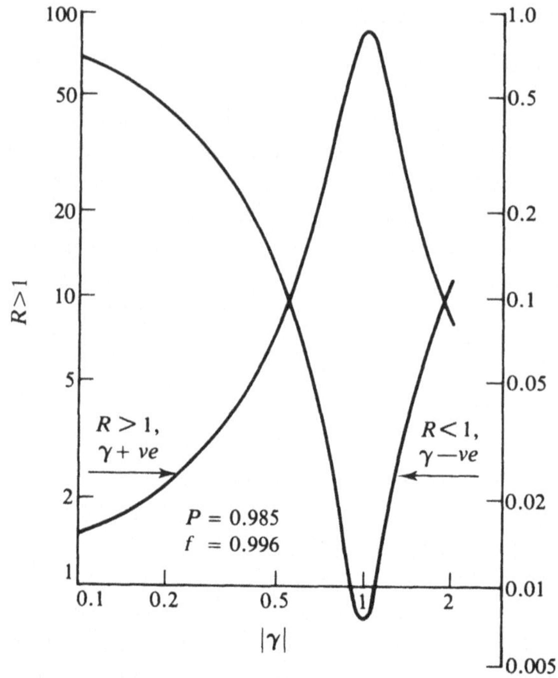

Fig. 4.1 The relationship between the flipping ratio R and the γ parameter for typical polarizing crystal and spin-flipper efficiencies in a polarized-beam diffractometer (from Brown and Forsyth 1964).

The relationships between R and γ given by this expression for typical values of P and f are illustrated in Fig. 4.1. The flipping ratio method is seen therefore to give highly precise results when $|\gamma| = |F_M/F_N|$ is either much less than or much greater than unity, since R is then relatively insensitive to instrumental imperfections. The most difficult cases to interpret and evaluate are when $|F_M| \simeq |F_N|$, since instrumental corrections then become very important, and sometimes a separate experiment may be needed to determine which of the γ roots is appropriate. In most cases of practical importance the appropriate $|\gamma|$, corresponding to an experimentally determined R, can be assigned on the basis of the $\boldsymbol{\kappa}$ dependence of $|\gamma|$, e.g. at high $\boldsymbol{\kappa}$, F_M is small and $|\gamma|$ is always less than unity.

The third instrumental parameter which influences the observed flipping ratio is the higher-order wavelength contamination in the polarized monochromatic beam. The effect arises from two causes: the lower polarizing efficiencies of higher order reflections, and the fact that action of the spin-flipper is often neutron-wavelength-dependent. A method of applying these corrections for the (400) reflection in $Co_{0.92}$

$Fe_{0.08}$ and the (222) reflection in the Heusler alloy Cu_2MnAl has been developed by Boucherle, and is described by Forsyth (1979).

4.1.2 Extinction

There are two types of extinction: primary and secondary. The former is prevalent in near-perfect single crystals, and occurs because the amplitude of the incident neutron wave is progressively reduced as it undergoes successive Bragg reflections from a large number of crystal planes. The atoms near the bottom of a large mosaic block are therefore much less illuminated than those near the top. Its effect on the flipping ratio is negligible compared with other errors and, as demonstrated by Mook (1966), it can be ignored in comparison with secondary extinction.

Secondary extinction occurs in crystals which consist of numerous small perfect mosaic blocks which are slightly misoriented with respect to each other; the reduction in the beam strength within a single mosaic block is then relatively small. This extinction therefore only becomes appreciable deep within the crystal and at those blocks which have the same orientation as earlier blocks. The mosaic nature of a crystal is generally described by an angular distribution function $W(\Delta)$, which is defined such that $W(\Delta)\,d\Delta$ is the fraction of blocks which have their normals at angles between Δ and $(\Delta + d\Delta)$ from the crystal plane. This distribution function is usually well described by the normalized Gaussian,

$$W(\Delta) = (1/\eta\sqrt{2\pi})\exp[-\Delta^2/(2\eta^2)], \qquad (4.1.6)$$

where η represents the standard deviation of the mosaic spread. It is clear that in samples which exhibit secondary extinction the possibility exists that the effective intensities of the scattered (+) and (−) spin neutrons can be influenced by the mosaic structure, and this necessitates a correction to the measured flipping ratio.

The effect of secondary extinction in polarized neutron measurements has been discussed by Brown (1970). There are two extreme types of behaviour, in the first of which the flipping ratio varies across the rocking curve for a particular reflection, and in the second the flipping ratio remains constant over the rocking curve. The former result is relatively easily understood within the mosaic model for the crystal as exemplified by equn (4.1.6). On the lower slope of the rocking curve the extinction is extremely small, since there are only a few mosaic blocks in the correct orientation for reflection; the measured flipping ratio then has essentially no extinction correction. Near the peak of the rocking curve the secondary extinction is expected to be at its maximum, and is also larger for the neutron spin state with the larger scattering cross-section. The net effect is that the observed flipping ratio at the peak is reduced by the extinction and can approach unity. This type of behaviour was observed in iron by Shull and Mook (1966).

In another class of crystal the flipping ratio remains unchanged over the rocking curve, as found for example by Brown *et al.* (1968) for MnSi. The explanation here is that the angular spread of the scattered beam due to particle size dominates that due to the mosaic spread. In the model used by these authors to describe the secondary extinction in MnSi, an effective attenuation coefficient μ was used with

$$\mu = \mu_0 + gQ, \tag{4.1.7}$$

where μ_0 is the true attenuation coefficient, and g is an extinction constant. Q is the crystallographic quantity

$$Q = \lambda^3 N^2/(v_0^2 \sin 2\theta), \tag{4.1.8}$$

where N is the number of unit cells per unit volume, and v_0 is the unit cell volume. If we assume that the same g value applies for both (+) and (−) spin reflections then the flipping ratio becomes

$$R = \frac{\tfrac{1}{2}(1+P)(1+\gamma^2+2\gamma)\exp(-gQ_+t) + \tfrac{1}{2}(1-P)(1+\gamma^2-2\gamma)\exp(-gQ_-t)}{\tfrac{1}{2}[1+P(1-2f)](1+\gamma^2+2\gamma)\exp(-gQ_+t) + \tfrac{1}{2}[1-P(1-2f)](1+\gamma^2-2\gamma)\exp(-gQ_-t)}, \tag{4.1.9}$$

where t is the effective path length in the crystal and

$$Q_\pm = [\lambda^3/(v_0^2 \sin 2\theta)]\,|F_N \pm F_M|^2. \tag{4.1.10}$$

The first term in the numerator of equn (4.1.9) is the product of the (+) spin neutron intensity and the appropriate cross-section, and the second term gives the same product for (−) spin neutrons. The relationship may be simplified to

$$R = \frac{(1+\gamma^2)(\cosh x - P\sinh x) + 2\gamma(P\cosh x - \sinh x)}{(1+\gamma^2)[\cosh x - P(1-2f)\sinh x] + 2\gamma[P(1-2f)\cosh x - \sinh x]}, \tag{4.1.11}$$

where

$$x = (2\lambda^3 g t N^2 \gamma)/(v_0^2 \sin 2\theta). \tag{4.1.12}$$

For situations in which the secondary extinction is sufficiently small that the effective path lengths are the same for both (+) and (−) spin neutrons, then equn (4.1.11) can be solved self-consistently for γ by successive approximations. The results of applying this procedure to form factor measurements in MnSi are shown in Fig. 4.2; the upper points denote the extinction-corrected values.

A simplifying assumption in the above analysis is that the neutron

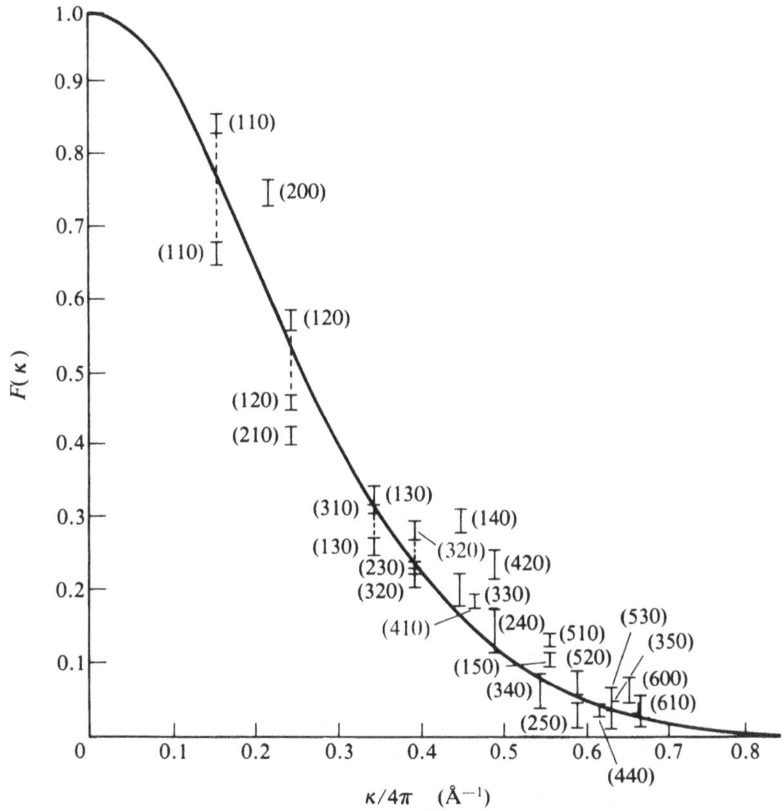

Fig. 4.2 The effect of extinction on the form factor measurements for Mn in MnSi. The lower values are those obtained with no extinction corrections, and the solid curve is the theoretical form factor for Mn^{2+} (from Brown 1970).

maintains its spin state within the crystal, i.e. that there is no spin-flip scattering. Magnetic spin-flip scattering is only possible when the direction of the scattering vector is no longer exactly perpendicular to the neutron polarization or sample magnetization direction, as is the case when the detector is tilted to collect flipping ratio data from higher layers in reciprocal space. The effective beam polarization can then decrease as it traverses the crystal, though the effect is seldom large when the extinction is mainly secondary. Formal treatments of this effect on the flipping ratio are presented in the papers by Forsyth (1979) and Moon et al. (1972). Delapalme (1979) has reviewed several extinction models, and has shown how polarized neutrons can provide a sensitive test for the validity of the various approaches.

The x parameter in equn (4.1.12) is a measure of the magnitude of the extinction effect and is seen to reduce at shorter wavelengths and smaller crystal thicknesses. Flipping ratio measurements as a function of wavelength and crystal thickness thus provide an experimental means of assessing the importance of extinction in single-crystal specimens. It is sometimes possible to reduce the extinction effect to an acceptably low level by taking data at short neutron wavelengths. A polarized-beam single-crystal diffractometer should therefore ideally be placed on a reactor hot-source beam tube.

4.1.3 Depolarization

The depolarization of neutron beams as they pass through magnetic materials is a phenomenon which can be used to investigate magnetic domain structures, and is described in Section 6.2.1. The magnitude of the effect can be characterized by a depolarization parameter D_m given by

$$2D_m = (1/t)\ln(P/P'), \tag{4.1.13}$$

where P is the incident beam polarization, and P' is the reduced value following transmission through a specimen of thickness t. D_m^{-1} is the mean free path for spin reversal in the specimen.

The effect of depolarization on the flipping ratio in polarized beam diffractometry can be modelled by assuming that D_m is a constant for all incident beam trajectories in the sample. If we consider first the polarized beam intensity reflected from a small element of crystal volume dV with internal incident flight path d, then this is given by

$$I \propto 1 + \gamma^2 + 2P\gamma \exp(-2D_m d)\, dV. \tag{4.1.14}$$

There is a corresponding expression with $P(1-2f)$ replacing P for the spin-flipped reflected beam intensity. By defining a new depolarization factor D', given by the integral

$$D' = \int \exp(-2D_m d)\, dV \tag{4.1.15}$$

taken over the whole crystal volume, we obtain an extinction-free flipping ratio

$$R = (1 + \gamma^2 + 2PD'\gamma)/[1 + \gamma^2 + 2P(1-2f)D'\gamma] \tag{4.1.16}$$

Note that D' is different for each reflection, and is sample-geometry-dependent. An approximate corrected form of the flipping ratio which includes the effect of extinction may be obtained by replacing P in equn (4.1.11) by PD'.

4.2 MAGNETIC BRAGG DIFFRACTION

Essentially all of the experiments to be discussed in this section have involved making flipping ratio measurements from magnetized single-crystal specimens. Polarized neutron powder diffraction has potential advantages such as (a) no extinction, (b) ease of sample preparation, and (c) a simpler data collection procedure; despite these it is not often attempted. In addition to the poorer peak-to-background ratios and reflection overlap difficulties associated with the powder diffraction method, there are also sample-dependent effects which can strongly influence the flipping ratios. The first of these is neutron depolarization, which occurs due to the varying B fields in strongly magnetic materials but can also exist in paramagnetic powders. As an illustration of this phenomenon we cite the observation by Moon and Koehler (1975) of an applied field-dependent flipping ratio in a $^{160}Gd_2O_3$ powder specimen. Another effect, attributed to the alignment of individual crystallites by an applied magnetic field, has been reported by Aldred *et al.* (1975) for a $NpFe_2$ powder specimen. The result is that the Bragg peak flipping ratios are either enhanced or diminished, and the effect is expected to be particularly marked in highly anisotropic specimens. The polarized neutron powder method is therefore limited to weakly magnetic specimens with low anisotropy, such as paramagnets containing transition metal or lanthanide ions, paramagnetic molecules, and free radicals.

Before presenting typical polarized-beam results obtained with magnetized single crystals, it is worthwhile recalling the significance of the $\gamma(hkl)$ values measured for the different Bragg reflections. These give the unit cell structure factor ratios $F_M(\boldsymbol{\kappa})/F_N(\boldsymbol{\kappa})$, and sometimes (see for example Fig. 4.12) the data is simply presented in this form. In the majority of cases, however, a separate structural determination of $F_N(\boldsymbol{\kappa})$ is made using unpolarized neutrons, and this allows $F_M(\boldsymbol{\kappa})$ or the magnetic scattering amplitudes $p(hkl)$ to be evaluated. The $\boldsymbol{\kappa}$-dependence of the magnetic form factor can be plotted in several ways: as $p(hkl)/b(hkl)$ versus $\boldsymbol{\kappa}$, in terms of atomic Bohr magnetons (β) multiplied by $F(\boldsymbol{\kappa})$, or by using magnetization data to normalize $F(\boldsymbol{\kappa})$ to unity at $\boldsymbol{\kappa}=0$. The fundamental quantity measured is always the magnetic form factor $F(\boldsymbol{\kappa})$ and, since this is the Fourier transform of the magnetization density in the unit cell, it is frequently used to provide magnetization density distributions within the unit cell; see equn (2.2.30).

4.2.1 Ferromagnetic metals and compounds of the 3d group

Some of the earliest data obtained using the polarized neutron diffraction technique were in the ferromagnetic transition elements Fe (Shull and Yamada 1962), Co (Moon 1964), and Ni (Mook 1966). The Ni magnetic form factor and the magnetization density distribution observed in the (001) plane of its unit cell are shown in Figs 2.7 and 2.8. The Fourier maps of all these metals revealed for the first time the presence of a negatively polarized moment, which was distributed more or less uniformly throughout the unit cells. In order to accommodate this in the comparison of measured and calculated form factors, the experimental data were fitted to the expression

$$F(\boldsymbol{\kappa}) = (2/g)[(1+\alpha)F_S(\boldsymbol{\kappa}) - \alpha\delta(\boldsymbol{\kappa})] + [(g-2)/g]F_L(\boldsymbol{\kappa}) + F_{\text{core}}(\boldsymbol{\kappa}), \quad (4.2.1)$$

where $F_S(\boldsymbol{\kappa})$ and $F_L(\boldsymbol{\kappa})$ are the spin and orbital contributions, and $F_{\text{core}}(\boldsymbol{\kappa})$ is due to spin polarization effects of core electrons within the 3d shell. The α parameter, which models the fraction of the total spin moment which is distributed smoothly over the unit cell, is assumed to be constant and the $\alpha\ \delta(\boldsymbol{\kappa})$ term only becomes finite for the (000) reflection. The magnetic electron distributions in these atoms are not spherically symmetrical, thus the form factors are not a smoothly varying function of $\boldsymbol{\kappa}$. For example, in a cubic crystal field, as occurs in Fe and Ni, the five-fold degenerate 3d electron orbitals split into triply degenerate orbitals T_{2g}, and doubly degenerate orbitals E_g. The scattering observed from these two sets of orbitals is different, and Weiss and Freeman (1959) showed that the spin part of the form factor in a cubic field may be expressed by

$$F_S(\boldsymbol{\kappa}) = \langle j_0 \rangle + (\tfrac{5}{2}\Gamma - 1)A_{hkl}\langle j_4 \rangle \quad (4.2.2)$$

where $\langle j_0 \rangle$ represents the spherical part of the spin distribution, and $\langle j_4 \rangle$ is the aspherical part. Γ is an asphericity parameter giving the fraction of 3d electrons in E_g orbitals and equals 0.4 for spherical symmetry. A_{hkl} depends on the direction examined in the crystal and in a cubic material it is given by

$$A_{hkl} = \frac{h^4 + k^4 + l^4 - 3(h^2k^2 + h^2l^2 + k^2l^2)}{[h^2 + k^2 + l^2]^2}. \quad (4.2.3)$$

The effect of asphericity in Ni is readily apparent in Fig. 2.7, where four pairs of reflections with the same $\boldsymbol{\kappa}$ values are observed to have significantly different form factor values; a similar effect was observed in Fe and is shown in Fig. 4.3. Measurements of this type indicated that there is a preference for E_g symmetry in Fe, T_{2g} symmetry in Ni, and nearly spherical symmetry in Co.

Initial comparisons of theory and experiment using equn (4.2.1) used component form factors evaluated using Hartree–Fock calculations, e.g. those of Watson and Freeman (1961). The only adjustable parameters in the data-fitting were therefore α and Γ, and excellent fits were obtained for the three ferromagnetic transition metals Fe, Co and Ni with $\alpha = +0.10$, $+0.18$, and $+0.19$ respectively. These positive α values therefore supported the concept that the uniformly distributed magnetization component in the unit cells of these metals was negative. Figure 4.4 illustrates the three-dimensional picture of the regions of reverse magnetization deduced by Shull and Mook (1966) for the unit cell of Fe.

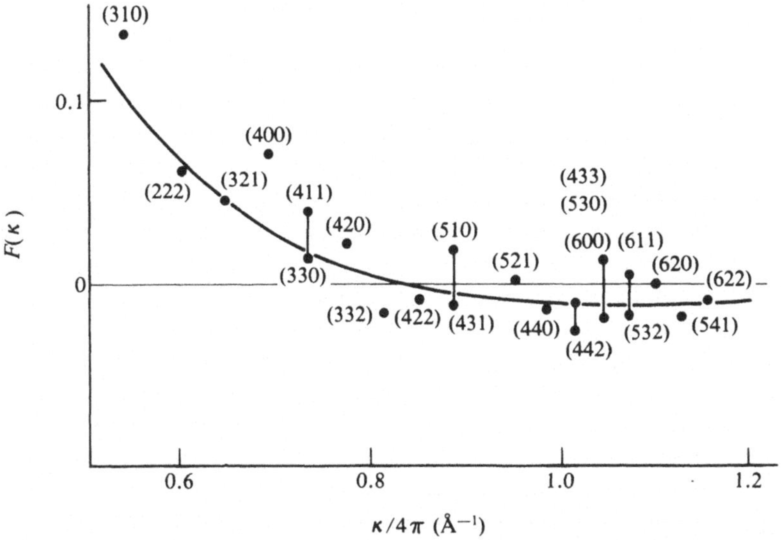

Fig. 4.3 Magnetic form factor data observed by Shull and Yamada (1962) for Fe, showing the different values for five pairs of reflections with the same κ value.

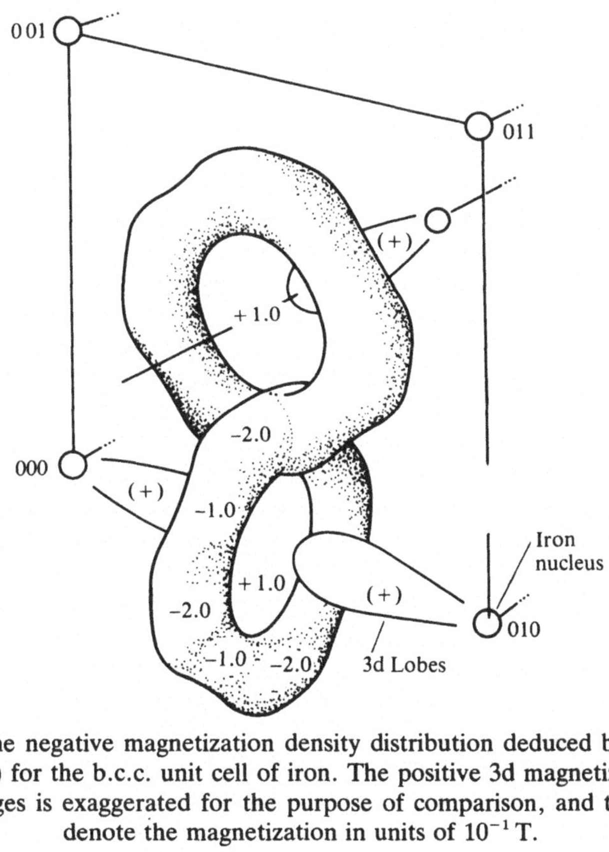

Fig. 4.4 The negative magnetization density distribution deduced by Shull and Mook (1966) for the b.c.c. unit cell of iron. The positive 3d magnetization along the cube edges is exaggerated for the purpose of comparison, and the numbers denote the magnetization in units of 10^{-1} T.

Further evidence for the existence of a uniform negative spin component was provided following the introduction by Moon (1971) of the direct Fourier analysis method, which gave very similar results to those obtained using the above procedure. The significance of the α parameter has, however, been questioned, and in fact van Laar *et al.* (1979), using a sophisticated technique for the Fourier analysis of the experimental data, found no evidence for a finite α value. The current status of this controversy has been summarized by Moon (1982), who concludes that its resolution must await further theoretical developments.

Following the investigations of the pure ferromagnetic metals, there have been a series of measurements on their ordered alloys with (a) other 3d metals, e.g. $FeNi_3$ (Cable and Wollan 1973), (b) 4d elements, e.g. $FePd_3$ (Shirane *et al.* 1964), (c) 5d elements, e.g. $CoPt_3$ (Menzinger and Paoletti 1966), (d) rare-earth metals, e.g. $TbNi_2$ (Givord *et al.* 1976), and (e) non-magnetic elements, e.g. Fe_3Si (Moss and Brown 1972). A review which gives many more examples as well as a critical evaluation of the results has been published by Menzinger and Sacchetti (1979). Van Laar *et al.* (1980) have made a comprehensive analysis of the magnetic form factors in cubic binary 3d–3d metal alloys using their modified Fourier method.

4.2.2 Paramagnetic metals

One of the first observations of an induced moment form factor in an aligned paramagnetic transition metal was that in vanadium by Shull and Ferrier (1963), who found excellent agreement with the free atom form factor given by Freeman and Watson (1961). This was a very suitable experiment for the polarized-beam technique because of the low nuclear coherent scattering amplitude in vanadium; the results were consistent with a 37% 3d spin and 63% 3d orbital contribution to the induced moment.

A more recent measurement is that by Stassis *et al.* (1975) on chromium, where the results were also explained in terms of a free-ion form factor with a 40% 3d spin and 60% 3d orbital contribution to the induced moment. Figure 4.5 shows a comparison of measured and calculated magnetic scattering amplitudes for Cr. Stassis (1979) concluded that the electronic wave functions at the Fermi levels in V and Cr metals are only slightly more extended than in the free ions, and this, together with the relatively small spin contributions to the induced moments, explains the success of the ionic model in fitting the experimental data.

However, for some transition metals like scandium and lutetium, in which the Fermi levels are close to the bottom of the d-band, greater deviations from the free-ion form factor calculations are both expected

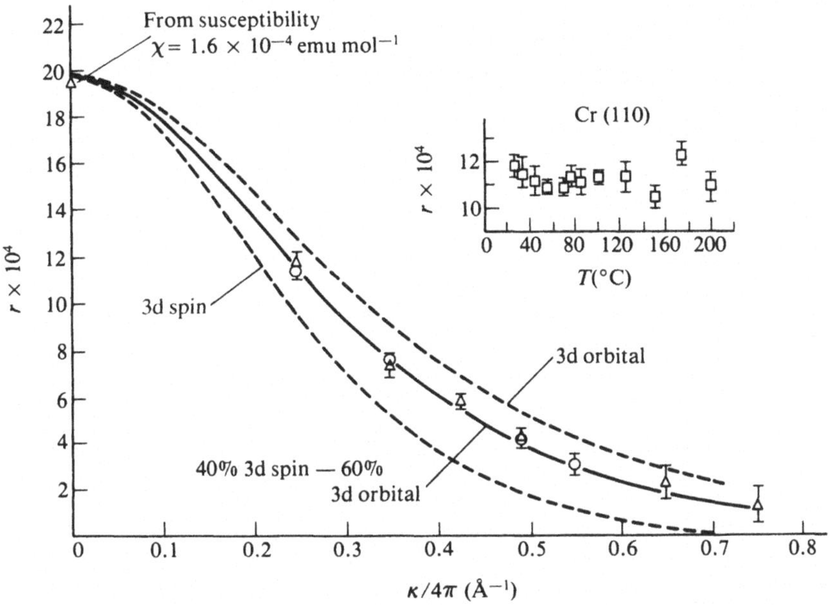

Fig. 4.5 Comparison of the measured induced moment form factor for Cr with atomic calculations; ○, 100°C; △, room temperature; $H = 22\,200$ Oe (from Stassis *et al.* 1975).

and observed. A comparison of the experimental results for scandium with both a free-ion Hartree–Fock calculation and a more sophisticated 'APW' calculation introduced by Gupta and Freeman (1976) is shown, following Koehler and Moon (1976), in Fig. 4.6. Similar results have been reported by Stassis *et al.* (1977) for lutetium, and their significance is discussed by Stassis (1979).

Moon (1982) has listed the paramagnetic metals which have been examined by the polarized neutron diffraction technique; these are Ti, Y, Zr, Nb, Pd, Pt and α-U in addition to the four metals discussed above. All these measurements owe their success to the development of high-field superconducting magnets, and even then the induced moment may be only several milli-Bohr magnetons. Numerous corrections, which can often be ignored in ferromagnetic samples, must be included in the data analysis; they include (i) nuclear polarization, (ii) the neutron spin–neutron orbit interaction (Shull 1963), (iii) diamagnetic scattering (see Section 4.2.8), (iv) sample purity, and (v) magnetic field effects on the neutron path and velocity (Moon *et al.* 1976). Moon (1982) has argued that the intrinsic limit to the accuracy of form factor determinations in paramagnetic transition metals is set by systematic errors due to

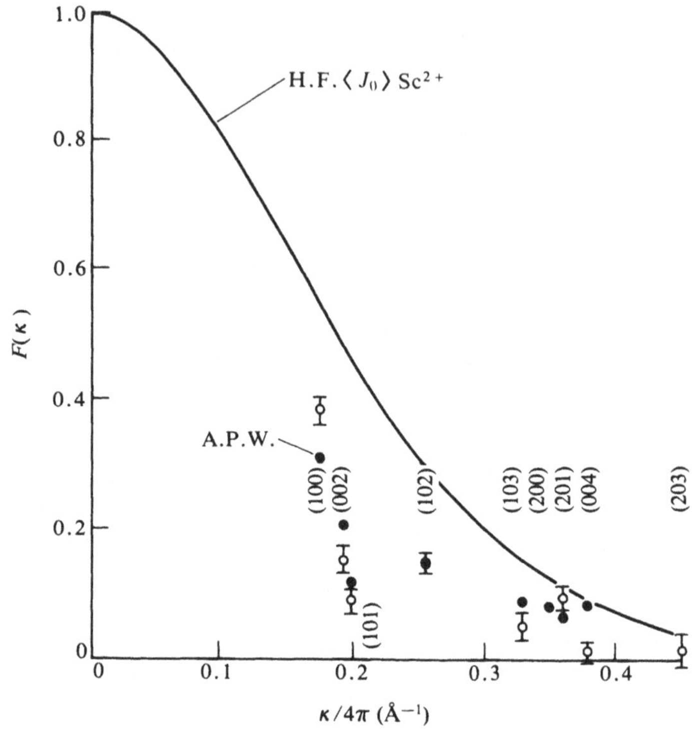

Fig. 4.6 Comparison of the measured form factor of Sc (○) with atomic (continuous line) and APW (●) calculations (from Koehler and Moon 1976).

(v). The residual flipping ratio for an elemental paramagnet is given by

$$r = R - 1 = (1.93\chi B F(\boldsymbol{\kappa}))/b, \qquad (4.2.4)$$

where χ is the susceptibility in e.m.u./mole, B is the applied field in tesla units, and b is the nuclear scattering amplitude in units of 10^{-12} cm. In order to determine $F(\boldsymbol{\kappa})$ to an accuracy within $\pm 2.5\%$, r must be measured to an accuracy

$$\Delta r = (4.83 \times 10^{-2} \chi B)/b. \qquad (4.2.5)$$

Figure 4.7 shows these required Δr values for the transition metal paramagnets, together with the number of counts needed to give a 2.5% statistical error in $\Delta r/r$ or $\Delta F(\boldsymbol{\kappa})/F(\boldsymbol{\kappa})$ when the applied field $B = 6$ T. The broken line gives the estimated systematic error Δr due to magnetic field effects, so that for the elements below the line it is impossible to attain the $\pm 2.5\%$ accuracy in $F(\boldsymbol{\kappa})$ merely by accumulating counts. These conclusions remain the same at higher applied fields, where the effect is to move both the points and the broken line towards higher values. The

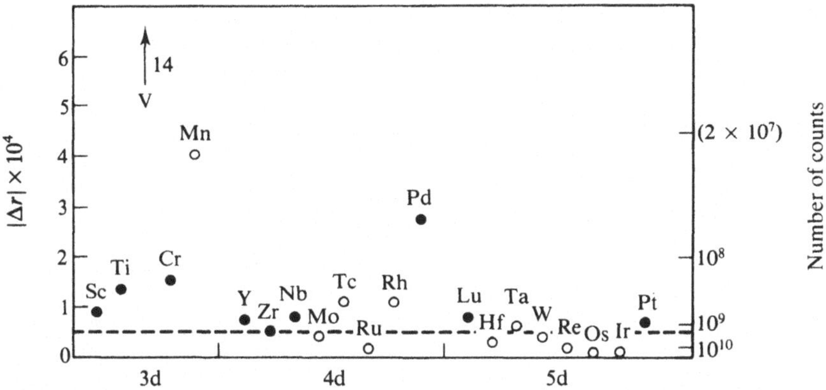

Fig. 4.7 Required errors in the reduced flipping ratio r which allows the normalized form factors for the paramagnetic transition metal elements to be determined to a 2.5% accuracy in an applied field $B = 6\,\mathrm{T}$. ● Indicates elements investigated up to 1982. The required detected counts are shown in the right, where $|\Delta r|$ is purely statistical, and the dashed line is the systematic error due to magnetic field effects (from Moon 1982).

solid circles show the results for those elements which had been investigated by the polarized beam method up to 1982.

4.2.3 Lanthanide metals and compounds

The orbital moment of the atoms in the 4f series is unquenched, and this can make a significant contribution to the magnetization density. The effective form factor $F(\boldsymbol{\kappa})$ is therefore in general a weighted combination of the spin form factor $F_S(\boldsymbol{\kappa})$ due to a pure spin moment, and the orbital form factor $F_L(\boldsymbol{\kappa})$ due to a pure orbital moment. The weighting factors depend on the total, spin and orbital angular momentum quantum numbers J, S and L, and $F(\boldsymbol{\kappa})$ is given (see for example Bacon (1975)), by

$$F(\boldsymbol{\kappa}) = \frac{C_L F_L(\boldsymbol{\kappa}) + C_S F_S(\boldsymbol{\kappa})}{3J(J+1) + S(S+1) - L(L+1)}, \tag{4.2.6}$$

with the orbital coefficient

$$C_L = J(J+1) + L(L+1) - S(S+1) \tag{4.2.7}$$

and the spin coefficient

$$C_S = 2\{J(J+1) + S(S+1) - L(L+1)\}. \tag{4.2.8}$$

The orbital magnetization density is usually more localized than the spin density, see Fig. 2.4, hence $F_L(\boldsymbol{\kappa})$ often extends to greater $\boldsymbol{\kappa}$ values

TABLE 4.1 Spin, orbital and total angular momentum quantum numbers for rare-earth tripositive ions with their orbital and spin form factor weighting coefficients (see equns (4.2.6) to (4.2.8)

Rare-earth ion	S	L	J	$(C_L/C)^a$	$(C_S/C)^a$
Pr^{3+}	1	5	4	1.50	−0.50
Nd^{3+}	3/2	6	9/2	1.75	−0.75
Sm^{3+}	5/2	5	5/2	6.00	−5.00
Gd^{3+}	7/2	0	7/2	0	1
Tb^{3+}	3	3	6	0.33	0.67
Dy^{3+}	5/2	5	15/2	0.50	0.50
Ho^{3+}	2	6	8	0.60	0.40
Er^{3+}	3/2	6	15/2	0.67	0.33

[a] Note: $C = 3J(J+1) + S(S+1) - L(L+1)$.

than $F_S(\boldsymbol{\kappa})$; expressions for these two form factors have been derived by Blume *et al.* (1962). The resultant form factor $F(\boldsymbol{\kappa})$ has a strong dependence on the weighting coefficients C_L and C_S. This is particularly evident when they have opposite signs, and these coefficients are shown for some of the rare-earth metals in Table 4.1.

The 4f shell in gadolinium is half-filled and the orbital moment is zero. The magnetic form factor therefore gives a direct determination of the 4f electron distribution, and this was one of the first rare-earth metals studied by the polarized beam diffraction technique. The results of Moon *et al.* (1972) on metallic Gd are shown in Fig. 4.8. At the time of these measurements only non-relativistic Hartree–Fock calculations of the form factor were available, and these were in very poor agreement with the observations. This stimulated further theoretical work and a relativistic Hartree–Fock–Slater calculation was carried out by Davis and Cooke (1972) followed by a Dirac–Fock calculation by Freeman and Desclaux (1972); the latter in particular gave excellent agreement with the data. This advance also revealed that there is an additional contribution to the scattering at low $\boldsymbol{\kappa}$ values which is due to the polarization of the conduction electrons. Conduction electron form factors were obtained for Gd by Moon *et al.* (1972) by separating the localized 4f electron contribution from the measured form factor.

Measurements on erbium which, unlike Gd, possesses an orbital magnetic moment were performed by Stassis *et al.* (1976). In these experiments the coherent paramagnetic scattering amplitude was determined with the applied field both along and perpendicular to the crystal c-axis; the results together with Dirac–Fock calculations for free Er^{3+} are shown in Fig. 4.9(a). The discrepancies between theory and experiment at low $\boldsymbol{\kappa}$ were similar in magnitude to those observed for Gd, and were also attributed to conduction electrons. The $\boldsymbol{\kappa}$-dependence of the

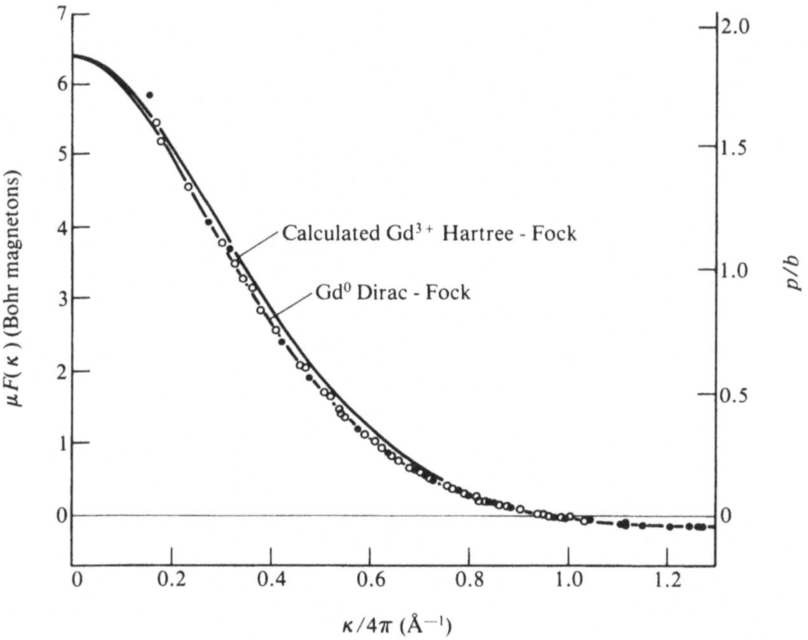

Fig. 4.8 Polarized-beam magnetic moment measurements on ^{160}Gd at 96 K by Moon *et al.* (1972), showing a comparison with relativistic Hartree–Fock and Dirac–Fock calculations (●, (*hk*0); ○, (0*kl*)).

conduction electron form factor extracted from this data is shown in Fig. 4.9(b).

The power of the relativistic Dirac–Fock (DF) method for calculating precise form factors has been supported by polarized-beam measurements in two ionic compounds, $Tb(OH)_3$ and Gd_2O_3, in which the complications due to conduction electron polarization effects are absent. Excellent agreement was obtained between DF calculations and the form factors measured by Lander *et al.* (1973) on $Tb(OH)_3$, and by Moon and Koehler (1975) on Gd_2O_3.

Other polarized neutron experiments on pure 4f metals include those by Stassis *et al.* (1978) on γ-Ce; Lebech *et al.* (1979) on Pr and Nd; Koehler and Moon (1972) on Sm; and Felcher *et al.* (1976) on Ho.

There have been numerous studies of rare-earth–metal compounds, see for example the discussion by Boucherle *et al.* (1982) on the results for the $ReAl_2$ series. These clearly show that the form factors for those Re atoms with an orbital moment are more expanded in reciprocal space than in the case of Gd for which $L = 0$. Another interesting observation is that the form factors are more expansive in the first half of the series. This follows from Table 4.1 since, when $J = L - S$, the total magnetization density results from a cancellation of the more localized orbital and

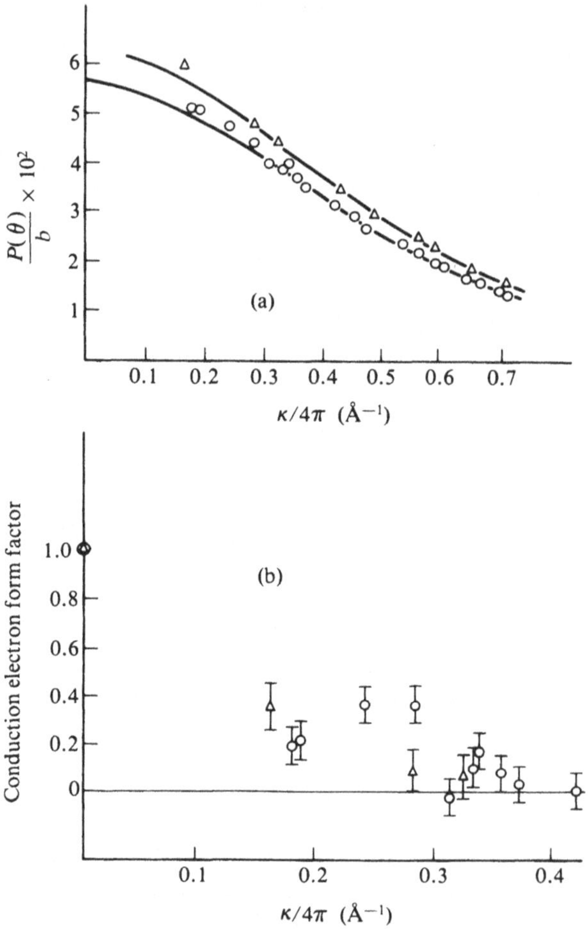

Fig. 4.9 (a) Polarized-beam form factor measurements on Er by Stassis *et al.* (1976) with a comparison of Dirac–Fock calculations (full curves). (b) The conduction electron form factor in Er extracted from the data. △, $H \parallel c$; ○, $H \perp c$; $H = 22\,000$ Oe.

the more expansive spin density. We give as our example the extreme case of Sm in $SmCo_5$, for which $C_L/C = 1.75$, $C_S/C = -0.75$, and where at low temperatures the orbital moment dominates over the spin moment. The samarium magnetic scattering amplitudes observed by Givord *et al.* (1979) at 4.2 K are shown in Fig. 4.10(a); the maximum in $F(\boldsymbol{\kappa})$ at non-zero $\boldsymbol{\kappa}$ results from the magnitudes and the opposite signs of C_L and C_S when substituted into equn (4.2.6). At higher temperatures there is mixing of higher-state J multiplets with the ground-state $J = 5/2$ multiplet, and since these excited levels have a dominant spin character this causes the peak shape of the Sm form factor to become more

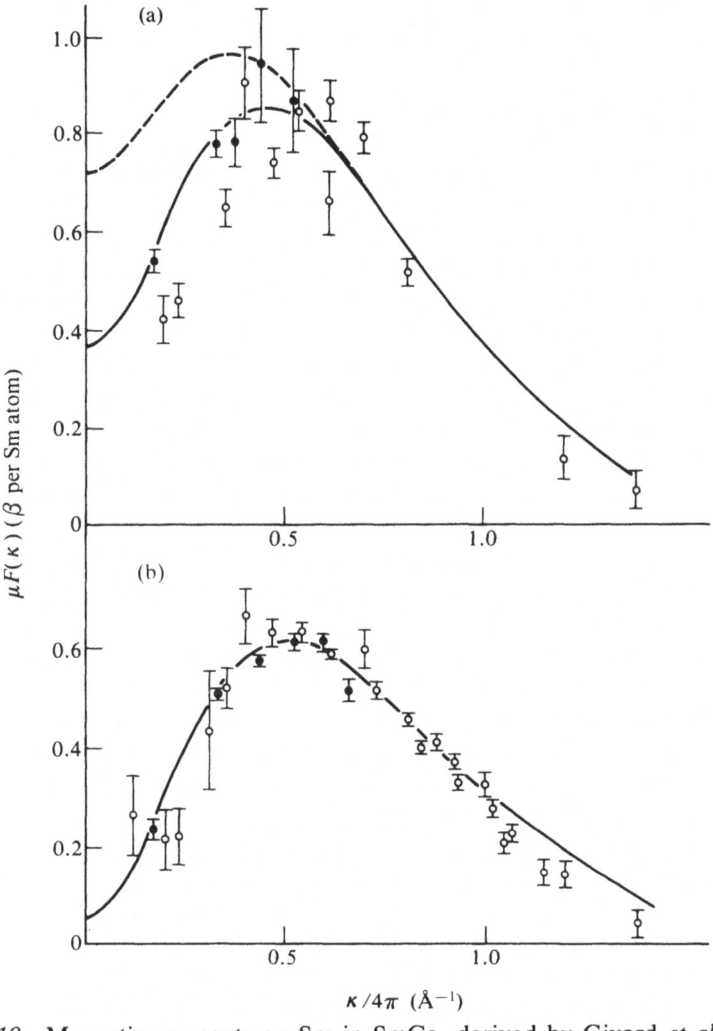

Fig. 4.10 Magnetic moments on Sm in $SmCo_5$ derived by Givord *et al.* (1979) from polarized neutron measurements at (a) 4.2 K and (b) 300 K; the broken line corresponds to the saturated state $|5/2, 5/2|$ and the full lines to a complete energy level scheme.

pronounced. The magnetic amplitudes observed by Givord *et al.* (1979) at 300 K are shown in Fig. 4.10(b).

The results and interpretation of magnetization density measurements in a wide range of rare-earth compounds have been reviewed by Boucherle *et al.* (1982). There is, however, a class of rare-earth compounds, the mixed-valence materials, in which the magnetic properties cannot be explained by assigning integral valence numbers to the rare-earth ions. We now describe some of the polarized neutron experiments performed on intermediate valence compounds.

4.2.4 Mixed-valence materials

These materials do not exhibit the normal behaviour associated with a localized ionic moment, i.e. they do not order magnetically at low temperatures, neither do their susceptibilities diverge as $T \to 0$. Their unusual properties are thought to arise from the coexistence at the Fermi level of ionic 4f levels and a broad s–d conduction band; electron transfer between the 4f levels and the conduction band can then cause the ionic valence to fluctuate between integral values. In the polarized neutron experiments performed on these materials, an applied field at the sample induces a magnetization whose form factor is determined in the usual way. This elastic measurement senses a time-average of the induced magnetization and gives information on the mixed-valence ground state.

Two of the earliest mixed-valence materials examined by polarized neutrons were the samarium compounds SmS and $Sm_{0.76}Y_{0.24}S$, see Moon *et al.* (1978). The former has a susceptibility at atmospheric pressure which is characteristic of Sm^{2+}, but it transforms to a mixed-valence state, with apparent valence +2.8, at a pressure ~6 kbar. Polarized-beam measurements gave a Sm^{2+} form factor both above and below the phase transition, and there was no evidence of a Sm^{3+}-like form factor such as that shown in Fig. 4.10. It was concluded that the Curie terms in the Sm^{3+} susceptibility make no contribution to the mixed-valence phase. A similar result was observed for $Sm_{0.76}Y_{0.24}S$. At room temperature and atmospheric pressure this material has an apparent valence +2.6 which changes to +2.3 as the temperature is lowered. The observed form factor, which is shown in Fig. 4.11, is essentially temperature-independent and very close to that calculated for Sm^{2+}.

The polarized beam method has also been applied to the investigation of two cerium-based mixed-valence alloys: $Ce_{0.74}Th_{0.26}$ (Moon 1979), and $CeSn_3$ (Stassis *et al.* 1979). The γ-phase of $Ce_{0.74}Th_{0.26}$ measured at 180 K gave the form factor expected for the Ce^{3+} ion, however the form factor observed at 50 K for the mixed-valence α-phase could only be satisfactorily fitted by adding a spatially diffuse induced magnetization component to the $4f^1$ susceptibility of Ce^{3+}. It was tempting to associate this second term with conduction electron effects, since its magnitude was very similar to that observed by Moon *et al.* (1972) for metallic Gd. Thermal expansion and compressibility measurements indicate that $CeSn_3$ is also a mixed-valence compound, however its low-temperature magnetic suceptibility, which rapidly increases below 40 K, is very uncharacteristic of these systems. Polarized-beam measurements of the induced-moment form factor over the temperature range 300–40 K, at

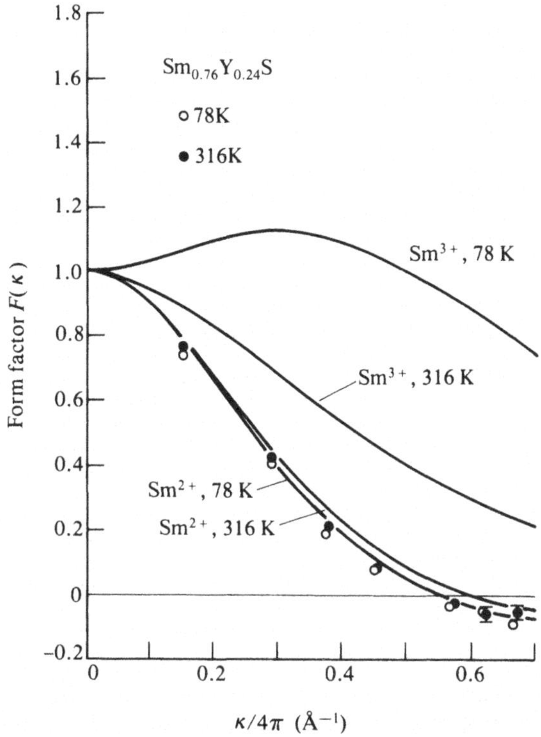

Fig. 4.11 Magnetic form factors in $Sm_{0.76}Y_{0.24}S$ measured by Moon *et al.* (1978), (○, 78 K; ●, 316 K), compared with ionic calculations (curves).

which the susceptibility is roughly constant, were in excellent agreement with that expected for the Ce^{3+} free ion. However, at temperatures below 40 K large deviations from the 4f form factor were observed, and a typical set of results is shown in Fig. 4.12. This data could be fitted by augmenting the 4f contribution with a large 5d contribution with E_g symmetry. The net susceptibility obtained in this way was still only approximately 50% of the bulk susceptibility, so it appears that there is a large positive contribution to the low-temperature susceptibility of $CeSn_3$ which is not sensed by the neutron measurement.

A further polarized neutron magnetic form factor investigation of $CeSn_3$ was reported by Boucherle *et al.* (1986). Measurements were made on two samples with different stoichiometries: $CeSn_{2.9}$ and $CeSn_3$. The 4f-type localized moment was observed to be the same in both compounds, however the 3d contributions were much weaker than had been observed by Stassis *et al.* (1979), being 25% of the total magnetization in $CeSn_3$, and zero in $CeSn_{2.9}$. The $CeSn_{2.9}$ result was attributed to the presence of Ce_2Sn_5 impurities in ordered regions of the sample, as well as

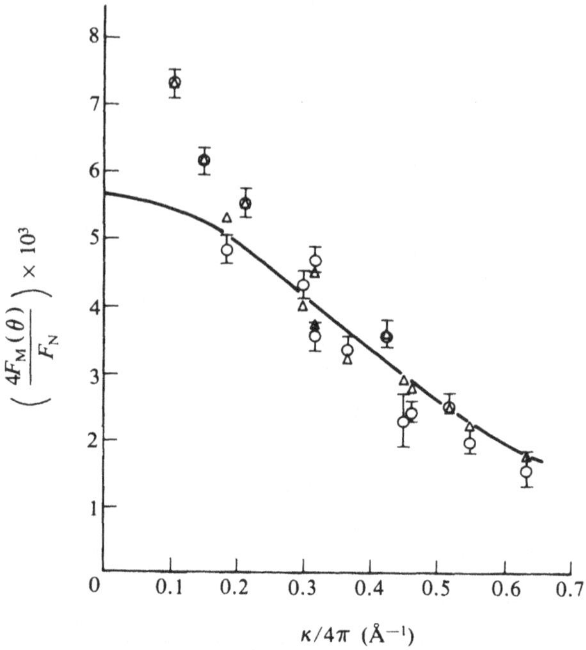

Fig. 4.12 The magnetic form factor of $CeSn_3$ at 4.2 K, $H = 4.25$ T, from Stassis *et al.* (1979). The full curve shows the 4f electron contribution, open circles are the measurements, and open triangles are calculated values assuming a 62% 4f contribution with a 38% 5d contribution with E_g symmetry.

disordered Ce^{2+} paramagnetic impurities. The intrinsic properties of $CeSn_3$ are only achieved when both these impurities are absent, illustrating the importance of the coherence between all the Ce atoms in producing a large 5d contribution to the magnetization.

4.2.5 Actinide compounds

The ground-state properties and even the ionization states of many of the compounds of the actinide elements U, Np and Pu remain unknown. This uncertainty in the outer electronic configurations of these solids is not surprising, since the free ions have 5f, 6d and 7s outer orbitals which are very close in energy. Magnetization density measurements using polarized neutrons therefore represents an important method of investigating the electronic structure of these materials.

Some of the uranium compounds which have been studied include US by Wedgwood (1972), USb by Lander (1976), UGe_2 by Ballou *et al.* (1982), and UNi_2 by Blaise *et al.* (1982). The experiments all involved comparing measured form factors with calculations, and are typified by the first measurements made by Wedgwood (1972) on US. These results showed that the magnetic moment distribution was associated primarily with the 5f shell of the uranium ion, that there was a small negative moment due to the uranium 7s or 6d electrons and a negligible moment on the sulphur ion. The experimental data were compared with various theoretical form factor models. The best agreement was obtained for a U ion designated $5f^2$, $^3H_4$, Γ_1, in which the exchange field was comparable in magnitude to the crystal field, and gave an ordered magnetic moment of 1.7β per uranium atom. Figure 4.13 shows a comparison of the measured data with four of the models used in this investigation. The effect of covalency in the simple ionic compound UCl_4 has been observed by measuring the magnetization distribution, see Lander *et al.* (1985).

Neptunium form factors have been measured in $NpAs_2$ by Amoretti *et al.* (1982) and by Blaise *et al.* (1982), in NpO_2 also by Blaise *et al.* (1982), and in the transition metal alloys $NpFe_2$, $NpCo_2$ and $NpMn_2$ (see for example Aldred *et al.* (1975)). The most detailed theoretical discussion of Np form factors is that given by Amoretti *et al.* (1982), who analysed their $NpAs_2$ results for the three ionic states Np^{3+}, Np^{4+} and Np^{5+} with both a free-ion wave function and a crystal field wave function for the 5f electrons. Their results were compatible with both 3+ and 4+ valencies.

A polarized neutron study of plutonium in the form of a single crystal

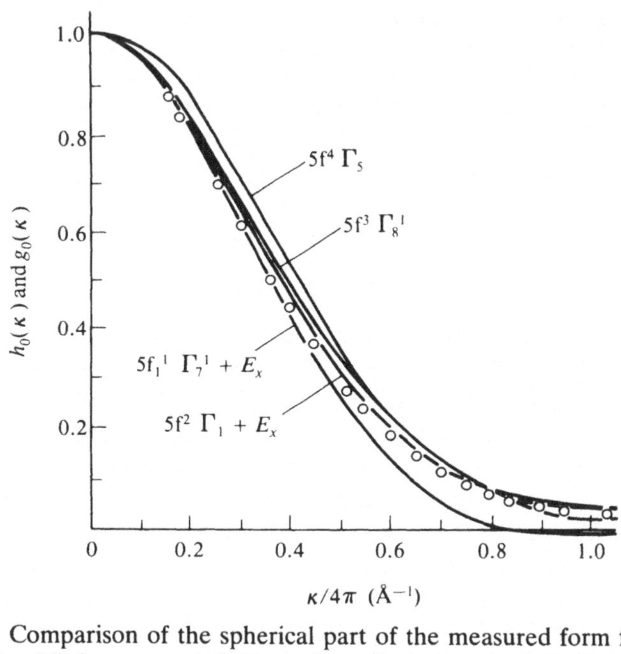

Fig. 4.13 Comparison of the spherical part of the measured form factor data in US ($h_0(\boldsymbol{\kappa})$) with four crystal field model calculations ($g_0(\boldsymbol{\kappa})$) (from Wedgwood 1972).

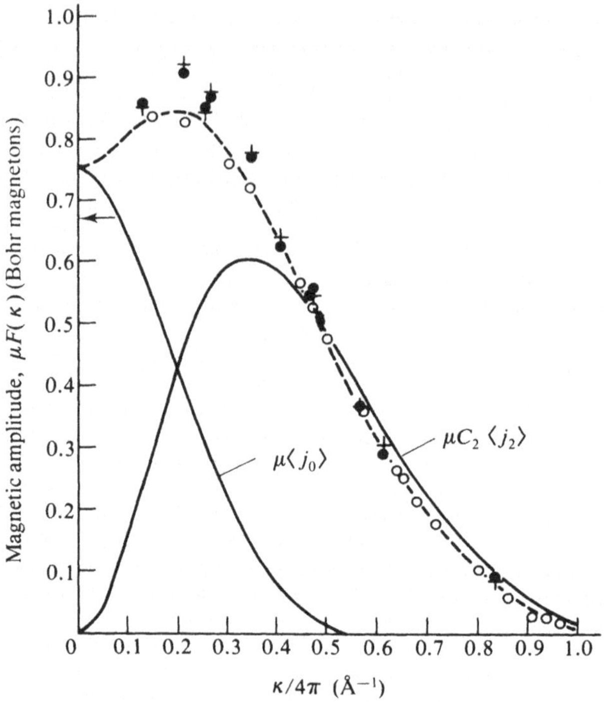

Fig. 4.14 Measurements of $\mu F(\boldsymbol{\kappa})$ for Pu in ferromagnetic PuSb at 10 K. The broken curve represents a best-fit for reflections in the scattering plane. Reflections out of the scattering plane are shown as solid circles, and their calculated amplitude values as crosses. The solid lines are calculated contributions from the spin ($\mu\langle j_0\rangle$) and orbital ($\mu C_2\langle j_2\rangle$) moments. The bulk magnetization measured for PuSb gives a zero-κ amplitude shown by the arrow (from Lander *et al.* 1984).

of ferromagnetic ^{239}PuSb has been reported by Lander *et al.* (1984); their measurements of $\mu F(\boldsymbol{\kappa})$, where μ is the magnetic moment, are shown in Fig. 4.14. The most striking feature of these results, which are reminiscent of those observed for Sm^{3+}, is that they peak at non-zero $\boldsymbol{\kappa}$. The resulting magnetic amplitude is essentially a combination of a factor $\mu\langle j_0\rangle$ due to a spin moment, and a factor $\mu C_2\langle j_2\rangle$ due to an orbital moment; values for these quantities taken from relativistic Dirac–Fock wave functions are also included in the figure. The observed magnetic form factor arises because the coefficient C_2 is large, and the function itself is zero at $\boldsymbol{\kappa}=0$. If the orbital moment were absent, C_2 would be ~ 0, and therefore the observation of the maximum is an unambiguous signature of the trivalent ground state of Pu in ferromagnetic PuSb.

4.2.6 Covalency

The application of polarized neutrons to the study of metal–ligand covalent bonding by polarized neutron diffraction has been reviewed by Tofield (1975), and by Forsyth (1979). Any deviations from a pure ionic bond involving magnetic ions results in the production of an electron spin density which can be observed directly in a Fourier map, or by examining details in the magnetic form factor. We saw in Section 2.1.2, for example, that a ligand spin density results in a sharply peaked form factor, since this density is far-removed from the metal site. There are three broad classes of magnetic substances whose covalency have been investigated by polarized neutron diffraction, viz. salts and oxides of the 3d transition metals, six-coordination complexes of the 3d and 5d transition metals, and one case involving an actinide (5f) metal. Results will be presented for a typical example from each of these classes.

Some of the polarized-beam covalency measurements of 3d metal oxides and salts include those by Alperin (1962) on NiO, Rakhecha and Satya Murthy (1978) on Fe_3O_4, Nathans *et al.* (1963) on MnF_2, Brown and Forsyth (1967) on $MnCO_3$, and Brown and Forsyth (1977) on $FeCO_3$. We take as an example the study by Nathans *et al.* (1963) of the antiferromagnetic salt MnF_2. This has the rutile structure in which the magnetic moments of the corner and body-centre atoms are oppositely directed, and this gives reflections of three different types: (a) $(0kl)$ reflections with $(k+l)$ odd which are purely magnetic, (b) reflections with $(h+k+l)$ odd which are a mixture of magnetic scattering and nuclear scattering from the fluorine atoms, and (c) reflections with $(h+k+l)$ even which are purely nuclear but may contain a small magnetic contribution if the magnetic moment distribution around the Mn atoms is non-spherical. The antiferromagnetically coupled ions are not related by a translation vector of the crystal unit cell, and the Bragg peak intensities may be neutron-polarization-dependent in a sample which is not equidomained; see Alperin *et al.* (1961).

Figure 4.15 shows the magnetic structure factors derived by Nathans *et al.* (1963) for MnF_2 using the polarized-beam method. The most striking observation was that reflections of type (c) did in fact display a magnetic component, i.e. they gave flipping ratios which differed from unity, and this was attributed to a non-spherical moment distribution around the Mn atoms. The results were later shown by Hubbard and Marshall (1965) to be consistent with a model in which the effect of covalency is to place a spin on the F^- ions which is parallel to that of its nearest magnetic atom neighbour. In MnF_2 each F^- ion is bonded to three Mn atoms, two with one spin direction and the other with opposite spin, and therefore the net spins on the F^- ions are in the same direction as the majority spin direction of its three Mn neighbours.

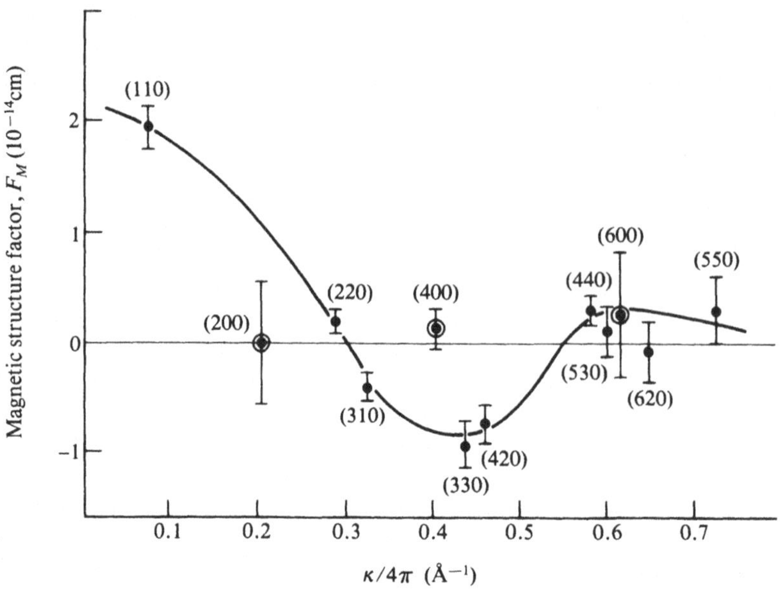

Fig. 4.15 Magnetic structure factors measured by Nathans *et al.* (1963) for MnF_2; the points with the dotted circles are those corresponding to Bragg reflections where the magnetic scattering is expected to be identically zero.

There have been several investigations of 3d and 5d transition metal ions in which there is six-fold coordination of ligand atoms or molecules. Representative examples include K_2NaCrF_6 (Wedgwood 1976), K_2IrCl_6 (Lynn *et al.* 1976b), $Cs_2KFe(CN)_6$ (Brown *et al.* 1982b), and $Ni(NH_3)_4(NO)_2$ (Chandler *et al.* 1982). We now discuss Wedgwood's results for K_2NaCrF_6. This is a cubic insulating compound in which the Cr^{3+} ions are at the centre of an octahedron of F^- ions, and the orbital magnetic moment is completely quenched. It has a large unit cell, thus the $(CrF_6)^{3-}$ clusters can be considered to be essentially isolated, and the alkali metal ions can be ignored in any discussion of its magnetism. The material is paramagnetic down to 1.5 K, and the spin-only aligned moment per Cr atom under these conditions is 1.08β.

Form factor data were collected up to a scattering vector value $\kappa = 9.12$ Å^{-1}, and these were Fourier-transformed in the usual way to yield spin density maps. Figure 4.16 shows the averaged spin density profile through the (001) plane, where the average is taken over a small cube with dimensions $\delta = 0.4$ Å. The T_{2g} symmetry around the Cr ion is clearly visible, though the nodes expected in the X and Y directions through the Cr atom are smeared out owing to the instrumental resolution. The other significant features of this map are the small islands of spin density which are centred just off the centre of the F sites, and

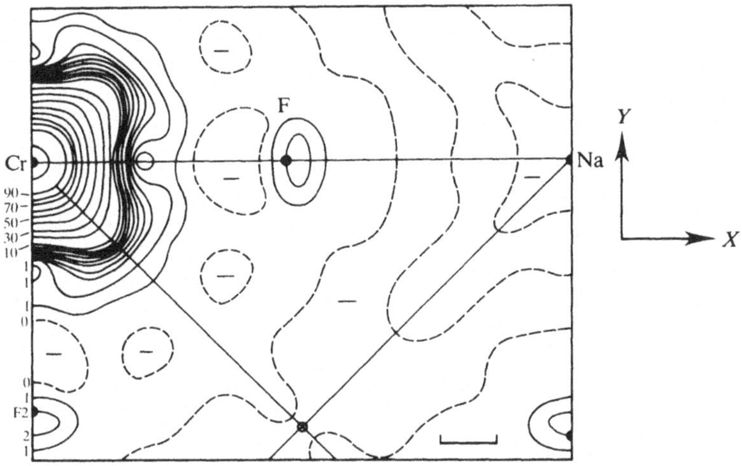

Fig. 4.16 Spin density distribution in the (001) plane of K_2NaCrF_6 through a Cr site. The contours are in units of 0.01 $\beta/Å^3$ (from Wedgwood 1976).

this directly confirms that covalent bonding spin transfer occurs in this compound. The off-centre distribution at the F atoms arises because the unpaired spin is associated with antibonding orbital wave functions which have a negative overlap region between the Cr^{3+} and F^- ions, and this pushes out the maximum on the ligand spin density. These antibonding orbitals between the metal 3d electrons and ligand 2p electrons can be expressed in terms of a single covalency parameter A_π, and a simple molecular orbital model analysis by Wedgwood gave $A_\pi = 0.24$. A single A_π value was not, however, consistent with all the reflection data collected, and it was concluded that a spin-polarized Hartree–Fock calculation would be needed to give a more satisfactory agreement with the complete set of measurements.

3d electron systems have extensive orbitals which are sensitive to the atom's environment, and are therefore ideally suited to covalency studies. 5f electron systems are also potentially amenable to such investigations since their wave functions are significantly less localized than in 4f systems, even though they are not as extensive as in 3d atoms. The first direct observation of covalency in an actinide system was that of Lander *et al.* (1985) in the ionic compound UCl_4. The structure of the single crystal shows that there is an eight-fold coordination of Cl atoms around each U atom; four of the U–Cl bonds have lengths 2.65 Å, and the other four have lengths 2.88 Å. The polarized neutron experiment gave a magnetization density distribution which revealed deviations from that expected for a single-ion crystal-field model. This was convincingly demonstrated by evaluating difference Fourier maps derived from the

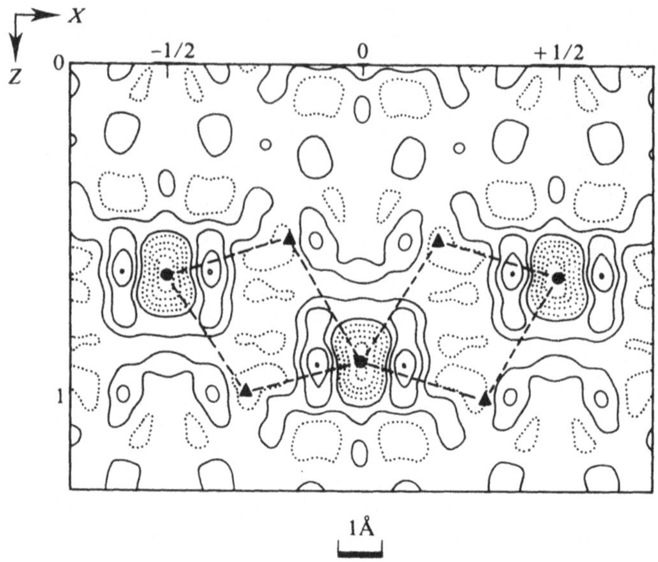

Fig. 4.17 Difference magnetization density distribution in a plane containing a U atom, two short-bond Cl atoms, and two long-bond Cl atoms in UCl_4. ●, U; ▲, Cl; full lines, positive contours; dotted lines, negative contours (from Lander *et al.* 1985).

differences between the observed and calculated magnetic structure factors for a range of reflections. Figure 4.17 shows such a map through a plane containing a U atom, two 2.65 Å bond Cl atoms, and two 2.88 Å bond Cl atoms. The contours in this section were drawn at intervals of $1.5 \times 10^{-3}\ \beta/\text{Å}^3$ with the first contour at $\pm 0.75 \times 10^{-3}\ \beta/\text{Å}^3$, and with a significance level $\sim 1 \times 10^3\ \beta/\text{Å}^3$. It was concluded that within the limits of the observations there was no transferred spin at the Cl sites, however there was a small depletion of the 5f density near the U site and a small positive increase at ~1 Å from the U site, mainly in the direction of the shorter U–Cl bonds.

4.2.7 Free radicals

In addition to its uses in examining d and f electron magnetism, the polarized neutron diffraction method can also reveal the distribution of magnetism in p electron systems. Free radicals are molecules containing one unpaired electron spin which is often delocalized over several atoms, and the spin densities at the atoms are commonly determined by observing nuclear hyperfine interactions in electron spin resonance (ESR) spectra. The ESR method, however, relies on the use of empirical coupling constants, and only gives a single-valued electron spin density at the atom, whereas, as we have seen, the polarized neutron method gives the distribution of the spin density over the whole molecule.

The first polarized neutron determination of the spin density in a free radical was that by Brown *et al.* (1979), who examined the nitroxide biradical di(2,2,6,6-tetramethyl-4-piperidinyl-1-oxyl) suberate. It was known that the unpaired spin in this radical is almost entirely concentrated in an antibonding p_π orbital constructed to a good approximation from 2p orbitals on the nitrogen and oxygen atoms of the N—O group, and the aim of the experiment was to obtain the precise distribution between these two atoms. An unpolarized beam measurement was performed to give the nuclear structure factors F_N for a set of Bragg reflections for which $\gamma = F_M/F_N$ had been determined by the polarized-beam technique. The latter results were obtained at 1.5 K in a 1.65 T applied field, and under these conditions the induced moment in the paramagnetic crystal was 1.61 β/molecule. Owing to the complexity of the crystal structure, it did not prove possible to measure a sufficiently large set of F_M values to allow a classical Fourier synthesis of the spin density, the main difficulty being that many of the peaks had very weak nuclear components. Instead it was found more useful to analyse the data in terms of a multipole expansion around the atomic centres, and this provided an analytical description of the spin density distribution. The spin density maps in the (100) plane through the N—O groups, evaluated both by including and omitting the carbon chain residue of the biradical, were almost identical, thus showing that the unpaired spin was indeed confined to the N—O group. It also appeared from the results that there was an equal sharing of this spin between the N and O atoms.

The multipole expansion method can also be applied to the analysis of the flipping ratios obtained from non-centrosymmetric structures, and it was used by Boucherle *et al.* (1982) in their study of the free radical diphenylpicrylhydrazyl (DPPH) whose structure is shown in Fig. 4.18.

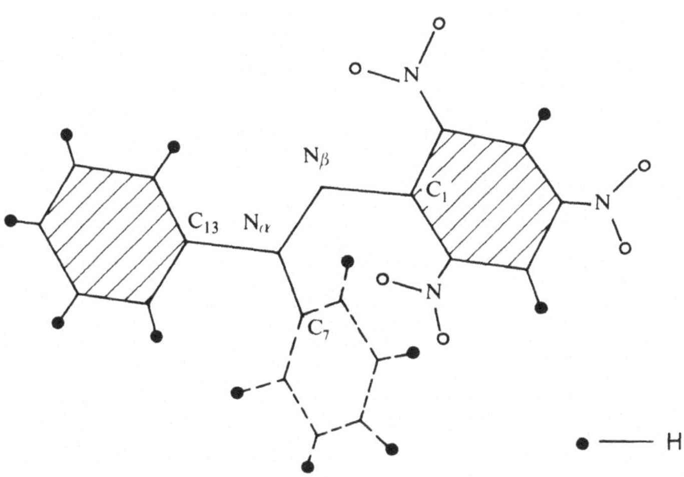

Fig. 4.18 Structure of DPPH.

This was examined in the crystallized form DPPH-C_6H_6, with the paramagnetic moments again oriented at low temperature in an external magnetic field of 4.65 T. The spin density determined for the hydrazyl group and projected into the plane containing N_α—N_β and the bisector of $C_7N_\alpha C_{13}$ is shown in Fig. 4.19(a); this represents approximately 70% of th spin density of the whole free radical molecule, and is evenly separated into two 2p-type atomic orbitals. The projection of the spin density into the phenyl ring plane containing C_{13} is shown in Fig. 4.19(b), and shows that the density alternates in sign between neighbouring carbon atoms.

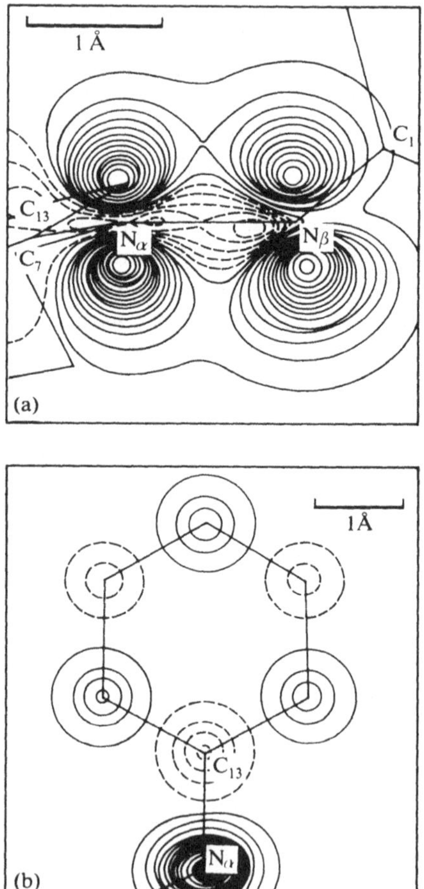

Fig. 4.19 Projections of the spin densities in DPPH: (a) in a plane perpendicular to the trigonal N_α plane and containing N_α—N_β; and (b) in the phenyl ring plane containing C_{13}. Adjacent contour lines are separated by 0.03 $\beta/\text{Å}^2$ (from Boucherle *et al.* 1982a).

Further examination of other projections showed that the two phenyl rings in DPPH are not in fact completely identical.

4.2.8 Diamagnets

An external magnetic field B induces diamagnetic currents in an atom, and their interaction with the neutron magnetic moment was first formulated by Stassis (1970). For a spherical atom, the diamagnetic scattering amplitude may be writen as

$$p_{\mathrm{d}}(\boldsymbol{\kappa}) = -\gamma_{\mathrm{n}}\chi_{\mathrm{d}}\left(\frac{eB}{\hbar c}\right)F_{\mathrm{d}}(\boldsymbol{\kappa})\boldsymbol{q}_{\mathrm{m}}\cdot\hat{\boldsymbol{\sigma}}, \tag{4.2.9}$$

where χ_{d} is the atomic diamagnetic susceptibility, $\hat{\boldsymbol{\sigma}}$ is the Pauli spin matrix, $\boldsymbol{q}_{\mathrm{m}}$ is the magnetic interaction vector defined as

$$\boldsymbol{q}_{\mathrm{m}} = \tilde{\boldsymbol{\kappa}}(\tilde{\boldsymbol{\kappa}}\cdot\tilde{\boldsymbol{h}}) - \tilde{\boldsymbol{h}}, \tag{4.2.10}$$

$\tilde{\boldsymbol{h}}$ and $\tilde{\boldsymbol{\kappa}}$ are unit vectors in the magnetic field and scattering vector directions, and $F_{\mathrm{d}}(\boldsymbol{\kappa})$ is the normalized diamagnetic form factor for the atom,

$$F_{\mathrm{d}}(\boldsymbol{\kappa}) = \frac{-3}{\langle r^2\rangle}\frac{1}{\kappa}\left(\frac{\mathrm{d}F_{\mathrm{c}}(\boldsymbol{\kappa})}{\mathrm{d}\kappa}\right). \tag{4.2.11}$$

$F_{\mathrm{c}}(\boldsymbol{\kappa})$ in this expression is the normalized charge form factor for the atom, and $\langle r^2\rangle$ is the second moment of the atomic charge distribution. The relation shows that $F_{\mathrm{d}}(\boldsymbol{\kappa})$ falls off with increasing κ much more rapidly than the charge form factor $F_{\mathrm{c}}(\boldsymbol{\kappa})$, and this can be qualitatively understood since the current density distribution probed by the neutron moment is much more extended in space than the charge distribution. Typical values of $p_{\mathrm{d}}(\boldsymbol{\kappa})$ in a magnetic field ~2 T are in the region $\sim 10^{-15}$–10^{-17} cm atom^{-1}, and the measurement of such small amplitudes is just within the sensitivity of the polarized-beam diffraction technique. We have already remarked that in the measurement of induced-moment form factors in paramagnetic metals, the diamagnetic scattering correction is an important one.

Bismuth, which has an anomalously high diamagnetic susceptibility, attributed to five valence electrons outside a core of 78 electrons, was the first diamagnetic material to be studied by polarized neutron scattering. These measurements were carried out by Collins (1979), and are reported by Moon (1982). The residual flipping ratios measured with a 4.25 T field applied along the binary a axis of the crystal are shown in Fig. 4.20. The two significant features to note are that the outer electron contribution is only evident for the first Bragg peak, and that the $\kappa = 0$ point, obtained from the bulk susceptibility, is far off-scale.

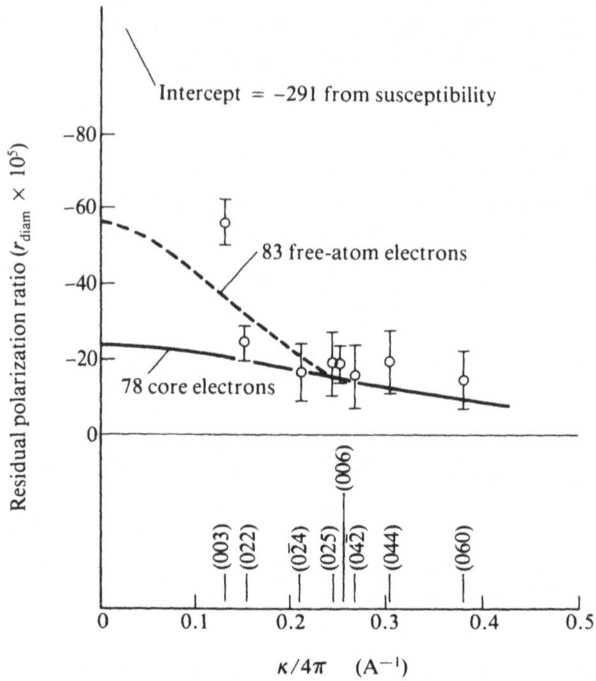

Fig. 4.20 Residual flipping ratios measured by Collins (1979) for diamagnetic bismuth at 290 K; H = 4.25 T parallel to binary (*a*) axis (from Moon 1982).

4.3 AMORPHOUS MAGNETIC SOLIDS

Despite the lack of any long-range atomic order, the magnetic atoms in amorphous materials do possess a local environment, and it is of interest to determine how the magnetism is distributed in the locality of the magnetic atom. The most definitive method for studying such effects is to use polarized neutron diffraction. Our first example is amorphous Fe prepared by decomposing $Fe(CO)_5$ under an electric discharge. The difference cross-section for measurements with both (+) and (−) spin neutrons at a polarized-beam diffractometer for a material with only one type of magnetic atom is

$$\Delta\left(\frac{d\sigma}{d\Omega}\right)^{\text{Pol}} = \left(\frac{d\sigma}{d\Omega}\right)_{+}^{\text{Pol}} - \left(\frac{d\sigma}{d\Omega}\right)_{-}^{\text{Pol}}$$
$$= 2b\,|r_0|\,\mu F(\boldsymbol{\kappa})S(\boldsymbol{\kappa}), \tag{4.3.1}$$

where $S(\boldsymbol{\kappa})$ is the monatomic interference function given in equn (2.1.46). The difference spectrum for amorphous Fe, as measured on the D5 polarized beam instrument at Grenoble, has been published in a review paper of Schweizer (1982). The magnetic scattering amplitude $\mu F(\boldsymbol{\kappa})$ was calculated by dividing this data by $2b\,|r_0|\,S(\boldsymbol{\kappa})$, where $S(\boldsymbol{\kappa})$ was measured by using X-rays, and the result is shown in Fig. 4.21. The $\mu F(\boldsymbol{\kappa})$ values at $(\kappa/4\pi) > 0.45\ \text{Å}^{-1}$ agree well with the theoretical form factor curve for Fe^{2+}, showing that the magnetic electrons in amorphous Fe have the same radial distribution as in crystalline Fe. The experimental points observed at lower $\boldsymbol{\kappa}$ values were attributed to CO impurities in the sample. These contribute terms such as $b_C r_0 \mu F(\boldsymbol{\kappa}) S_{\text{Fe–C}}$ and $b_O r_0 \mu F(\boldsymbol{\kappa}) S_{\text{Fe–O}}$ to the neutron cross-section. The X-ray measurement is insensitive to CO impurities, and does not show these features.

The application of polarized neutrons to determine the partial structure factors of magnetic atoms in amorphous binary ferromagnets was first reported by Blétry and Sadoc (1975). The (+) and (−) spin cross-sections, which contain the partial structure factors S_{AA}, S_{BB} and S_{AB}, are given in equns (2.2.31) and (2.2.32), and it was shown that polarized neutron diffraction can be used to determine these three factors without either varying the binary alloy composition or using special isotopes. In the experiment two scattering patterns are recorded with (+) and (−) spin neutrons incident on the sample in the usual $\tilde{\boldsymbol{P}} \cdot \tilde{\boldsymbol{\kappa}} = \tilde{\boldsymbol{M}} \cdot \tilde{\boldsymbol{\kappa}} = 0$

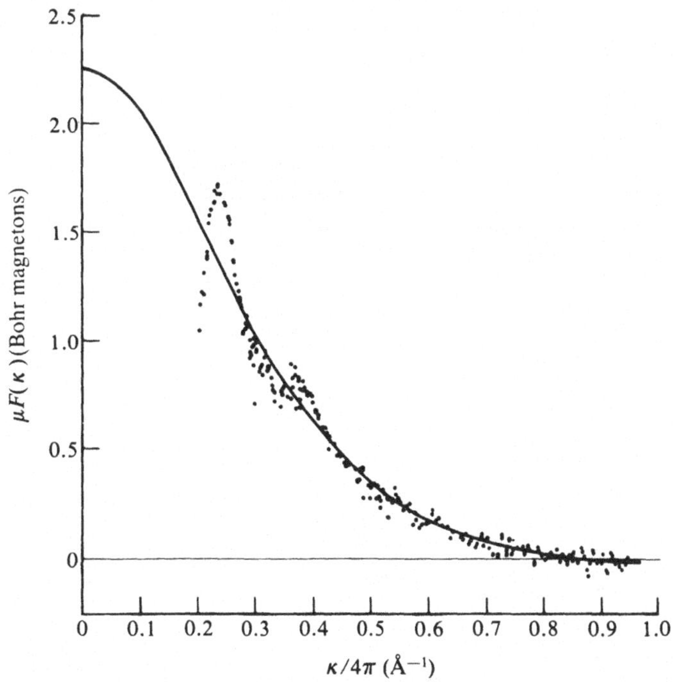

Fig. 4.21 Magnetic form factor for amorphous iron measured with polarized neutrons (·) and compared with a theoretical form factor for Fe^{2+} (full curve) (from Schweizer 1982).

geometry, and a third pattern is obtained with either polarization state, but with the magnetization parallel to the scattering vector. For their first experiment, Blétry and Sadoc determined the three partial structure factors in a Co–P alloy, and showed how to correct the measured data for Laue diffuse, incoherent, and multiple scattering. Data analysis showed the cobalt magnetic form factor $F(\boldsymbol{\kappa})$ to be approximately 15% wider (at half-maximum height) than the normal Co^{2+} form factor, which indicates a greater localization of the magnetic moment in the amorphous alloy.

A similar experiment was carried out on the ferromagnetic metallic glass $Fe_{83}B_{17}$ by Guoan *et al.* (1982). Two of the partial structure factors $S_{Fe\text{-}Fe}(\boldsymbol{\kappa})$ and $S_{Fe\text{-}B}(\boldsymbol{\kappa})$ were evaluated and are shown in Fig. 4.22; the small value of the coefficient $(c_B b_B)^2$ of $S_{BB}(\boldsymbol{\kappa})$, however, did not allow the third partial structure factor to be determined with sufficient accuracy. The iron form factor used in the data analysis was modified from the normal form factor to include a width parameter, and a broadening in $F(\boldsymbol{\kappa})$ of almost exactly the same magnitude as observed by Blétry and Sadoc for Co–P was deduced. Both results, therefore, point to a significant localization of the magnetic electrons in these amorphous

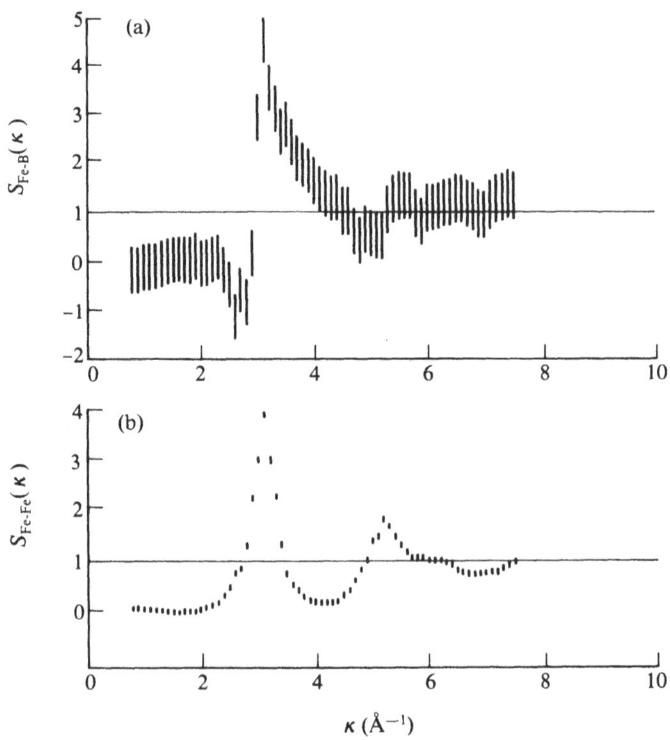

Fig. 4.22 The partial structure factors: (a) $S_{Fe\text{-}B}(\boldsymbol{\kappa})$ and (b) $S_{Fe\text{-}Fe}(\boldsymbol{\kappa})$ in $Fe_{83}B_{17}$ determined by polarized neutron diffraction (from Guoan *et al.* 1982).

binary ferromagnets, whereas the normal distribution was observed in monoatomic amorphous iron.

4.4 POLARIZED-BEAM DIFFUSE SCATTERING IN FERROMAGNETIC ALLOYS

The determination of the diffuse magnetic scattering cross-sections in binary alloys gives information on the spatial fluctuations of the magnetic moments from site to site. We now discuss the measurement of these cross-sections with polarized neutrons and magnetized specimens. In a directly analogous way to magnetic diffraction in single crystals, such experiments are much more sensitive than their field-switching unpolarized beam counterparts. The parameter usually measured is the difference cross-section for experiments with (+) and (−) spin incident neutrons, and the simplest expression for this is given by equn (2.2.35):

$$\Delta\left(\frac{\mathrm{d}\sigma}{\mathrm{d}\Omega}\right)^{\mathrm{Pol}} = \left(\frac{\mathrm{d}\sigma}{\mathrm{d}\Omega}\right)^{\mathrm{Pol}}_{+} - \left(\frac{\mathrm{d}\sigma}{\mathrm{d}\Omega}\right)^{\mathrm{Pol}}_{-}$$
$$= 2\,|r_0|\,c(1-c)(b_{\mathrm{A}}-b_{\mathrm{B}})(F_{\mathrm{A}}(\boldsymbol{\kappa})\mu_{\mathrm{A}}-F_{\mathrm{B}}(\boldsymbol{\kappa})\mu_{\mathrm{B}}), \quad (4.4.1)$$

which is valid in the usual $\tilde{\boldsymbol{\kappa}} \cdot \tilde{\boldsymbol{\eta}} = -1$ geometry, and within the limits of the assumptions of the Shull–Wilkinson model for the alloy (Shull and Wilkinson 1955). This equation is a special case of the more general expression, given for example by Rainford (1982),

$$\Delta\left(\frac{\mathrm{d}\sigma}{\mathrm{d}\Omega}\right)^{\mathrm{Pol}} = 2\,|r_0|\,[\tilde{\boldsymbol{M}} - (\tilde{\boldsymbol{M}} \cdot \tilde{\boldsymbol{\kappa}})\tilde{\boldsymbol{\kappa}}]I(\boldsymbol{\kappa}), \quad (4.4.2)$$

with

$$I(\boldsymbol{\kappa}) = \sum_{i,j} \exp[i\boldsymbol{\kappa} \cdot (\boldsymbol{R}_i - \boldsymbol{R}_j)]\langle(b_i - \bar{b})(F_j(\boldsymbol{\kappa})\mu_j - \overline{F(\boldsymbol{\kappa})\mu})\rangle, \quad (4.4.3)$$

where $\tilde{\boldsymbol{M}}$ is a unit vector in the direction of the magnetization of the sample, and the triangular brackets denote a configurational averaging. The experiments are always carried out with the magnetizing field normal to the scattering vector, hence the geometrical parameter is +1. In the Shull–Wilkinson (1955) model the constituent A and B atoms in the binary alloy are assumed not to experience local environment effects, hence any magnetic moment fluctuations must be a direct consequence of concentration fluctuations. In this case

$$\bar{\mu} = c\mu_{\mathrm{A}} + (1-c)\mu_{\mathrm{B}}, \quad (4.4.4)$$

and Rainford (1982) showed that equn (4.4.2) then reduces to equn (4.4.1). One of the essential requirements for this type of measurement is that there should be a good contrast between the two nuclear scattering amplitudes of the A and B atoms. By extrapolating the neutron measurements to $\boldsymbol{\kappa}=0$, and combining the result with bulk magnetization determinations of the average moment per atom $\bar{\mu}$, the individual atomic moments μ_A and μ_B can be found. Collins and Forsyth (1963) used this technique to determine the signs of the moment differences in Fe–Co and Fe–Ni alloys.

In dilute alloys, when for example c is small, it is reasonable to assume that all A atoms have the same moment, and to introduce a perturbation term to account for the change in the B atom moments due to the impurity A atoms. An exact solution to the diffuse cross-sections then becomes possible. The moment at the ith site is given by

$$\mu_i = \mu_B + (\mu_A - \mu_B)p_i + \sum_r g(r)p_{i+r}, \tag{4.4.5}$$

where $p_i = 1$ if site i is occupied by an A atom, or $p_i = 0$ if it is occupied by a B atom, and $g(r)$ is the disturbance function operative on the B atom moments at a distance r from an A atom. $I(\boldsymbol{\kappa})$ then becomes

$$I(\boldsymbol{\kappa}) = c(1-c)(b_A - b_B)[F_A(\boldsymbol{\kappa})\mu_A - F_B(\boldsymbol{\kappa})\mu_B + F_B(\boldsymbol{\kappa})G(\boldsymbol{\kappa})], \tag{4.4.6}$$

with

$$G(\boldsymbol{\kappa}) = \sum_r \exp(i\boldsymbol{\kappa}\cdot\boldsymbol{r})g(r), \tag{4.4.7}$$

and in the limit $\boldsymbol{\kappa}=0$,

$$I(\boldsymbol{\kappa}=0) = c(1-c)(b_A - b_B)\left(\frac{\mathrm{d}\bar{\mu}}{\mathrm{d}c}\right). \tag{4.4.8}$$

For polycrystalline samples

$$G(\boldsymbol{\kappa}) = \sum_s z_s g(r_s)\frac{\sin \kappa r_s}{\kappa r_s}, \tag{4.4.9}$$

where z_s is the coordination number of shell s. Equation (4.4.8) shows that the polarized-beam cross-section has a forward positive or negative peak with a magnitude which is proportional to $(\mathrm{d}\bar{\mu}/\mathrm{d}c)$. At large $\boldsymbol{\kappa}$ values $G(\boldsymbol{\kappa})$ approaches zero, and the diffuse cross-section approximates well to the Shull–Wilkinson result.

Figure 4.23 shows the data measured by Mezei (1976b) for an **Fe**Mn alloy containing 2.8 at% Mn. These were analysed by assuming $F_A(\boldsymbol{\kappa}) = F_B(\boldsymbol{\kappa}) = F_{Fe}(\boldsymbol{\kappa})$ in equn (4.4.6), i.e.

$$\Delta\left(\frac{\mathrm{d}\sigma}{\mathrm{d}\Omega}\right)^{\mathrm{Pol}} = 2\,|r_0|\,c(1-c)(b_{Mn} - b_{Fe})F_{Fe}(k) \times\left[(\mu_{Mn} - \mu_{Fe}) + \sum_s z_s\Delta\mu(\mathrm{Fe})_s\frac{\sin \kappa r_s}{\kappa r_s}\right]. \tag{4.4.10}$$

The measurements were made at the D7 diffuse scattering instrument at

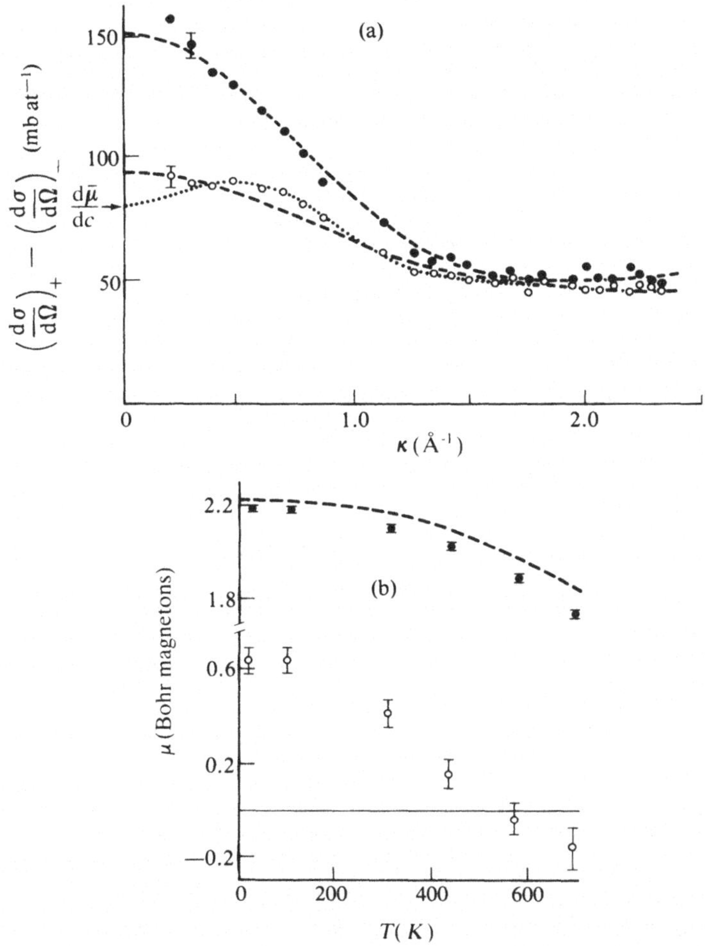

Fig. 4.23 (a) Polarized neutron difference cross-sections for Fe(2.8% Mn) at 648 K (●) and 305 K (○). (b) The temperature dependence of the magnetic moments on the Mn(ǫ) and nearest-neighbour Fe atoms (•), compared with the pure Fe moment μ_{Fe}, (broken curve) (from Mezei 1976b).

the Institut Laue–Langevin operating at $\lambda = 4.90$ Å with a resolution $\delta\lambda/\lambda \sim 8\%$; data were collected over a temperature range from 4.2 to 700 K. The results showed that the Mn moment changes by $0.80 \pm 0.1\beta$ between 0 and 700 K, and that the moment perturbation in the neighbouring Fe shells was strongly temperature-dependent. The Mn moment at zero degrees kelvin was found to be $0.64 \pm 0.15\beta$.

The analysis of the diffuse magnetic scattering from concentrated alloys becomes more complicated owing to the wider range of local environments, and because of the possibility that magnetic fluctuations can affect both the atomic constituents. A useful description of such alloys is that introduced by Marshall (1968), who assumed that the moment perturbations at a given site are linear in the possible chemical fluctuations in its proximity. If site i is occupied by a B atom, then

$$\mu_i = \bar{\mu}_{\mathrm{B}} + \sum_{r_s} g(r_s)(p_{i+r_s} - c), \tag{4.4.11}$$

whereas if site i is occupied by an A atom,

$$\mu_i = \bar{\mu}_{\mathrm{A}} + \sum_{r_s} h(r_s)(p_{i+r_s} - c). \tag{4.4.12}$$

The model therefore contains two moment disturbance functions: $g(r)$ for B atoms and $h(r)$ for A atoms. The polarized-beam difference cross-section parameter $I(\boldsymbol{\kappa})$ then becomes

$$I(\boldsymbol{\kappa}) = c(1-c)(b_{\mathrm{A}} - b_{\mathrm{B}})F(\boldsymbol{\kappa})S(\boldsymbol{\kappa})M(\boldsymbol{\kappa}), \tag{4.4.13}$$

where

$$M(\boldsymbol{\kappa}) = \mu_{\mathrm{A}} - \mu_{\mathrm{B}} + (1-c)G(\boldsymbol{\kappa}) + cH(\boldsymbol{\kappa}). \tag{4.4.14}$$

$G(\boldsymbol{\kappa})$ and $H(\boldsymbol{\kappa})$ are given by the Fourier transforms of $f(r)$ and $h(r)$, and the magnetic form factors for the A and B atoms are assumed to be the same. The $S(\boldsymbol{\kappa})$ term allows for deviations from complete randomness in the alloy, and for a polycrystalline specimen

$$S(\boldsymbol{\kappa}) = 1 + \sum_{s} z_s \alpha(r_s) \frac{\sin \kappa r_s}{\kappa r_s}, \tag{4.4.15}$$

where $\alpha(r_s)$ is the Cowley (1950) short-range order parameter for the sth coordination shell, see equn (2.1.60). $S(\boldsymbol{\kappa})$ has an important influence on the measured cross-sections, and magnetic information can only be deduced after this function has been evaluated accurately by determining the nuclear diffuse cross-section. Equation (4.4.14) shows that $M(\boldsymbol{\kappa})$ still allows $(\mu_{\mathrm{A}} - \mu_{\mathrm{B}})$ to be determined, as well as the combination of $(1-c)G(\boldsymbol{\kappa})$ and $cH(\boldsymbol{\kappa})$. Further, the polarized neutron forward scattering cross-section at $\boldsymbol{\kappa} \to 0$ remains proportional to $\mathrm{d}\bar{\mu}/\mathrm{d}c$.

Mirebeau *et al.* (1987) used Marshall's formalism to analyse polarized neutron diffuse scattering measurements on a series of ferromagnetic $Fe_{1-c}V_c$ alloys with $0.01 < c < 0.2$. Some of their results for the magnetic moment distribution function $-M(\boldsymbol{\kappa})$, see equn (4.4.14), are shown in Fig. 4.24. The results showed a constant negative moment of about -1β at the V sites up to V concentrations of approximately 10%, in contrast to earlier results of Kajzar and Parette (1980) in which a mean V moment of -3β was found for a 1 at% V alloy.

Marshall's theory is limited in the sense that it is only concerned with the chemical environment in the alloy, and more advanced models which take into account the magnetic environment, notably the model of

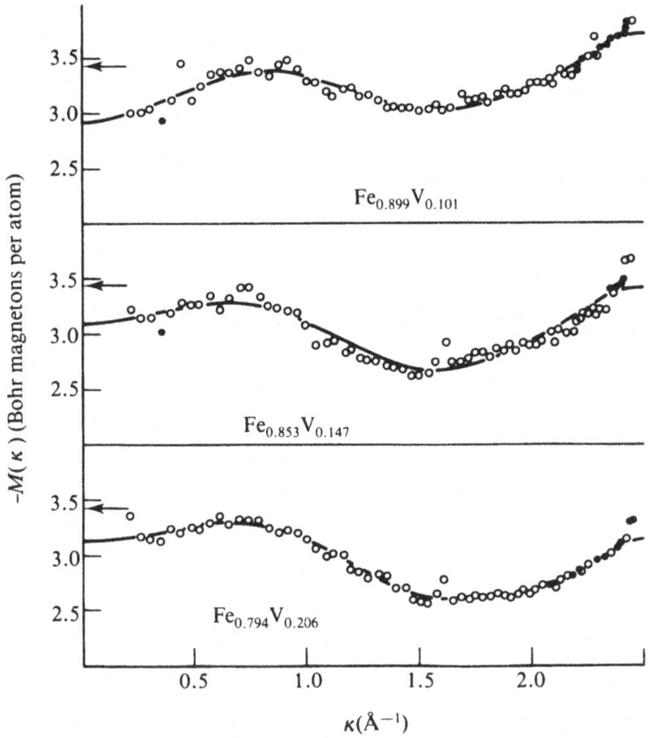

Fig. 4.24 Polarized neutron determination of the magnetic function $-M(\boldsymbol{\kappa})$ of equn (4.4.14) for $Fe_{1-c}V_c$ alloys by Mirebeau *et al.* (1987). The arrows show the values of $-d\mu/dc$ deduced from magnetization measurements.

Medina and Cable (1977), were introduced later. A more detailed discussion is beyond the scope of this book, and further information may be obtained in the reviews by Rainford (1982) and Hicks (1979b).

4.5 SCATTERERS CONTAINING POLARIZED NUCLEI

The purely nuclear scattering from specimens containing polarized nuclei is discussed in Section 2.2.1. The main result, see equn (2.2.15), for the coherent scattering cross-sections with (+) and (−) incident spin neutrons for a single polarizable nucleus is

$$\left(\frac{d\sigma}{d\Omega}\right)_{\text{coh}\,\pm} = (b \pm \tfrac{1}{2}\mathcal{B}iP_{\text{N}})^2 \qquad (4.5.1)$$

where b is the usual coherent scattering amplitude

$$b = [(i+1)b_+ + ib_-]/(2i+1) \qquad (4.5.2)$$

and

$$\mathscr{B} = [b_+ - b_-]/(i + \tfrac{1}{2}). \tag{4.5.3}$$

Let us consider the particularly simple example of a Bravais lattice containing one polarizable nucleus, and where the atom has no atomic moment. Equation (4.5.1) can also be expressed in terms of the nuclear ($\boldsymbol{P}_\mathrm{N}$) and neutron ($\boldsymbol{P}$) polarization vectors, see equn (2.2.18), as

$$\left(\frac{\mathrm{d}\sigma}{\mathrm{d}\Omega}\right)_{\mathrm{coh}} = b^2 + \frac{\mathscr{B}^2}{4} i^2 P_\mathrm{N}^2 + b\mathscr{B}i\boldsymbol{P}_\mathrm{N} \cdot \boldsymbol{P}. \tag{4.5.4}$$

An equivalent expression for non-Bravais crystal lattices is obtained by using the unit cell nuclear coherent and nuclear spin structure factors given by equns (2.2.21) and (2.2.22); however, the main features of the coherent scattering from assemblies of polarized nuclei are all evident in this simpler equation. First, just as for magnetic scattering, we see that it is possible to extract information on nuclear magnetic effects by using unpolarized neutrons ($\boldsymbol{P} = 0$). As examples we cite the experiments of Steiner (1985) on CoF_2 and $DyPO_4$. Secondly, and again in an analogous way to magnetic scattering, the most sensitive measurements of nuclear magnetism use polarized neutrons to exploit the interference term in equn (4.5.4). Most of the experiments to be discussed in this section are of this type, and involve purely nuclear scatterers. An extension of equn (4.5.4) which includes scattering due to atomic magnetism will be given when describing experiments involving polarized vanadium.

Next we restate the result for the nuclear spin (NS) incoherent cross-section which is given by equn (2.2.20),

$$\left(\frac{\mathrm{d}\sigma}{\mathrm{d}\Omega}\right)_{\mathrm{NS}} = \frac{\mathscr{B}^2}{4}[i(i+1) - i^2 P_\mathrm{N} - i\boldsymbol{P}_\mathrm{N} \cdot \boldsymbol{P}]. \tag{4.5.5}$$

It is now instructive to show how these neutron-spin-dependent coherent and spin incoherent cross-sections vary with the nuclear polarization P_N. We take as our example the scattering from protons where the $(i + \frac{1}{2})$ compound state scattering length is $b_+ = 1.04 \times 10^{-12}$ cm, and the $(i - \frac{1}{2})$ scattering length is $b_- = -4.7 \times 10^{-12}$ cm; the nuclear spin scattering parameter $\mathscr{B}$ is therefore $\mathscr{B} = 5.74 \times 10^{-12}$ cm, and this is the largest known $\mathscr{B}$ value for any nucleus. Figure 4.25 shows calculated values of the nuclear spin incoherent cross-section given by equn (4.5.5), and coherent scattering amplitudes given by

$$b(\pm) = [b^2 + (\mathscr{B}^2/4)i^2 P_\mathrm{H}^2 \pm b\mathscr{B}iP_\mathrm{H}]^{1/2} = [b \pm (\mathscr{B}/2)iP_\mathrm{H}]^{1/2} \tag{4.5.6}$$

for positively (+) and negatively (−) polarized neutrons, and for a range 0–100% in the proton polarization P_H. This figure shows that the coherent scattering lengths can be changed just by changing the proton polarization, and this effect can be exploited in diffraction experiments in

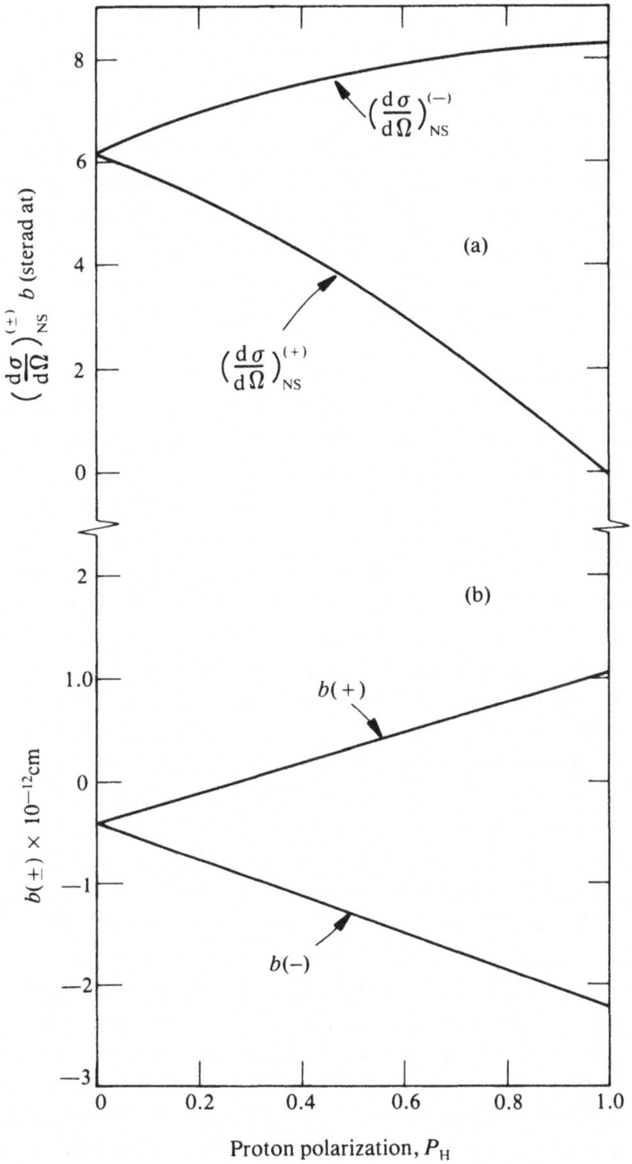

Fig. 4.25 Neutron spin (±) dependences of (a) the spin incoherent scattering cross-sections, and (b) the coherent scattering amplitudes, for protons as a function of the proton polarization.

an analogous way to heavy atom substitution with X-rays, or isotopic substitution in conventional neutron diffraction. One of the methods available for polarizing protons, the 'solid-state effect', can produce both negative and positive polarizations, thus this technique gives an even wider variation in $b(+)$ and $b(-)$ than is shown in the figure. Another important observation from Fig. 4.25 is the diminution of the (+) spin incoherent scattering as P_H increases. Neutron diffraction from hydrogenous systems suffers considerably from poor signal-to-noise ratios owing to the large spin-incoherence of hydrogen, and the use of polarized neutrons and polarized protons is one way of reducing the featureless background.

The viability of the polarized neutron–polarized nucleus diffraction method depends, see for example equn (4.5.6), on the nucleus having a large nuclear spin scattering parameter $\mathscr{B}$, and also on the experimental possibility of producing a large nuclear polarization P_N. The three main methods for generating nuclear polarizations: dynamic polarization using the 'solid-state effect', and static polarization using either the 'brute force' method or the hyperfine field interaction, have all been applied to produce polarized nuclei for neutron experiments. The principles underlying these methods are discussed in Section 3.1.2, and so it suffices here to make only a few additional remarks. The dynamic polarization method using spin resonance pumping, though technically difficult, is unique in that it can produce both positively and negatively polarized nuclei. It has been used to prepare polarized proton and polarized fluorine scattering samples. The hyperfine field method is experimentally more attractive, though in samples containing absorbing nuclear resonances, scattering experiments are only possible at neutron energies far removed from these resonances. Possible candidates, with their hyperfine fields as given by Heeringa (1985) in parentheses, are Ho (900 T), Dy (520 T) and Au (145 T). However, the method which has so far received least attention is the 'brute force' method. It requires very low temperatures and high magnetic fields, so that only the lowest magnetic substates of the nuclear spins are populated. With current technologies, applied fields $B \sim 10\,\mathrm{T}$ and temperatures $T \sim 10\,\mathrm{mK}$ (from dilution refrigerators) are readily achievable, and Heeringa (1985) has given the nuclear polarizations that can be reached for a range of nuclei under these experimental conditions. We reproduce his results for those nuclei pertinent to this discussion in Table 4.2. In the sixth column we calculate the product $\frac{1}{2}\mathscr{B}iP_N$ using $(b_+ - b_-)$ values given by Glattli *et al.* (1979); this is the nuclear spin coherent scattering length and its magnitude gives an indication of the feasibility of the appropriate polarized neutron-polarized nucleus scattering experiment.

We now describe some of polarized neutron–polarized nuclei scattering experiments from which results are available.

TABLE 4.2 Nuclear spin coherent scattering amplitudes for several nuclei polarized by the 'brute force' polarization method

Nucleus	Nuclear spin, i	Nuclear polarization, P_N[a]	$(b_+ - b_-)$ $\times 10^{-12}$ cm[b]	$\mathscr{B}$ $\times 10^{-12}$ cm	$b_{coh}^{NS} = \frac{1}{2}\mathscr{B}iP_N$ $\times 10^{-12}$ cm	$b_{coh}(P_N = 0)$ $\times 10^{-12}$ cm
^{1}H	1/2	0.77	+5.82	+5.82	+1.12	−0.374
^{7}Li	3/2	0.566	−0.45	−0.225	−0.096	−0.220
^{45}Sc	7/2	0.60	−1.36	−0.34	−0.357	1.23
^{51}V	7/2	0.631	+1.294	+0.324	+0.357	−0.0414
^{65}Cu	3/2	0.443	+0.37	+0.185	+0.061	1.00

[a] P_N values calculated by Heeringa (1985) for the 'brute force' method in a 10 T magnetic field at 10 mK.
[b] Values given by Glattli *et al.* (1979), and measured by the spin precession method.

4.5.1 Polarized protons

Protons have the largest $\mathscr{B}$ value of any nuclei, and the first demonstration of the neutron spin dependence of the coherent scattering from polarized protons was the diffraction experiment of Hayter *et al.* (1974). These workers measured the flipping ratios for a number of Bragg peaks from a lanthanum magnesium nitrate (LMN) single crystal in which the protons in the water of crystallization were dynamically polarized by using the 'solid-state effect'. The aim of the experiment was to evaluate the feasibility of using polarized neutron diffraction to determine hydrogen positions. In this diffraction experiment there are two unit cell structure factors, first the usual unpolarized beam structure factor

$$F_N(\boldsymbol{\kappa}) = \sum_N b_N \exp(i\boldsymbol{\kappa} \cdot \boldsymbol{R}_N), \tag{4.5.7}$$

and a polarized beam or nuclear spin structure factor

$$F_{NS}(\boldsymbol{\kappa}) = \sum_H \tfrac{1}{2}\mathscr{B}_H i_H P_H \exp(i\boldsymbol{\kappa} \cdot \boldsymbol{R}_H), \tag{4.5.8}$$

where for simplicity we have omitted the Debye–Waller factors, and the label H refers to protons. The flipping ratios in the polarized beam experiment, uncorrected for instrumental inefficiencies, are then

$$\mathrm{R}(\boldsymbol{\kappa}) = \frac{[F_N(\boldsymbol{\kappa}) + F_{NS}(\boldsymbol{\kappa})]^2}{[F_N(\boldsymbol{\kappa}) - F_{NS}(\boldsymbol{\kappa})]^2} \tag{4.5.9}$$

A flipping ratio of 8.5:1 was observed at the maximum of the (0 −3 12) reflection in LMN, and Fig. 4.26 shows the reduction of this flipping ratio after the dynamic microwave pumping was stopped. The observed time-dependence of the intensities gives a measure of the decay of P_H, and the nuclear spin relaxation time deduced from the data was 25 min. In a later publication, Leslie *et al.* (1980) summarized their results after

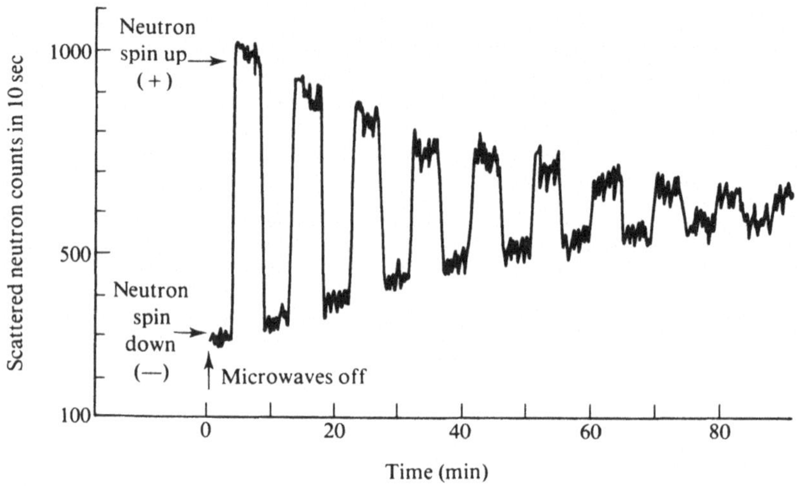

Fig. 4.26 Relaxation of (+) and (−) neutron spin intensities for the (0 −3 12) reflection in a dynamically polarized LMN crystal following the removal of the polarizing microwave radiation (from Hayter *et al.* 1974).

measuring 500 independent flipping ratios from 500 separate reflections. The data analysis method favoured by these authors was to use X-ray scattering to determine the positions of non-hydrogen atoms, and then to make an intelligent guess of the hydrogen positions which was then used in a least-squares refinement of the measured flipping ratios. The application of this procedure to the LMN data gave a final crystallographic R factor of 12.5%, and the hydrogen positions agreed well with those obtained from a conventional low-temperature unpolarized neutron diffraction study.

A similar experiment, which concentrated on making a careful measurement of the flipping ratio for the single (111) diffraction peak in a dynamically polarized single crystal of LiH, was performed by Abragam *et al.* (1972a). The aim of this measurement was to establish the viability of the method and apparatus used to determine the spin-dependent scattering parameter $\mathscr{B}$ for ^{19}F. The measured flipping ratio in the LiH experiment was $R = 1 - (5.7 \pm 0.8) \times 10^{-2}$, and was shown to be consistent with the known b and $\mathscr{B}$ values for H and Li.

The polarization of protons can also be used to provide contrast variation in neutron small-angle scattering experiments, and promises to be an important tool in macromolecule structure determinations. Stuhrmann *et al.* (1985) have started such a research programme by preparing a polarized-proton 'frozen-spin' target from the protein lysozyme dissolved in D_2O. This was doped with a deuterated paramagnetic free radical to provide the free electrons for the 'solid-state effect' dynamic polarization. Clusters of 700 protons defined by the lysozyme structure were polarized to a proton polarization $P_H = 0.75$.

4.5.2 Fluorine-19

Polarized neutron diffraction from a polarized single crystal of CaF_2 was used by Abragam *et al.* (1972a) to measure the spin-dependent scattering parameter $\mathcal{B}$ for ^{19}F. The flipping ratio equation (4.5.9) in the case $F_N \gg F_{NS}$ becomes

$$R = 1 + 4F_{NS}(\boldsymbol{\kappa})/F_N(\boldsymbol{\kappa}), \tag{4.5.10}$$

and in the CaF_2 experiment, for which $\mathcal{B}_F$ was expected to be very small, a detailed examination was made of the (200) reflection, since this has a low $F_N(\boldsymbol{\kappa})$ value. Abragam *et al.* (1979a) showed that the expected flipping ratio for this reflection is

$$R = 1 + \frac{4\mathcal{B}_F i P_F P(1-2f)}{\bar{b}_{Ca} - 2\bar{b}_F}, \tag{4.5.11}$$

where $i = \frac{1}{2}$ for ^{19}F, P_F is the fluorine polarization, P is the neutron polarizing efficiency of the $Co_{92}Fe_8$ polarizer used for the measurements, and f is the spin-flipper efficiency. The nuclear polarization P_F was set up using the 'solid-state effect', and values between -0.35 and $+0.40$ were achieved. The linear dependence of R and $\mathcal{B}_F$ was then used to give a $\mathcal{B}_F$ or $(b_+ - b_-)_F$ value of $-0.0135(2) \times 10^{-12}$ cm. This result was ten times smaller than a theoretical estimate of this quantity.

4.5.3 Vanadium-51

The first experiment which examined nuclear polarization effects by polarized neutron scattering was that by Shull and Ferrier (1963) on a single crystal of vanadium. This metal exhibits a weak paramagnetic susceptibility which is essentially temperature-independent, and this property allowed the electronic and nuclear polarization contributions to the measured flipping ratios to be distinguished. In the presence of magnetic scattering, equn (4.5.9) for the flipping ratio becomes

$$R = \frac{[F_N(\boldsymbol{\kappa}) + F_{NS}(\boldsymbol{\kappa}) + F_M(\boldsymbol{\kappa})]^2}{[F_N(\boldsymbol{\kappa}) - F_{NS}(\boldsymbol{\kappa}) - F_M(\boldsymbol{\kappa})]^2}, \tag{4.5.12}$$

where the magnetic structure factor $F_M(\boldsymbol{\kappa})$ is given by an expression such as eqn (2.2.23). If we again simplify this to the case of a single nucleus or atom with nuclear and magnetic scattering lengths b and p we obtain, for the $\tilde{\boldsymbol{P}} \cdot \tilde{\boldsymbol{\kappa}} = 0$ geometry,

$$R = \frac{b^2 + p^2 + (\mathcal{B}^2/4)i^2P_N^2 + 2bp + b\mathcal{B}iP_N + p\mathcal{B}iP_N}{b^2 + p^2 + (\mathcal{B}^2/4)i^2P_N^2 - 2bp - b\mathcal{B}iP_N + p\mathcal{B}iP_N}. \tag{4.5.13}$$

We note that only the fourth and fifth (interference) terms in this expression are neutron-spin-dependent. The magnetic scattering amplitude p contains the magnetic form factor and is given by equn (2.1.77). In the vanadium experiments the largest term in the numerator and denominator is b^2, followed by the fourth and fifth terms which are some two orders of magnitude smaller, and the other three terms are a further two orders of magnitude smaller again. We can therefore make the approximation

$$R = \left(\frac{b + p + (\mathcal{B}/2)iP_{\mathrm{N}}}{b - p - (\mathcal{B}/2)iP_{\mathrm{N}}}\right)^2 \simeq 1 + \frac{4[p + (\mathcal{B}/2)iP_{\mathrm{N}}]}{b}. \qquad (4.5.14)$$

The vanadium polarization was created by the 'brute force' method by cooling the specimen to a temperature of 2.81 K in an applied field $B = 1.32\,\mathrm{T}$. In the approximation $\mu_{\mathrm{N}}B/k_{\mathrm{B}}T \ll 1$, which is valid for this experiment, where μ_{N} is the vanadium nuclear moment, this polarization is

$$P_{\mathrm{N}} = \frac{1}{3}\frac{(i+1)}{i}\frac{\mu_{\mathrm{N}}B}{k_{\mathrm{B}}T}, \qquad (4.5.15)$$

with $i = 7/2$ for vanadium. Equation (4.5.14) therefore shows that the flipping ratio contains a temperature-independent electronic term, and a nuclear polarization term which is proportional to T^{-1}. The temperature dependence of the flipping ratio observed by Shull and Ferrier (1963) for the vanadium (110) reflection is shown in Figure 4.27. The magnitude of $\mathcal{B}$ was known from the spin incoherent scattering cross-section, and this experiment showed for the first time that its sign is positive. It should be remarked that the magnitude of P_{N} in this experiment was exceedingly small ($P_{\mathrm{N}} = 0.0004$), yet the great sensitivity of the polarized-beam technique allowed its effect to be detected very easily.

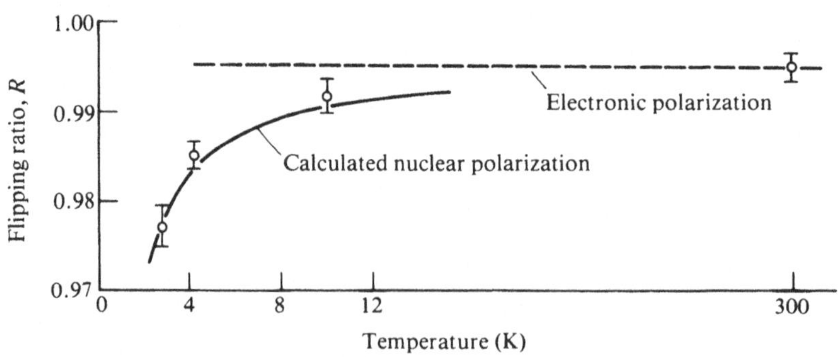

Fig. 4.27 Flipping ratios measured by Shull and Ferrier (1963) for the (110) vanadium reflection showing combined electronic and nuclear polarization at low temperature.

4.5.4 Cobalt-59

Ito and Shull (1969) used the polarized neutron diffraction method to investigate the effect of nuclear polarization on the coherent scattering in a single crystal of hexagonal cobalt. The study was restricted to the (220) and (140) Bragg reflections, and the ^{59}Co nuclear polarization was generated by the hyperfine field interaction in the ferromagnetic element, which was cooled to 2.23 K. The flipping ratios contained both electronic and nuclear magnetic terms, and the data were analysed using equn (4.5.14). From the observations the spin-state scattering amplitudes b_+ and b_-, corresponding to neutron–nucleus compound states with total angular momenta $(i + \frac{1}{2})$ and $(i - \frac{1}{2})$, were determined to be $(-0.380 \pm 0.054) \times 10^{-12}$ cm and $(+1.060 \pm 0.070) \times 10^{-12}$ cm respectively.

4.5.5 Holmium-165

Steiner (1985) has reported polarized-beam diffraction experiments on $HoVO_4$, which is a two-nucleus spin system in which both the ^{165}Ho and ^{51}V nuclei were polarized by the 'brute force' method. At temperatures above approximately 50 mK and in applied fields >1 T it was shown that the polarized beam scattering is determined essentially by the hyperfine field polarized ^{165}Ho nuclei with a $(b_+ - b_-)$ value of -0.35×10^{-12} cm. (Glattli *et al.* 1979). The result is that the (+) neutron spin scattered intensity is expected to decrease with temperature, corresponding to a build-up in the ^{165}Ho nuclear polarization. At temperatures below ~50 mK, however, the ^{51}V-system becomes polarized, and since ^{51}V has a large $(b_+ - b_-)$ value of opposite sign to that of ^{165}Ho $((b_+ - b_-)_V = +1.294 \times 10^{-12}$ cm), this has the effect of increasing the (+) neutron spin scattered intensity. Steiner (1985) shows calculated curves of the (+) polarized beam scattered intensities for the (200) Bragg reflection in $HoVO_4$ over a temperature range 1–1000 mK and in applied fields between 0.2 T and 4 T. These curves all show the minima predicted by the preceding argument, and this behaviour was also confirmed experimentally by Steiner.

4.6 POLARIZED NEUTRONS AND SURFACE MAGNETISM

In this section we make a diversion from conventional neutron diffraction to give examples of the application of polarized neutron optical reflectivity measurements to the study of surface magnetism. The sample property probed in these experiments is the variation of the coherent scattering density normal to the reflecting surface, and in the cases to be discussed here this contains a magnetic component. The fundamental principles underlying the polarized-beam critical reflection method have been covered in Section 3.3.5; it has similarities with polarized neutron diffraction in the sense that the basic measured parameter is a flipping ratio, which, after applying corrections, gives the ratio of (+) neutron spin and (−) neutron spin optical reflectivities. We now present examples of some of the experiments carried out by Felcher and co-workers at the Argonne National Laboratory's pulsed neutron source polarized-beam reflectometer, which is illustrated in Fig. 3.36.

4.6.1 Magnetism in iron oxide films

The depth profiles of the magnetization in sputtered films of Fe_3O_4 and its oxidized form γ-Fe_2O_3 were obtained by Parkin *et al.* (1986) using the polarized neutron reflection technique. These materials show great promise as high-density magnetic recording media, and the objectives of the measurements were to explain why the saturation magnetizations of such films are much lower than for the bulk materials, and also to ascertain whether the magnetization is uniform throughout the film. The iron oxide samples were prepared on substrates of polished silicon (111) single crystals, since silicon is a weak neutron reflector. Two identical Fe_3O_4 films of thickness 2600 Å were first deposited, and one of these was subsequently oxidized to γ-Fe_2O_3. Their spin-dependent neutron reflectivities measured in a 0.25 T magnetic field are shown in Fig. 4.28; the difference is striking and shows immediately that the magnetization profiles of the two samples are different. The behaviour observed for Fe_3O_4 is in fact characteristic of a film in which the magnetization density is constant, since equal spin reflectivities $\mathcal{R}_+ = \mathcal{R}_-$ are observed at wavelengths which can be precisely scaled according to

$$\lambda^+ = \lambda^-\{(b/v) - B[(b/v) + B]\}^{1/2}, \tag{4.6.1}$$

where the terms are as defined in equn (3.3.10). In contrast, the magnetization profile in the γ-Fe_2O_3 film is not uniform. Parkin *et al.* (1986) showed that the observation that the reflectivities are spin-independent at $\lambda < 4.8$ Å for the incidence angle $\theta = 0.548°$ of these measurements is consistent with the existence of a magnetic dead layer of thickness ~150 Å at the surface of the film. Attempts to substantiate this hypothesis using diffraction techniques were not, however, successful. Structural studies showed the top-most layers to be the same as in the bulk of the films, and also that the grain sizes were similar. It was surmised that the dead layer may simply be related to the many defects which are formed in the spinel structure during the oxidation process.

The (−) neutron spin reflectivity observed for Fe_3O_4 shows significant modulations due to interference between the front and back surfaces of the film as depicted in Fig. 2.18(a). A calculation of this thin-film interference using the optics theory described in Section 2.4.2 gave a film

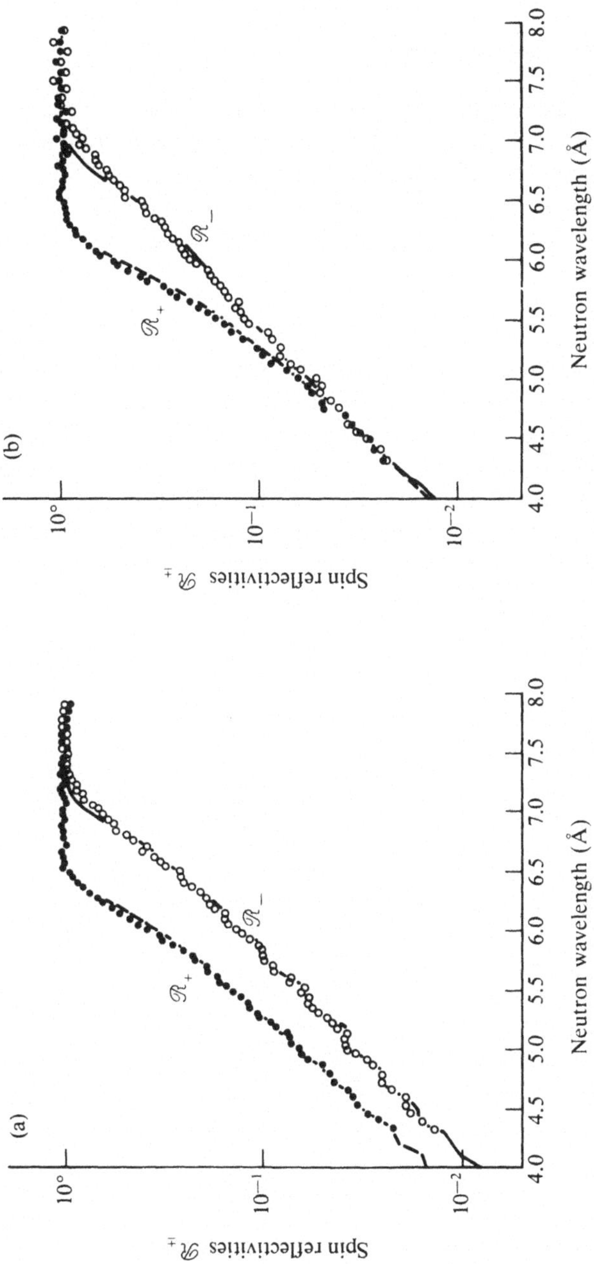

Fig. 4.28 **Neutron spin reflectivities for (a) Fe_3O_4 and (b) γ-Fe_2O_3 thin films measured by Parkin *et al.* (1986).**

thickness $d = 2700$ Å, in excellent agreement with the known film thickness.

4.6.2 Superconductor surfaces

When a type-II superconductor is subjected to an applied magnetic field H, then this is expelled from its bulk provided that its magnitude is less than a critical value H_{c1}. The depth variation of the magnetic induction B can be approximated as decreasing exponentially with distance z from the surface, and we may define a characteristic penetration depth Λ as follows:

$$B(z) = H \exp(-z/\Lambda). \tag{4.6.2}$$

The characteristic penetration distance Λ is typically a few hundred angströms and, through equn (3.3.9), it can be conveniently measured by the polarized neutron reflection method. Felcher *et al.* (1985) have reported measurements on niobium, pure lead, and an alloy $Pb_{0.992}Bi_{0.008}$; here we restrict ourselves to the Nb results.

Niobium is a type-II superconductor and exhibits complete field expulsion only below the critical field H_{c1}, above which a fluxoid lattice is

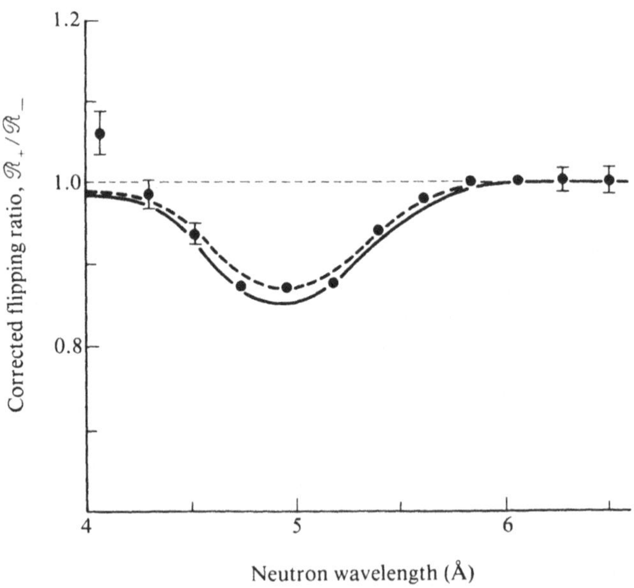

Fig. 4.29 'Corrected' flipping ratios observed for a superconducting Nb film at 4.6 K in an applied field of 0.05 T and with a reflection angle $\theta = 0.34° \pm 0.002°$. The lines are calculated fits to the data (broken curve $\Lambda = 480$ Å, full curve, $\Lambda = 380$ Å); from Felcher *et al.* (1985).

expected. Measurements on a sample having $H_{c1} = 0.1$ T were performed over a range of different applied fields and temperatures. Figure 4.29 shows a typical result of the 'corrected' flipping ratios $\mathcal{R}_+/\mathcal{R}_-$ observed as a function of neutron wavelength at fixed reflection angle $\theta = 0.34° \pm 0.02°$ and in an applied field $H = 0.05$ T. These results were compared with optics calculations, where it was assumed that the neutron-spin-dependent refractive index was determined by a field decay as given in equn (4.6.2). The data indicated that $\Lambda = 430 \pm 40$ Å at 4.6 K. A set of measurements taken at 7 K gave a best-fit value $\Lambda = 600 \pm 80$ Å, thus revealing the expected increase in the penetration depth with temperature. An extrapolation of all the results gave a zero-temperature penetration depth value $\Lambda(0) = 410 \pm 40$ Å, which was in excellent agreement with values derived from magnetic induction experiments.

5

Neutron polarization analysis experiments

This chapter is devoted to describing the results of neutron polarization analysis experiments. The principal objective is to give a representative account of the kinds of measurements which are possible, rather than to review the many experiments that have now been performed. We start by showing how the raw data collected in a polarization analysis experiment can be corrected to provide the required spin-dependent partial cross-sections. Corrections are frequently necessary for factors such as imperfect polarization, geometrical effects due to the relative directions of the scattering vector and polarization for magnetic scatterers, and sometimes multiple scattering in the sample. This is followed by a description of nuclear scattering experiments, where the aim is to separate the coherent and spin incoherent scattering cross-sections. The major part of the chapter, however, concerns the use of polarization analysis to isolate unequivocally the magnetic scattering, often in samples in which the nuclear scattering is appreciable. These experiments can be either elastic or inelastic.

5.1 CORRECTIONS TO POLARIZATION ANALYSIS DATA

The polarization analysis method in neutron scattering is intensity-limited, and the quality of the experimental data has not yet reached the degree of sophistication associated with polarized beam diffraction. It is, however, always necessary to correct the measured counts for instrumental inefficiencies, and frequently there is also a need to correct for geometrical effects depending on the directions of $\boldsymbol{P}$ and $\boldsymbol{\kappa}$, or for multiple scattering in the sample. We now discuss how to apply these corrections in the simplest polarization analysis experiment which separates the spin-flip (SF) and the non-spin-flip (NSF) scattering cross-sections. The experimental arrangement is typified by that shown in Fig. 2.9 without the need for the second spin-flipper. The two measured observables are the count rate with the spin-flipper in front of the sample 'off', C_{OFF}, and the count rate when this is activated, C_{ON}. There are two common approaches, one involves taking a 'corrected' difference of these count rates, and the other uses the flipping ratio. We first discuss the corrections for the instrument component inefficiencies.

5.1.1 Instrumental inefficiencies

Just as in polarized-beam diffraction, imperfect polarization and spin-flipping can have a serious effect on the detected intensities in polarization analysis experiments. The required spin-dependent scattering cross-sections can only be evaluated by correcting the raw data to account for these inefficiencies. Let us consider the measurement of the spin-dependent differential scattering cross-sections $(\mathrm{d}\sigma/\mathrm{d}\Omega)_{\mathrm{SF}}$ and $(\mathrm{d}\sigma/\mathrm{d}\Omega)_{\mathrm{NSF}}$; the extension to the double differential cross-sections normally measured on a triple-axis spectrometer is straightforward. The count rate measured with the spin-flipper 'off' is

$$C_{\mathrm{OFF}} = \phi_0 T\Omega N\varepsilon\left(\frac{1}{2}\right)\left\{\left(\frac{\mathrm{d}\sigma}{\mathrm{d}\Omega}\right)_T + PP_{\mathrm{A}}(1-D)\left[\left(\frac{\mathrm{d}\sigma}{\mathrm{d}\Omega}\right)_{\mathrm{NSF}} - \left(\frac{\mathrm{d}\sigma}{\mathrm{d}\Omega}\right)_{\mathrm{SF}}\right]\right\}, \tag{5.1.1}$$

where ϕ_0 is the incident beam flux at the sample, T is a composite transmittance factor for the polarizer, sample, and spin analyser, Ω is the effective analysed solid angle, N is the number of scattering atoms or nuclei, and ε is the detector efficiency. The product $PP_{\mathrm{A}}(1-D)$ contains the polarizing efficiency of the polarizer, analyser, and an instrumental depolarization term D due to inefficient spin guiding; we abbreviate this product to P^*, and assume that it can be calibrated using the methods described in Section 3.3.6. It is also convenient to collect together the instrumental factors $\phi_0 T\Omega N\varepsilon$, and we denote this by I. The count rate observed with the spin flipper 'on' then becomes

$$C_{\mathrm{ON}} = I\left(\frac{1}{2}\right)\left\{\left(\frac{\mathrm{d}\sigma}{\mathrm{d}\Omega}\right)_T - P^*E\left[\left(\frac{\mathrm{d}\sigma}{\mathrm{d}\Omega}\right)_{\mathrm{NSF}} - \left(\frac{\mathrm{d}\sigma}{\mathrm{d}\Omega}\right)_{\mathrm{SF}}\right]\right\}, \tag{5.1.2}$$

where $E = (2f-1)$, and f is the spin flipper efficiency as previously defined. From these equations we observe

$$C_{\mathrm{OFF}} - C_{\mathrm{ON}} = \left(\frac{1}{2}\right)IP^*(1+E)\left[\left(\frac{\mathrm{d}\sigma}{\mathrm{d}\Omega}\right)_{\mathrm{NSF}} - \left(\frac{\mathrm{d}\sigma}{\mathrm{d}\Omega}\right)_{\mathrm{SF}}\right] \tag{5.1.3}$$

and

$$C_{\mathrm{OFF}} + C_{\mathrm{ON}} = I\left\{\left(\frac{\mathrm{d}\sigma}{\mathrm{d}\Omega}\right)_{\mathrm{NSF}} + \left(\frac{\mathrm{d}\sigma}{\mathrm{d}\Omega}\right)_{\mathrm{SF}} + \frac{P^*}{2}(1-E)\left[\left(\frac{\mathrm{d}\sigma}{\mathrm{d}\Omega}\right)_{\mathrm{NSF}} - \left(\frac{\mathrm{d}\sigma}{\mathrm{d}\Omega}\right)_{\mathrm{SF}}\right]\right\}. \tag{5.1.4}$$

Hence the spin-flip and non-spin-flip cross-sections can be determined if P^* and f are known, and if I is calibrated using a standard scatterer such as vanadium.

In practice $|P^*|$ and $|f|\to 1$, and it is frequently a good approximation to assume that $E = 1$ in these equations. We then obtain, for the spin-flip scattering cross-section

$$\left(\frac{d\sigma}{d\Omega}\right)_{SF} = \frac{1}{2IP^*}[C_{OFF}(P^* - 1) + C_{ON}(1 + P^*)], \qquad (5.1.5)$$

or, in terms of the flipping ratio R_0 observed with the two polarizers and no scattering sample, i.e.

$$R_0 = (1 + P^*)/(1 - P^*), \qquad (5.1.6)$$

we have

$$\left(\frac{d\sigma}{d\Omega}\right)_{SF} = \frac{1}{I(R_0 - 1)}[C_{ON}R_0 - C_{OFF}]. \qquad (5.1.7)$$

A similar expression has been derived by Ziebeck and Brown (1980). In most instruments with high polarization, R_0 is large, so that $R_0/(R_0 - 1)$ approaches unity and $(R_0 - 1)^{-1}$ is small. The determination of $(d\sigma/d\Omega)_{SF}$ therefore depends less critically on determining C_{OFF} compared with C_{ON}, and it is statistically more advantangeous to use most of the measuring time with the spin-flipper 'on'. The optimum division of the measuring time is such that the ratio of time spent in determining C_{OFF} compared with C_{ON} should be $\sqrt{C_{OFF}}/(R_0\sqrt{C_{ON}})$. Schärpf (1985) used a further approximation, i.e. $(R_0 - 1) \sim R_0$, to give

$$\left(\frac{d\sigma}{d\Omega}\right)_{SF} = \frac{1}{I}\left[C_{ON} - \frac{C_{OFF}}{R_0}\right] \qquad (5.1.8)$$

in his treatment of the data measured at the diffuse scattering polarization analysis instrument D7 at Grenoble.

The corresponding equations for the non-spin-flip scattering cross-sections are

$$\left(\frac{d\sigma}{d\Omega}\right)_{NSF} = \frac{1}{2IP^*}[C_{OFF}(1 + P^*) + C_{ON}(P^* - 1)] \qquad (5.1.9)$$

$$= \frac{1}{I(R_0 - 1)}[C_{OFF}R_0 - C_{ON}], \qquad (5.1.10)$$

or, in the Schärpf approximation,

$$\left(\frac{d\sigma}{d\Omega}\right)_{NSF} = \frac{1}{I}\left[C_{OFF} - \frac{C_{ON}}{R_0}\right]. \qquad (5.1.11)$$

Note the symmetrical nature of equns (5.1.7) and (5.1.10). It is clear that the optimum time division when measuring non-spin-flip scattering should be such that the ratio of time spent in determining C_{ON} compared with C_{OFF} is $\sqrt{C_{ON}}/(R_0\sqrt{C_{OFF}})$.

An alternative method of analysing polarization analysis data is to use the flipping ratio R given by

$$R = C_{\mathrm{OFF}}/C_{\mathrm{ON}} \tag{5.1.12}$$

$$= \frac{1 + PP_{\mathrm{A}}(1-D)(1-2r)}{1 + PP_{\mathrm{A}}(1-D)(1-2f)(1-2r)}, \tag{5.1.13}$$

where r is the fraction of spin-flip scattering, i.e.

$$r = \left(\frac{\mathrm{d}\sigma}{\mathrm{d}\Omega}\right)_{\mathrm{SF}} \Big/ \left(\frac{\mathrm{d}\sigma}{\mathrm{d}\Omega}\right)_{\mathrm{T}} = \left(\frac{\mathrm{d}\sigma}{\mathrm{d}\Omega}\right)_{\mathrm{SF}} \Big/ \left[\left(\frac{\mathrm{d}\sigma}{\mathrm{d}\Omega}\right)_{\mathrm{NSF}} + \left(\frac{\mathrm{d}\sigma}{\mathrm{d}\Omega}\right)_{\mathrm{SF}}\right]. \tag{5.1.14}$$

A rearrangement of equn (5.1.13) gives

$$\left(\frac{\mathrm{d}\sigma}{\mathrm{d}\Omega}\right)_{\mathrm{SF}} = \frac{1}{2}\left(\frac{\mathrm{d}\sigma}{\mathrm{d}\Omega}\right)_{\mathrm{T}}\left(1 - \frac{(R-1)}{PP_{\mathrm{A}}(1-D)[1-R(1-2f)]}\right) \tag{5.1.15}$$

and this cross-section can only be evaluated from the flipping ratio measurement if the unpolarized-beam scattering cross-section $(\mathrm{d}\sigma/\mathrm{d}\Omega)_{\mathrm{T}}$ is determined separately. The effect of uncertainties in the instrumental parameter $PP_{\mathrm{A}}(1-D)$ and f on the error in the spin-flip fraction r has been analysed by Williams (1975).

5.1.2 Spin-dependent cross-sections on an absolute scale

The usual method for obtaining absolute cross-sections in neutron scattering experiments is to compare the specimen count rates with those observed for a standard vanadium sample whose scattering characteristics are well established. A vanadium normalization can be used, for example, to eliminate the need to evaluate explicitly the I term in equn (5.1.7), and to place $(\mathrm{d}\sigma/\mathrm{d}\Omega)_T$ in equn (5.1.15) on an absolute scale.

An alternative procedure which does not require the vanadium experiment, and which is appropriate for calibrating paramagnetic diffuse scattering in the presence of a nuclear Bragg peak, was introduced by Ziebeck and Brown (1980). The diffuse intensity depends on a large number of instrumental parameters, all of which can be eliminated by comparing this with the integrated intensity of the Bragg peak. Two measurements of the spin-flip scattering are necessary, one with the neutron polarization parallel ($\|$) and the other perpendicular ($\perp$) to the scattering vector $\boldsymbol{\kappa}$. On a triple-axis polarization analysis spectrometer this gives a difference cross-section

$$\left(\frac{\mathrm{d}^2\sigma}{\mathrm{d}\Omega\,\mathrm{d}E'}\right)_{\mathrm{SF}}^{\|} - \left(\frac{\mathrm{d}^2\sigma}{\mathrm{d}\Omega\,\mathrm{d}E'}\right)_{\mathrm{SF}}^{\perp} = \frac{r_0^2}{3}\frac{V}{v_0} N_{\mathrm{m}} S(\boldsymbol{\kappa}, \omega) \tag{5.1.16}$$

where V is the sample volume, v_0 is the unit cell volume, N_{m} is the number of magnetic atoms per unit cell, and $S(\boldsymbol{\kappa}, \omega)$ is the magnetic response function. The difference intensity in an experiment in which the analysis is restricted to a receiving slit aperture with length l, width W, at a distance r from the sample, is

$$I_\Delta = I_0 \frac{r_0^2}{3} \frac{V}{v_0} N_m S(\boldsymbol{\kappa}, \omega)\, \Delta E' \frac{lW}{r^2}, \tag{5.1.17}$$

and $\Delta E'$ is the energy width of the analysing crystal. The sample also gives a Bragg powder line with an integrated intensity of

$$I_B = I_0 \frac{\lambda^3 l}{8\pi r \sin 2\theta_B} \frac{m F_N(\boldsymbol{\kappa})^2}{v_0^2 \sin \theta_B} \frac{VW}{r(\delta 2\theta_B)}. \tag{5.1.18}$$

The new terms in this expression are $\delta 2\theta_B$, the step length in the scattering angle in units of π as the Bragg peak is scanned, and m, the multiplicity of the nuclear reflection. A combination of these last two equations gives

$$\frac{I_\Delta}{I_B} = r_0^2 \frac{8\pi}{3} \frac{N_m v_0}{\lambda^3} \delta 2\theta_B \frac{\sin \theta_B \sin 2\theta_B}{F_N(\boldsymbol{\kappa})^2 m} \int_0^{\Delta E'} S(\boldsymbol{\kappa}, \omega)\, \delta E' \tag{5.1.19}$$

from which the magnetic response function integrated over the energy width of the analyser can be determined. This absolute intensity calibration of the paramagnetic scattering in polarization analysis experiments has also been used by other workers, see for example Böni *et al.* (1985).

5.1.3 General formulae for the spin-dependent partial cross-sections

The introductory discussion of polarization analysis applications given in Section 2.3 was restricted to one-dimensional polarization analysis, and to magnetic experiments in the two geometries $\tilde{\boldsymbol{P}} \cdot \tilde{\boldsymbol{\kappa}} = 0$ and $\tilde{\boldsymbol{P}} \cdot \tilde{\boldsymbol{\kappa}} = 1$. The $\tilde{\boldsymbol{P}} \cdot \tilde{\boldsymbol{\kappa}} = 1$ condition is rarely strictly adhered to in inelastic experiments, since it can only be satisfied at one scattering angle, and for one value of the energy transfer. More general formulae are therefore needed which give the spin-dependent cross-sections in terms of the angle between the incident spin polarization vector and the scattering vector. 'Corrected' data can then be used from a number of detectors, and this is especially valuable in time-of-flight instruments. A further extension is needed to allow for the possibility of three-dimensional polarization analysis, and this is also included in the following description. Figure 5.1 shows the geometry of the general polarization analysis scattering experiment in which the sample scatters a polarized neutron beam with a cross-section $d^2\sigma/d\Omega\, dE'$, and the scattered neutron polarization $\boldsymbol{P}'$ is measured. In the classical one-dimensional polarization analysis experi-

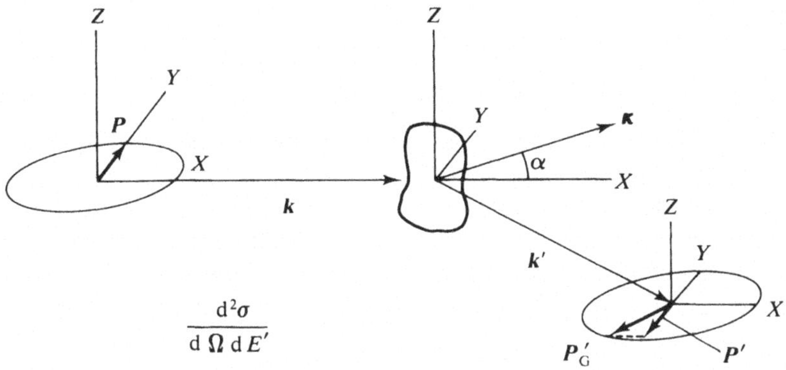

Fig. 5.1. Definition of the geometry for describing a generalized three-dimensional polarization analysis experiment. In the classical polarization analysis method only the component of $\boldsymbol{P}'$ which is parallel to $\boldsymbol{P}$ is measured.

ment the analysed component $\boldsymbol{P}'$ is always parallel to the incident polarization vector $\boldsymbol{P}$, but in the general three-dimensional experiment $\boldsymbol{P}'$ may be in any direction and is denoted by $\boldsymbol{P}'_G$. In the diagram (X, Y) is the scattering plane, and α is the angle between the longitudinal direction X and the scattering vector $\boldsymbol{\kappa}$.

For simplicity we consider the scattering from a paramagnetic sample with a magnetic cross-section abbreviated as σ_M, which also has a nuclear spin incoherent scattering cross-section σ_{NS}, and a combined coherent and isotopic incoherent nuclear scattering cross-section σ_N. The non-spin-flip (NSF) and spin-flip (SF) cross-sections in the one-dimensional experiment are

$$\sigma^Y_{NSF} = \tfrac{1}{3}\sigma_{NS} + \tfrac{1}{2}\sigma_M \cos^2\alpha + \sigma_N, \tag{5.1.20}$$

and

$$\sigma^Y_{SF} = \tfrac{2}{3}\sigma_{NS} + \tfrac{1}{2}\sigma_M(1 + \sin^2\alpha). \tag{5.1.21}$$

The partial cross-sections in the X and Z directions are

$$\sigma^X_{NSF} = \tfrac{1}{3}\sigma_{NS} + \tfrac{1}{2}\sigma_M \sin^2\alpha + \sigma_N, \tag{5.1.22}$$

$$\sigma^X_{SF} = \tfrac{2}{3}\sigma_{NS} + \tfrac{1}{2}\sigma_M(1 + \cos^2\alpha), \tag{5.1.23}$$

$$\sigma^Z_{NSF} = \tfrac{1}{3}\sigma_{NS} + \tfrac{1}{2}\sigma_M + \sigma_N, \tag{5.1.24}$$

$$\sigma^Z_{SF} = \tfrac{2}{3}\sigma_{NS} + \tfrac{1}{2}\sigma_M. \tag{5.2.25}$$

Three-dimensional polarization analysis is a particularly powerful method for isolating weak magnetic scattering intensities, and Schärpf (1985) has exploited this at the D7 multidetector polarization analysis instrument at Grenoble. It can be seen from the above equations that either of the following combinations of the NSF and SF partial cross-sections,

$$2\sigma^Z_{\rm NSF} - \sigma^Y_{\rm NSF} - \sigma^X_{\rm NSF} = \tfrac{1}{2}\sigma_{\rm M}, \tag{5.1.26a}$$

$$\sigma^X_{\rm SF} + \sigma^Y_{\rm SF} - 2\sigma^Z_{\rm SF} = \tfrac{1}{2}\sigma_{\rm M}, \tag{5.1.26b}$$

give the magnetic scattering directly without the need to specifically introduce the angle α.

This is obviously extremely useful in a multidetector instrument like D7, in which each detector corresponds to a different $\boldsymbol{\kappa}$ value. If the partial cross-sections are only measured in two dimensions, for example σ^X and σ^Z, then α must be known in order to determine $\sigma_{\rm M}$, e.g.

$$\sigma^X_{\rm SF} - \sigma^Z_{\rm SF} = \sigma_{\rm M}\cos^2\alpha. \tag{5.1.27}$$

A special case of the use of three-dimensional polarization analysis to measure the intermediate quasielastic scattering function $S(\boldsymbol{\kappa}, t)$ in isotropic magnetic systems, developed by Mezei and Murani (1979) for studying a spin glass system, will be discussed in Section 6.1.4.

5.1.4 Multiple scattering corrections

Neutrons which undergo multiple spin-flip scattering in a sample have the effect of changing the neutron polarization incident at the spin analyser compared with the case in which the scattering is entirely single. We have seen that each scattering event changes the neutron polarization by a factor $(1-2r)$, in an analogous way to the action of a spin-flipper; it is thus straightforward to calculate the effect of multiple scattering on the flipping ratio. This correction is particularly important in samples with large spin-flip scattering cross-sections. Following Williams (1973), we may replace the single-scattering term $(1-2r)$ in equn (5.1.13) by $(1-2r_{\rm m})$ with

$$(1-2r_{\rm m}) = F_1(1-2r) + F_2(1-2r)^2 + F_3(1-2r)^3 \ldots, \tag{5.1.28}$$

where F_1, F_2 are the fractions of the detected neutrons that have been scattered to first, second, etc., order. The experiment therefore determines $r_{\rm m}$, and by independently calculating the fractions F_1, F_2, etc., the spin-flip fraction r for single-scattering can be found.

A detailed experimental evaluation of multiple scattering effects in vanadium has been reported by Freeman *et al.* (1978). Their formulation uses the inverse of equn (5.1.12) to define the flipping ratio, though in the following we shall retain our current definition. From equns (5.1.13) and (5.1.28) we obtain

$$R^{-1} = 1 - \left(\frac{2fPP_{\rm A}(1-D)(1-2r)[F_1 + F_2(1-2r)\cdots F_m(1-2r)^{m-1}]}{1 + PP_{\rm A}(1-D)(1-2r)[F_1 + F_2(1-2r)\cdots F_m(1-2r)^{m-1}]}\right), \tag{5.1.29}$$

where the multiple scattering is summed to mth order, that is

$$F_m = N_m \Big/ \sum_m N_m \tag{5.1.30}$$

and N_m is the number of analysed and detected neutrons which are scattered m times. If the scattering is weak, it is a good approximation to assume a constant value for the ratio

$$N_{m+1}/N_m = c, \tag{5.1.31}$$

and in such cases equn (5.1.29) reduces to

$$R^{-1} = 1 - \left(\frac{2fPP_A(1-D)(1-c)(1-2r)}{1-c(1-2r)+PP_A(1-D)(1-c)(1-2r)}\right). \tag{5.1.32}$$

However, for some strong scatterers, when the experimental geometry is such that the average path length for the primary scattering is different from that for higher orders, this simplification is not valid, and a two-term expansion is required. By defining

$$N_2/N_1 = b \tag{5.1.33}$$

and

$$N_{m+1}/N_m = c \quad \text{for} \quad m > 1, \tag{5.1.34}$$

we obtain,

$$N_m = N_1(1+b-c)/(1-c). \tag{5.1.35}$$

The flipping ratio becomes

$$R^{-1} = 1 - 2f\zeta/(\zeta+1), \tag{5.1.36}$$

where

$$\zeta = \frac{PP_A(1-D)(1-2r)[1-\delta(1-2r)](1-c)}{(1-\delta)[1-c(1-2r)]} \tag{5.1.37}$$

and $\delta = c - b$.

These relationships were used by Freeman *et al.* (1978) to analyse the results of flipping ratio measurements from vanadium obtained at the D5 polarization analysis instrument at Grenoble. The aim of the experiment was to compare experimentally the measured flipping ratios with those calculated using Monte-Carlo predictions of the multiple scattering fractions F_m, and it was therefore important to carry out accurate calibrations of the efficiencies of the polarizers and the spin-flipper. Vanadium was chosen as the scatterer because it has a well-characterized cross-section, contains predominantly one isotope, and has $r = 0.663$. Varying degrees of multiple scattering were achieved by using three vanadium plate arrangements (designated A, B and C) with either different thicknesses or in different geometries, and a 20 mm cylindrical

TABLE 5.1 Experimental and theoretical flipping ratios for four vanadium sample geometries (from Freeman *et al.* 1978)

Vanadium sample geometry	Experimental flipping ratio, R^{-1}	Theoretical flipping ratio, R^{-1} (calc)
A	0.576 ± 0.008	0.591 ± 0.001
B	0.68 ± 0.02	0.678 ± 0.002
C	0.69 ± 0.04	0.703 ± 0.008
D	0.624 ± 0.005	0.628 ± 0.001

sample (D). Details are provided in the original reference. It was concluded that, for three of the sample geometries, the two-term expression, equn (5.1.37), gave a closer correspondence between theory and experiment than that assuming a constant multiple-scattering ratio. The experiments demonstrated that polarization analysis can provide a stringest test of the multiple scattering in a sample for which the spin-flip scattering is large. The experimental and theoretical flipping ratios for the four measurements are shown in Table 5.1. It should be noted that the observed R^{-1} values are significantly different from 0.5, which is the ratio expected for single scattering from a purely spin incoherent scatterer and an instrument with 100% efficient components.

Later Monte Carlo simulations of polarization analysis experiments with vanadium and weaker magnetic scatterers (see Mayers and Cywinski (1985)) showed that the constant multiple-scattering fraction approximation, equn (5.1.31), is perfectly adequate for describing most experiments of practical importance. These authors investigated both slab and cylindrical specimen geometries, and confirmed the importance of both sample size and geometry in determining the numbers of multiply scattered neutrons that are detected.

5.2 POLARIZATION ANALYSIS OF NUCLEAR SCATTERING

The experiments described in this section all concern the separation of coherent and isotopic incoherent scattering from nuclear spin incoherent scattering. Examples will be given of elastic and inelastic measurements, both of which have the same neutron spin dependences. Useful specimens all contain a large spin incoherent scatterer, i.e. a nucleus with a large nuclear spin scattering parameter $\mathscr{B}$, and in this sense they resemble the nuclear scatterers that can be studied by polarized neutron diffraction. Polarization analysis does not, however, require the nucleus to be polarized and this means that the experiment is far less complex. Furthermore, the technique also provides the opportunity to investigate liquids. It is essentially impossible to generate suitably large nuclear polarizations in liquids, and hence polarized-beam diffraction is inapplicable in this case.

5.2.1 Hydrogenous and deuterated materials

Heavy water

The coherent (5.6 barn at^{-1}) and spin incoherent (2 barn at^{-1}) cross-sections for deuterium are comparable in magnitude, and this means that deuterated compounds are well suited to the polarization analysis method. One of the first experiments performed was the separation of the coherent and spin incoherent differential cross-sections in D_2O by Dore *et al.* (1976). By substituting

$$\left(\frac{d\sigma}{d\Omega}\right)_{\mathrm{NSF}} = \left(\frac{d\sigma}{d\Omega}\right)_{\mathrm{coh}} + \left(\frac{1}{3}\right)\left(\frac{d\sigma}{d\Omega}\right)_{\mathrm{NS}} \tag{5.2.1}$$

and

$$\left(\frac{d\sigma}{d\Omega}\right)_{\mathrm{SF}} = \left(\frac{2}{3}\right)\left(\frac{d\sigma}{d\Omega}\right)_{\mathrm{NS}} \tag{5.2.2}$$

into equns (5.1.1) and (5.1.2), and assuming (i) the flipper efficiency $f = 1$, and (ii) multiple scattering corrections are negligible, we obtain for the coherent and nuclear spin incoherent cross-sections

$$\left(\frac{d\sigma}{d\Omega}\right)_{\mathrm{coh}} = \frac{1}{4I}\left(\frac{3(C_{\mathrm{OFF}} - C_{\mathrm{ON}})}{P^*} + (C_{\mathrm{OFF}} + C_{\mathrm{ON}})\right) \tag{5.2.3}$$

and

$$\left(\frac{d\sigma}{d\Omega}\right)_{\mathrm{NS}} = \frac{3}{4I}\left((C_{\mathrm{OFF}} + C_{\mathrm{ON}}) - \frac{(C_{\mathrm{OFF}} - C_{\mathrm{ON}})}{P^*}\right), \tag{5.2.4}$$

where P^* is the effective polarization term $PP_{\mathrm{A}}(1 - D)$. The coherent and spin incoherent intensities measured by Dore *et al.* (1976) for liquid D_2O on the D5 instrument at Grenoble are shown in Fig. 5.2. The characteristic diffraction pattern for D_2O is observed in the coherent part, whereas the incoherent component is, as expected, relatively featureless.

Polymers

Equivalent experiments to separate the coherent and spin incoherent differential cross-sections from various forms of the polymer poly(methyl methacrylate), both partially deuterated and undeuterated, have been carried out by Schärpf (1985) at the D7 diffuse scattering polarization analysis instrument at Grenoble. Both the first and second peaks in $S(\boldsymbol{\kappa})$ were observed for the deuterated specimens, and the first peak was seen in the hydrogenous specimen. This latter result is remarkable, since the ratio of the coherent to spin incoherent cross-sections for hydrogen is only about 2.2%.

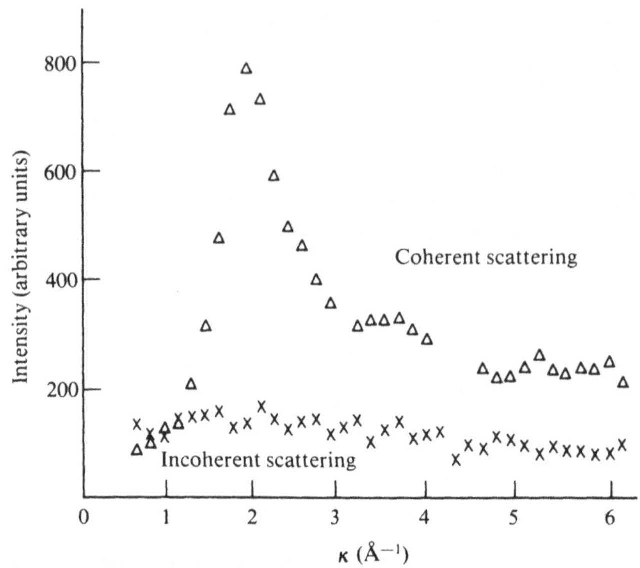

Fig. 5.2. The separation of the coherent (△) and spin incoherent (×) scattering from liquid D_2O using polarization analysis ($\lambda = 0.84$ Å) (from Dore *et al.* 1976).

5.2.2 Structure and orientational motion in C_2Cl_6

A clear separation of the coherent and spin incoherent elastic and quasielastic cross-sections in the plastic crystal C_2Cl_6 has been reported by Gerlach *et al.* (1982). The carbon atoms in the molecule are essentially pure coherent scatterers, whereas the chlorine atoms scatter both coherently ($\sigma_{coh} = 11.59$ barn at^{-1}) and incoherently ($\sigma_{inc} = 5.47$ barn at^{-1}). Approximately 70% of σ_{inc} results from a nuclear spin incoherence. Figure 5.3(a) shows the diffraction spectrum measured with $\lambda = 4.85$ Å neutrons at the D7 instrument in Grenoble over a $\boldsymbol{\kappa}$ range between 0.1 Å^{-1} and 2.2 Å^{-1}. The data points represent the spin-flip (SF) scattering, which equals two-thirds of the spin incoherent scattering, and the solid line is the non-spin-flip (NSF) scattering which contains the coherent, isotopic incoherent, and one-third of the spin incoherent scattering. For the special case of chlorine, the total incoherences in the SF and NSF scattering are roughly equal, and a subtraction of these patterns gives the coherent diffuse intensity shown (with diffraction peaks removed) in Fig. 5.3(b). These data are in agreement with measurements made with unpolarized neutrons, and can be explained as being due to the scattering from individual orientationally disordered molecules.

A study of the dynamics of this material was made by measuring time-of-flight spectra of the spin-flip scattering. The results, transferred to

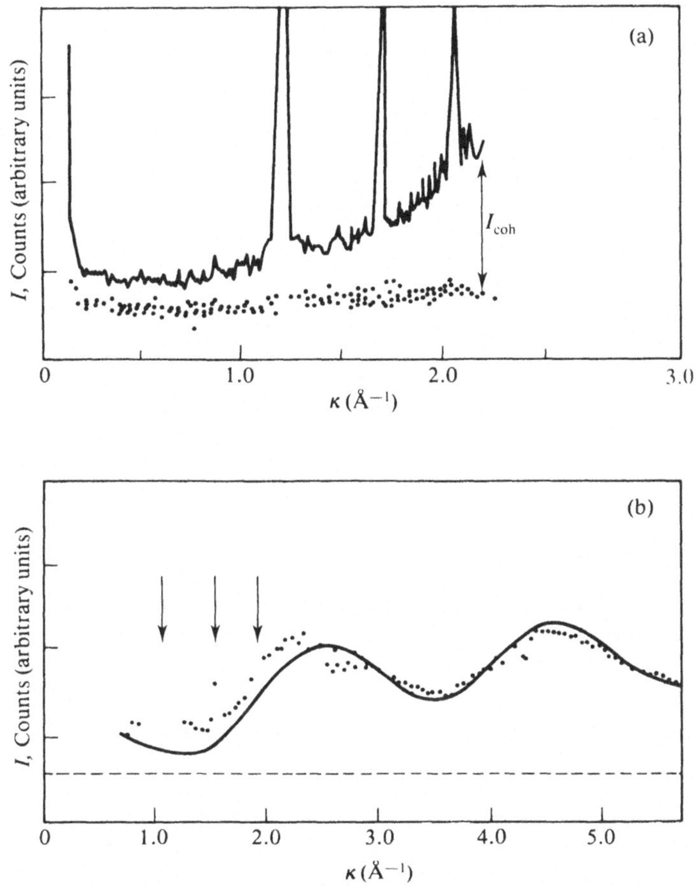

Fig. 5.3. Elastic polarization analysis measurements on C_2Cl_6. (a) The powder diffraction pattern, in which the full line is non-spin-flip scattering and the data points are the spin-flip scattering. (b) The diffuse intensity evaluated from the experiment (data points) compared with a measurement using unpolarized neutrons (full line); the arrows correspond to the **κ** values of the Bragg peaks shown in (a) (from Gerlach *et al.* 1982).

an energy scale, are shown at five **κ** values in Fig. 5.4. Since all this scattering is incoherent it was first used to evaluate the elastic incoherent structure factor (EISF), and from its **κ**-dependence it was deduced that the C_2Cl_6 molecule makes discrete jumps between a finite number of preferred orientations given by the crystal symmetry. The quality and statistical accuracy of the quasielastic scattering data were not, however, sufficiently adequate to permit the jump rates in the molecule to be deduced.

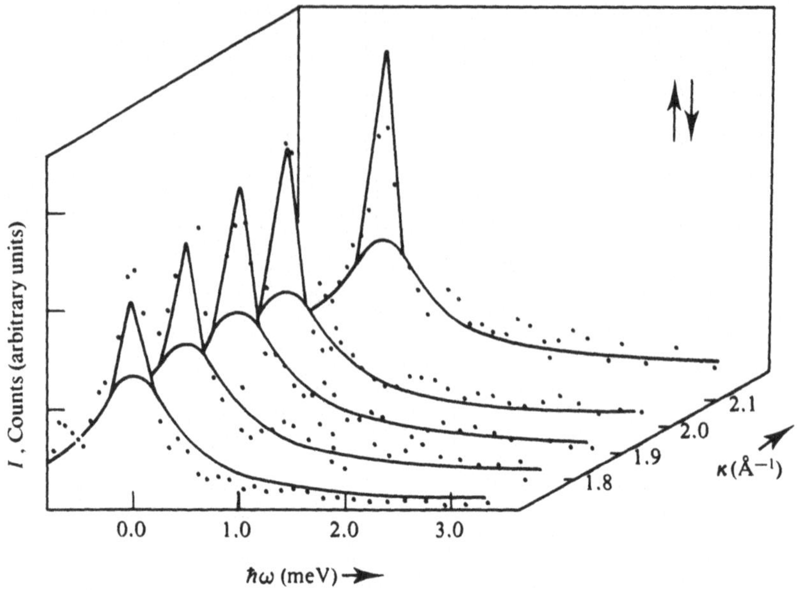

Fig. 5.4. Spin-flip scattering in C_2Cl_6 measured by time-of-flight polarization analysis and shown as a function of energy transfer and **κ** (from Gerlach *et al.* 1982).

5.2.3 Dynamic structure factors in liquid sodium

Schärpf (1985) has published results of a full polarization analysis inelastic scattering experiment on liquid sodium. Sodium is an ideal nuclear scatterer for polarization analysis measurements because the coherent and nuclear spin incoherent cross-sections are comparable: $\sigma_{coh} \sim 1.66$ barn at^{-1}, $\sigma_{NS} \sim 1.57$ barn at^{-1}. The experiment was carried out using the cold neutron polarization analysis instrument D7 at the Institut Laue–Langevin, Grenoble. The incoherent and coherent scattering laws $S_{inc}(\boldsymbol{\kappa}, \omega)$ and $S_{coh}(\boldsymbol{\kappa}, \omega)$ were successfully separated over a range of $\hbar\omega$ from zero to ~10 meV, and at **κ** values between 2 and 4 Å^{-1}. The former scattering law, see Section 2.1.1, is related to the self-correlation function $G_s(\boldsymbol{r}, t)$, whereas the latter gives the total correlation function $G(\boldsymbol{r}, t)$.

5.3 POLARIZATION ANALYSIS OF ELASTIC MAGNETIC SCATTERING

5.3.1 Bragg scattering in crystalline solids

Separation of nuclear and magnetic Bragg peaks

Polarization analysis provides a direct method of separating the nuclear and magnetic contributions in powder diffraction peaks. The method is of most value for antiferromagnetic materials, which can only be studied with unpolarized neutrons if measurements are taken both above and below the Néel point. The isothermal separation of the nuclear and magnetic Bragg peaks provided by polarization analysis evades any uncertainties that may arise owing to structural changes at or below the Néel point. The most useful geometry for these experiments is with the polarization vector parallel to the scattering vector, since (for identical polarizers) the flipper 'on' data corresponds to the spin-flip magnetic scattering and the flipper 'off' data gives the non-spin-flip nuclear scattering. Figure 5.5 shows the separation of the nuclear and magnetic

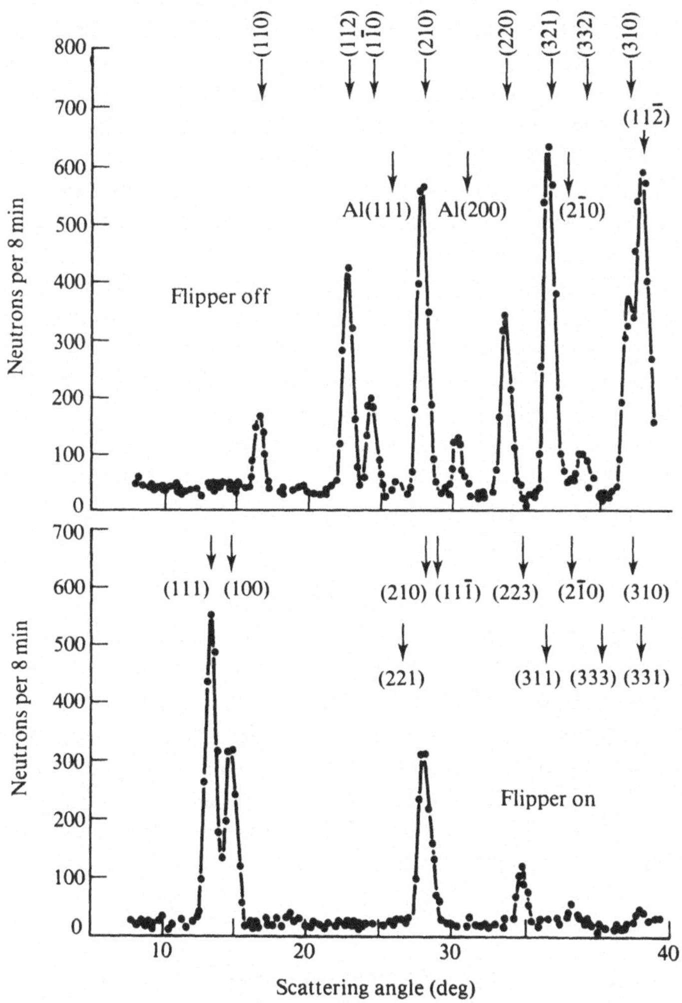

Fig. 5.5. Isothermal separation of nuclear (flipper off) and magnetic (flipper on) powder diffraction peaks from α-Fe_2O_3 by polarization analysis in the $\tilde{P}\cdot\tilde{\kappa}=1$ geometry (from Moon *et al.* 1969a).

Bragg peaks in haematite (α-Fe_2O_3) powder observed by Moon *et al.* (1969a). Aluminium peaks from the sample holder showed up in the non-spin-flip diffraction pattern, but not in the spin-flip pattern. The same procedure was used by Jacobson and Fender (1975) to separate the nuclear and magnetic powder diffraction peaks in $BiFeO_3$, which is both antiferromagnetic and ferroelectric.

Antiferromagnetism in Ti_2O_3 *and* V_2O_3

Moon *et al.* (1969b) used the polarization analysis of powder diffraction peaks to search for the occurrence of antiferromagnetism in Ti_2O_3. Previous unpolarized-beam measurements had shown that the structure of this oxide was either of the α-Fe_2O_3 type or of the Cr_2O_3 type. The room temperature polarization analysis measurements of Moon *et al.* (1969b) in the $\tilde{\boldsymbol{P}} \cdot \tilde{\boldsymbol{\kappa}} = 1$ geometry gave no evidence for either of these structures. A careful examination of the magnetic scattering in the (111), (110) and (100) reflection peaks gave what was essentially a null-result, and a maximum room temperature Ti moment of 0.03β was deduced from the spin-flip intensities.

Moon (1970) used the same technique to conclude that V_2O_3 is antiferromagnetic. The temperature dependences of the magnetic reflections in this oxide are very small and they disappear at 170 K owing to a structural change. The oxide is therefore not suited to an unpolarized-beam study, especially since the low-temperature structure is monoclinic. Moon used polarization analysis to establish that the antiferromagnetic structure consists of alternating ferromagnetic layers, and was able to assign a magnetic moment of 1.2β per vanadium atom.

Spin directions in chromium

Measurements of the flipping ratios in a polarization analysis experiment on a single crystal of chromium in an applied field of 1.5 T were used by Werner *et al.* (1969) to investigate the spin magnetization along different directions in the crystal. With the field applied in a vertical (V) direction perpendicular to the plane of scattering, the non-spin-flip (NSF) and spin-flip (SF) cross-sections are related to the magnetizations in the vertical ($M_V(\boldsymbol{\kappa})$) and longitudinal ($M_L(\boldsymbol{\kappa})$) directions (L is mutually orthogonal to V and the direction of $\boldsymbol{\kappa}$) by:

$$\sigma_{NSF} \propto |M_V(\boldsymbol{\kappa})|^2 = |M(\boldsymbol{\kappa})|^2 \overline{\cos^2 \beta} \qquad (5.3.1)$$

and

$$\sigma_{SF} \propto |M_L(\boldsymbol{\kappa})|^2 = |M(\boldsymbol{\kappa})|^2 \overline{\sin^2 \beta}\,. \qquad (5.3.2)$$

In these expressions it is assumed that the orientations of the domains fluctuate owing to thermal excitations, and β is defined as the angle between the net magnetization direction, i.e. $M(\boldsymbol{\kappa})$, and V for a given domain. The ratio $(\sigma_{SF}/\sigma_{NSF})$ therefore gives a measure of the magnetization components that are perpendicular and parallel to the neutron polarization direction. The purpose of the experiment was to determine whether the spin axis in the transversely polarized spin density wave state was confined to the cube edges since, with the $\langle 110 \rangle$ crystal axis aligned vertically, a $(\sigma_{SF}/\sigma_{NSF})$ ratio of unity would then be expected. This was found not to be the case, and the results at 298 K showed that it was almost as easy to align the spins into a $\langle 110 \rangle$ direction as into a $\langle 100 \rangle$ direction. Measurements at 200 K showed that the magnetic anisotropy increased with decreasing temperature. Observations of the temperature and magnetic field dependences of the moment direction were discussed in the context of a thermal activation model for the domains as expressed by equns (5.3.1) and (5.3.2), and a domain wall pinning mechanism where $\overline{\cos\beta}^2$ and $\overline{\sin\beta}^2$ replace $\overline{\cos^2\beta}$ and $\overline{\sin^2\beta}$ respectively in these equations. The results were shown to be more consistent with the domain wall pinning mechanism.

Ambiguous magnetic structure in neodymium

Unpolarized neutron single-crystal diffraction studies of the high-temperature (7.5 K − 19 K) phase of Nd revealed satellites at reciprocal lattice positions $(h \pm \kappa, 0, 0)$. Their origin was difficult to understand other than that were clearly related to a complexity in the magnetic structure. Two models were invoked to explain their existence: the so-called 'single **κ**' or a 'triple-**κ**' models, with **κ** defining the modulation vector. In the former, three domains are required with a periodic modulation along one of the three equivalent propagation directions in the double hexagonal close-packed structure. In the triple-**κ** model there is only one domain with periodic modulations along all three of the equivalent directions. The single-**κ** model with equally populated domains gives exactly the same scattered neutron intensities as a triple-**κ** model, and they cannot be differentiated by unpolarized neutron diffraction. This ambiguity was resolved for neodymium by Moon *et al.* (1979) using polarization analysis. The results of their measurements of the non-spin-flip (flipper 'off') and spin-flip (flipper 'on') intensities for the satellite at (7/8, 0, 0), and for two crystal orientations related by a 90° rotation about the scattering vector, are shown in Fig. 5.6. The scattering clearly changes from being completely non-spin-flip in orientation I to completely spin-flip in orientation II. The directions of the spectrometer scans are shown as the broken lines in the diagram. Moon *et al.* (1979) concluded that the satellites were purely magnetic in origin, and that they were due to modulated moments which were oriented perpendicular to

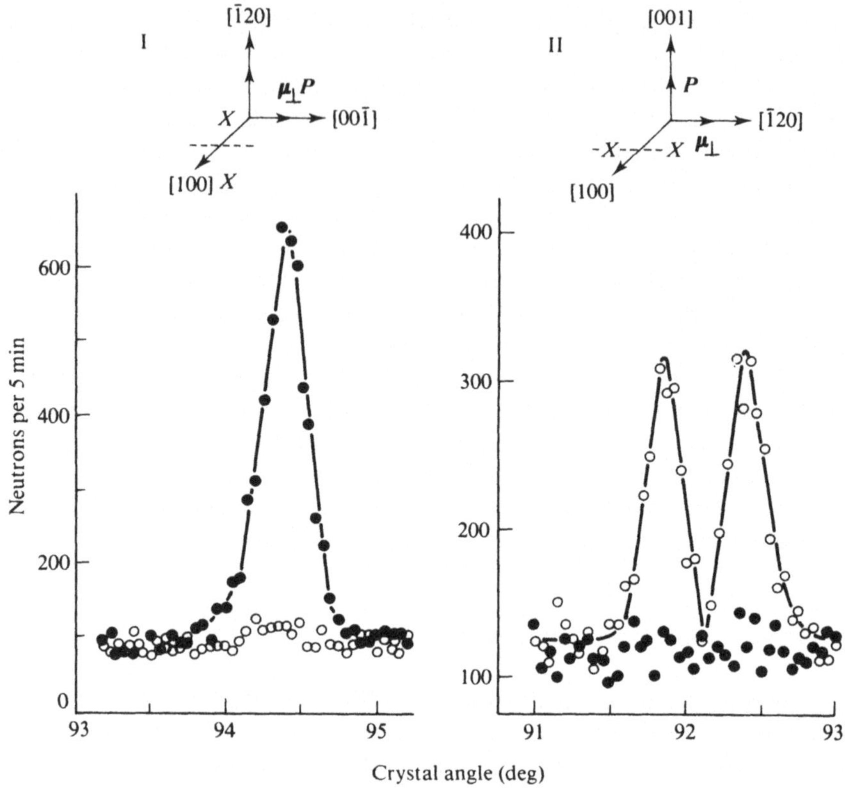

Fig. 5.6. Polarization analysis of satellite peaks from Nd at (7/8, 0, 0) measured at 9.9 K. The dashed lines in the geometry diagrams show the directions of the two spectrometer scans in reciprocal space, and the crosses show the positions of the satellites; ●, flipper off; ○, flipper on (from Moon *et al.* 1979).

both the [100] and [001] axes. This result is only possible for the 'single $\boldsymbol{\kappa}$' magnetic structure model.

The 'central' peak in praseodymium

Burke *et al.* (1981) used the polarization analysis method to examine the broad central peak in paramagnetic praseodymium which is observed below $T \sim 10$ K. This peak arises from elastic scattering centred at a wave vector $\boldsymbol{\kappa}_1 = 0.11\boldsymbol{\tau}_{110}$, though its origin is unclear. The investigation by Burke *et al.* was able to establish that a set of peaks measured at reciprocal lattice positions $(\boldsymbol{\kappa}_1, 0, 2m+1)$, with $m = 0, 1, 2, 3$, were all completely magnetic in character. Variable-temperature measurements were performed between 1.2 K and 25 K, and the results were discussed in relation to several mechanisms that have been proposed to explain this anomalous scattering.

5.3.2 Liquids and amorphous solids

Liquid oxygen

Polarization analysis was used by Deraman *et al.* (1985) to separate the nuclear and paramagnetic scattering in liquid oxygen. The experiment was performed at a liquid temperature of 82 K at the D5 polarization analysis triple-axis spectrometer at Grenoble with the wavelength set at 0.84 Å. The polarization direction was arranged to be parallel to the scattering vector, thus the non-spin-flip scattering corresponded to nuclear scattering and the spin-flip scattering gave the magnetic scattering (oxygen has essentially zero spin incoherence). The count rates observed for these scattering processes are shown as a function of scattering angle in Fig. 5.7.

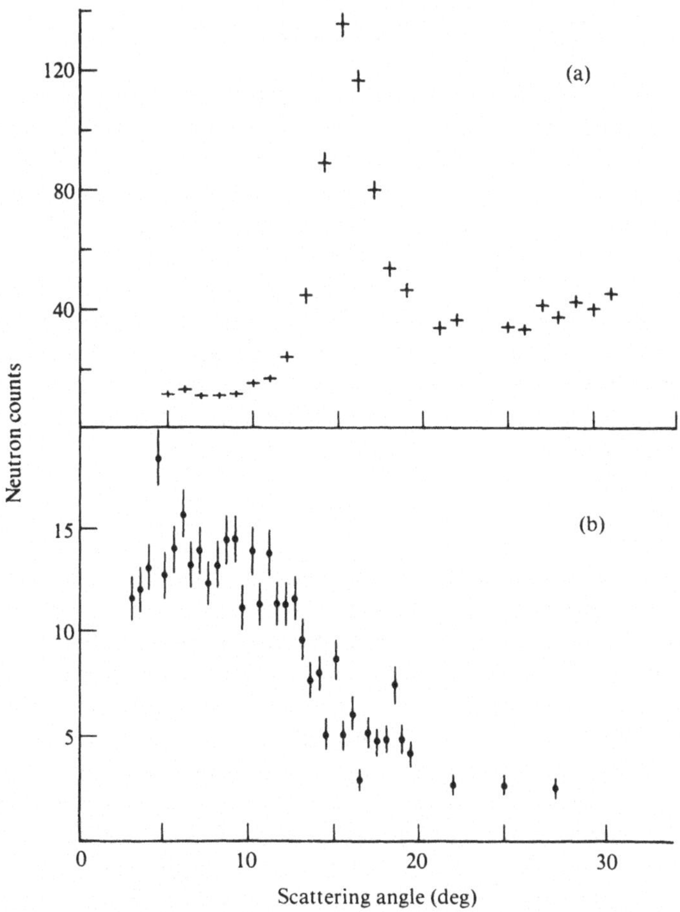

Fig. 5.7. Separation of nuclear (a) and magnetic (b) scattering in liquid oxygen by polarization analysis (from Deraman *et al.* 1985).

The interest in the experiment was mainly in the magnetic scattering, where the appropriate cross-section in the absence of any correlation between the magnetic moments is

$$\left(\frac{d\sigma}{d\Omega}\right)_M = \frac{2}{3} r_0^2 \langle F(\boldsymbol{\kappa})^2 \rangle \frac{\boldsymbol{\mu}^2}{4}, \tag{5.3.3}$$

and for oxygen, $\boldsymbol{\mu} = 2\boldsymbol{S}$, $\mu^2 = 4S(S+1) = 8\beta^2$. This averaging of the isolated molecule magnetic form factor $F(\boldsymbol{\kappa})$ is no longer applicable if there is a correlation between the moments, and a magnetic structure factor $S_M(\boldsymbol{\kappa})$ must be introduced such that

$$\left(\frac{d\sigma}{d\Omega}\right)_M = \frac{2}{3} r_0^2 \cdot \frac{1}{4} \langle F(\boldsymbol{\kappa})^2 \rangle S_M(\boldsymbol{\kappa}), \tag{5.3.4}$$

with

$$S_M(\boldsymbol{\kappa}) = 1 + \frac{4\pi}{\kappa} \int_0^\infty \rho_{mol}[g_M(r) - 1] r \sin \kappa r \, dr. \tag{5.3.5}$$

Here $g_M(r)$ is the magnetic pair correlation function and $\rho_{mol} = \rho_{nucl}/2$ is the molecular number density. A direct Fourier inversion of equn (5.3.4) gives a convolution expression containing $g_M(r)$, and the significant result in the experiment of Deraman *et al.* was that this function displayed a distinct negative correlation in the distance range 2.5–4.5 Å. The molecular centres correlation function in liquid oxygen, $g_N(r)$, has a prominent peak at 3.5 Å, and it was therefore concluded that the negative contribution in $g_M(r)$ must be a result of interactions between nearest-neighbour molecules. To summarize, this polarization analysis experiment allowed the observation of a strong short-range antiferromagnetic ordering in liquid O_2 at a tempeature which is four times the ordering temperature.

Amorphous $ErCo_2$

In common with most amorphous rare earth–transition metal alloys the Co moments in $ErCo_2$ are all parallel and coupled ferromagnetically, and the erbium moments are on average oppositely directed to the 3d moments of Co, though they do have a local anisotropy axis; such random ferrimagnetic structures are termed 'sperimagnetic'. Boucher *et al.* (1979) carried out a polarization analysis experiment on $ErCo_2$ with the object of looking for correlations between the directions of the easy axes at the Er sites. The usual $\tilde{\boldsymbol{P}} \cdot \tilde{\boldsymbol{\kappa}} = 1$ geometry was used with the sample magnetization directed parallel to $\boldsymbol{\kappa}$. Figure 5.8 shows (a) the non-spin-flip scattering, (b) the total spin-flip scattering, and (c) the magnetic spin-flip scattering observed after correcting for the Er form factor. The latter form factor correction was justified since the Er

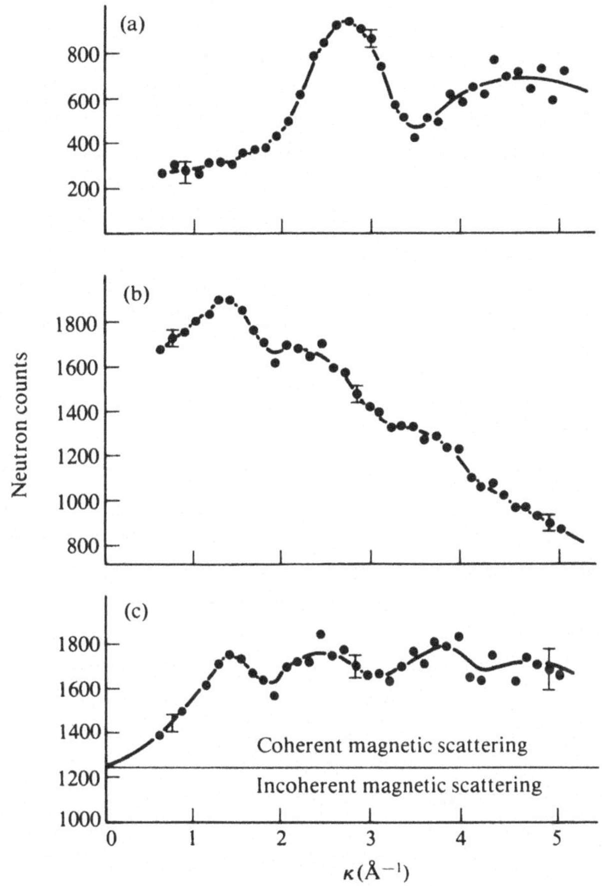

Fig. 5.8. Polarization analysis of neutron scattering from amorphous $ErCo_2$ showing (a) the non-spin-flip scattering, (b) the total spin-flip scattering, and (c) the magnetic spin-flip scattering corrected for the Er magnetic form factor (from Boucher *et al.* 1979).

magnetic moments ($\sim 9\beta$ at^{-1}) are much greater than the Co moments ($\sim 1\beta$ at^{-1}). The magnetic scattering in Fig. 5.8(c) contains an incoherent part arising from a random orientation of perpendicular magnetization components $M_\perp$, and a coherent part in which the oscillations show the neighbouring atom moments to be correlated. A Fourier inversion of the coherent magnetic scattering produced a magnetic correlation function $g_M(r)$ which led the authors to postulate a model in which the Er magnetic moments of neighbouring atoms tended to be parallel, with the coexistence of ferromagnetic and antiferromagnetic configurations. The ordering between Er first-neighbour moments can therefore be ferromagnetic, antiferromagnetic, or random, depending on the directions of the local easy-magnetization axes. This implies the existence of small subdomains in the material where the Co moment direction is fixed, but the Er magnetization changes from one subdomain to another.

Further polarization analysis experiments by Givord *et al.* (1985) on $ErCo_2$ were interpreted as revealing the presence of domains with magnetization directions which rotate with respect to each other, and this violates the cylindrical symmetry of the sperimagnet. The net magnetization due to these domains was determined to be at an angle ~18° to the average magnetization direction in the sample. This behaviour was attributed to the competition at low temperatures between the ferromagnetic 3d exchange interactions and the rare-earth ion crystalline electric field anisotropy.

5.3.3 Paramagnetic form factors

The easiest method for measuring paramagnetic form factors is to magnetize the specimen in a strong applied field, and to observe the flipping ratios for a series of Bragg reflections using the classical polarized neutron diffraction method; examples of such measurements are given in Section 4.2.2. For some samples, however, it is possible that the application of the field may modify the form factor, and in such cases polarization analysis provides a unique way of determining the zero-field form factor. Examples in which the field perturbation may be important are mixed-valence systems, in which there may be a preferential alignment of one of the ionic moments, and some metals and alloys in which the field polarizes the conduction electrons.

The first demonstration of the use of polarization analysis method for determining paramagnetic form factors was that of Moon *et al.* (1969a) on MnF_2; see Fig. 2.11. The spin-flip counts measured in the $\tilde{\boldsymbol{P}} \cdot \tilde{\boldsymbol{\kappa}} = 1$ geometry over a range of $\boldsymbol{\kappa}$ values gave, after normalization with the counts from a standard scatterer, the paramagnetic cross-section. The manganese magnetic form factor was then obtained by using equn (2.1.86):

$$\left(\frac{d\sigma}{d\Omega}\right)_{\text{para}} = \tfrac{2}{3} r_0^2 F(\boldsymbol{\kappa})^2 S(S+1), \qquad (5.3.6)$$

with $S = \frac{5}{2}$ for Mn^{2+}.

Another early paramagnetic form factor determination using exactly the same method was that by Moon *et al.* (1972) for gadolinium in Gd_2O_3. Their separation of the nuclear coherent and paramagnetic scattering for the isotopically enriched oxide powder $^{160}Gd_2O_3$ is shown in Fig. 5.9. A significant difference was observed between this oxide form factor and that observed for Gd metal, the former being larger towards higher $\boldsymbol{\kappa}$. The conclusion was that the oxide spin density was more contracted than the metallic 4f density, though the reason for this was not understood.

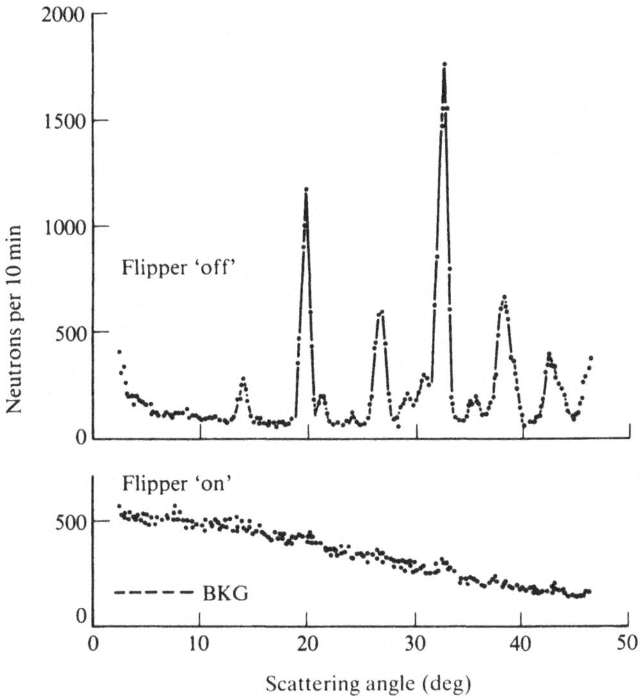

Fig. 5.9. Separation of coherent nuclear Bragg peaks from the paramagnetic diffuse scattering (flipper 'on' data) in $^{160}Gd_2O_3$ powder (from Moon *et al.* 1972).

5.3.4 Spin glasses

We now discuss some of the many 'quasielastic' polarization analysis experiments which have been carried out on Cu–Mn alloys, which are often regarded as prototype spin glasses. One of the characteristics of these materials is that below a concentration-dependent and fairly well defined transition temperature T_g, the spins are believed to freeze in random orientations without long-range order. In Cu–Mn alloys, spin glass order persists up to 72 at% Mn, with a maximum in T_g in the region of 50 at% Mn. Neutron polarization analysis of the diffuse scattering in these alloys is an ideal method for studying the static spin correlations, and such measurement have been reported by Ahmed and Hicks (1975), Davis *et al.* (1980), and Cable *et al.* (1984).

Neutron scattering from Cu–Mn alloys shows that they give a diffuse peak in the vicinity of the $(1, 0, \frac{1}{2})$ reciprocal lattice point, though early measurements reached conflicting conclusions about the nature of this peak, and of the sign of the spin pair correlation. This was resolved by the polarization analysis results of Davis *et al.* (1980), who showed that

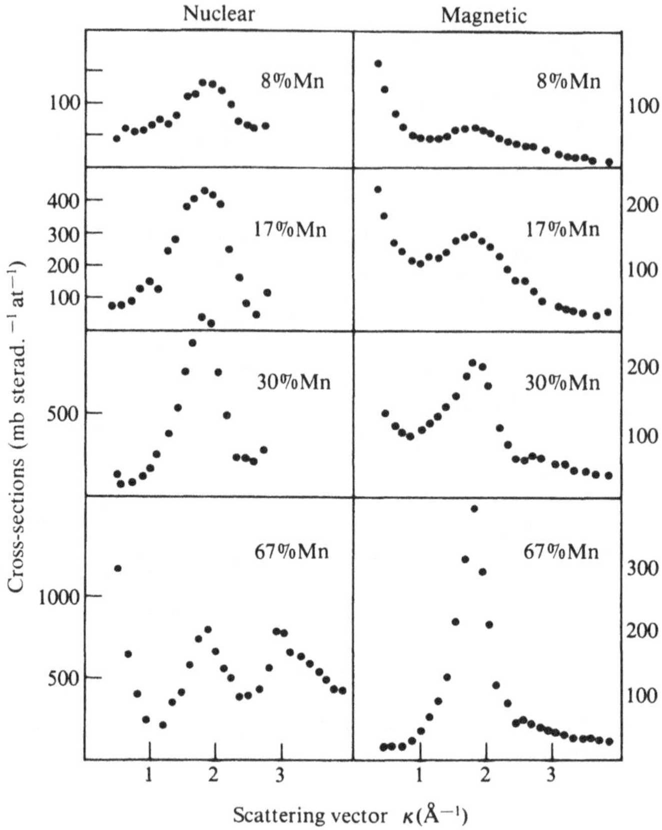

Fig. 5.10. Nuclear and magnetic scattering from Cu–Mn alloys separated by polarization analysis (from Davis *et al.* 1980).

this peak has both nuclear and magnetic components, and that the nearest-neighbour magnetic short-range parameters are negative, whereas the second-neighbour parameters are positive. The results of some of their measurements are shown in Fig. 5.10. The magnetic scattering, which was isolated by using polarization analysis in the $\tilde{\boldsymbol{P}}\cdot\tilde{\boldsymbol{\kappa}}=1$ geometry, was analysed using the expression

$$\left(\frac{d\sigma}{d\Omega}\right)_{SF}=r_0^2F_{Mn}(\boldsymbol{\kappa})^2\left(S(S+1)+\sum_i \beta_i\langle \boldsymbol{S}_0\cdot\boldsymbol{S}_i\rangle\frac{\sin \kappa R_i}{\kappa R_i}\right), \quad (5.3.7)$$

with

$$\beta_i=c+(1-c)\alpha_i, \quad (5.3.8)$$

where $\langle \boldsymbol{S}_0\cdot\boldsymbol{S}_i\rangle$ are the static spin pair correlation functions with the angular brackets denoting a thermal averaging, c is the concentration of

Mn atoms, and α_i is the Cowley (1950) short-range order parameter for the ith shell.

A further polarization analysis study of the spin structure of CuMn spin glasses was published by Cable *et al.* (1984), who made measurements on single-crystal samples. They separated the nuclear and magnetic scattering in the same way as Davis *et al.*, and their data for a 25 at% Mn alloy are presented as intensity contours in the (010) reciprocal lattice plane in Fig. 5.11. The nuclear scattering appears as broad diffuse peaks at

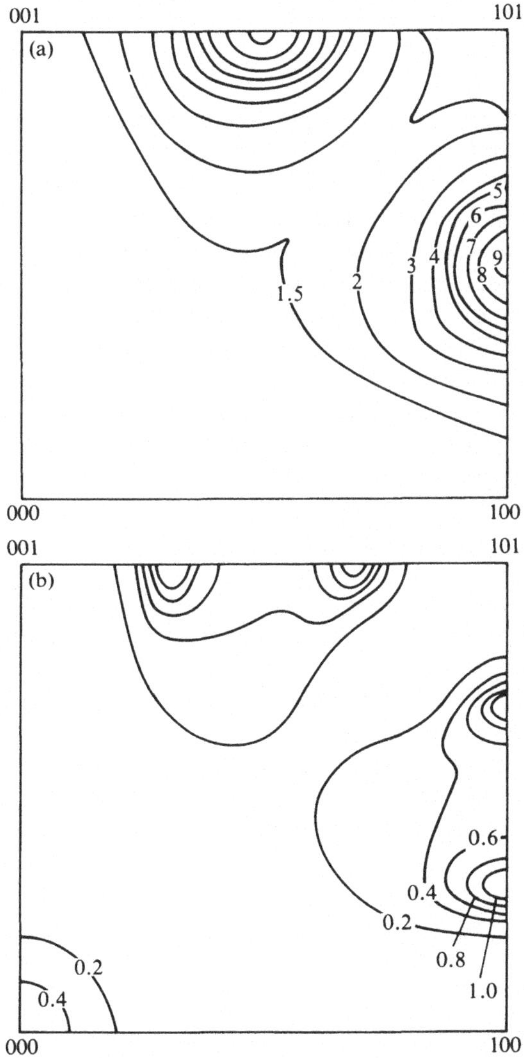

Fig. 5.11. Polarization analysis separation of (a) nuclear, and (b) magnetization densities for Cu at% Mn alloy at 10 K (from Cable *et al.* 1984).

$(1, 0, \frac{1}{2})$ and other symmetry-related positions. The magnetic scattering, see Fig. 5.11(b), has intensity spread around the $(0, 0, 0)$ and $(1, 0, \frac{1}{2})$ positions, but also has prominent and relatively sharp features at $(1, 0, \frac{1}{2} \pm \delta)$ and other symmetry-related positions. The positions of these peaks indicate that the spin correlations with which they are associated have a period which is long compared with the unit cell dimension ($a_0 \sim 3.7$ Å), and their widths suggest that this modulation is effective over appreciable distances. The period of the modulation was found to depend on the Mn concentration, being $\sim 6a_0$ for 5 at% Mn and $\sim 3a_0$ for 25 at% Mn. The results shown in Fig. 5.11 were measured at 10 K, and the peaks were observed to become broader and less intense at higher temperatures.

A Fourier analysis of the non-spin-flip or nuclear scattering data of Fig. 5.11(a) was carried out to obtain the atomic short-range order parameters. The results showed that Mn atoms prefer Cu nearest-neighbours and Mn second-neighbours. The first-neighbour interaction is antiferromagnetic while the second-neighbour interaction is ferromagnetic, therefore the atomic short-range order appears to cause the creation of ferromagnetic regions with dimensions determined by the range of this order. These small ferromagnetic regions coexist with the much larger modulated regions discussed above. Cable *et al.* (1984) concluded that the interactions between these ferromagnetic and long-modulation regions are a key element in understanding the complicated magnetic behaviour of spin glass systems.

5.3.5 Magnetic defects in antiferromagnets

Neutron polarization analysis is the only technique by which a direct unambiguous separation of the magnetic defect and nuclear disorder scattering in an antiferromagnet can be achieved; neither the field switching technique with unpolarized neutrons or the polarization reversal method with polarized beams are applicable for antiferromagnets. The magnetic defect scattering from impurity atoms A in an antiferromagnetic host B, where all the atomic moments are parallel or antiparallel to a unique direction $\tilde{\boldsymbol{\eta}}$, has a cross-section (see Rainford (1982)):

$$\left(\frac{d\sigma}{d\Omega}\right)_M = \left(\frac{r_0}{2}\right)^2 [1 - (\tilde{\boldsymbol{\kappa}} \cdot \tilde{\boldsymbol{\eta}})^2] T(\boldsymbol{\kappa}),$$

with

$$\tau(\boldsymbol{\kappa}) = S(\boldsymbol{\kappa} - \boldsymbol{\tau}) c(1-c)[F_A(\boldsymbol{\kappa})\mu_A - F_B(\boldsymbol{\kappa})\mu_B + F_B(\boldsymbol{\kappa}) G(\boldsymbol{\kappa} - \boldsymbol{\tau})]^2. \quad (5.3.9)$$

In this expression, μ_A and μ_B are the sublattice moments of species A and B, c is the concentration of the impurity atom A, and $\boldsymbol{\tau}$ is a vector such that $\exp(i\boldsymbol{\tau}\cdot\boldsymbol{R}_i) = +1$ for $\boldsymbol{R}_i$ on the 'up' sublattice or $\exp(i\boldsymbol{\tau}\cdot\boldsymbol{R}_i) = -1$ for $\boldsymbol{R}_i$ on the 'down' sublattice. It is usual to make the approximation $F_A(\boldsymbol{\kappa}) = F(\boldsymbol{\kappa})$, and to write

$$M(\boldsymbol{\kappa}) = \mu_A - \mu_B + G(\boldsymbol{\kappa}-\boldsymbol{\tau}). \qquad (5.3.10)$$

Hence,

$$\left(\frac{d\sigma}{d\Omega}\right)_M = \left(\frac{r_0}{2}\right)^2 [1-(\tilde{\boldsymbol{\kappa}}\cdot\tilde{\boldsymbol{\eta}})^2] S(\boldsymbol{\kappa}-\boldsymbol{\tau})c(1-c)F(\boldsymbol{\kappa})^2 M(\boldsymbol{\kappa})^2, \qquad (5.3.11)$$

were $M(\boldsymbol{\kappa})$ is the Fourier transform of the spatial distribution of the magnetic defect. It should be emphasised that these relationships are only strictly valid for a collinear magnetic structure.

The power of the polarization analysis method for determining the defect scattering in antiferromagnets lies in the fact that the magnetic cross-section can be measured directly as the spin-flip cross-section in the $\tilde{\boldsymbol{P}}\cdot\tilde{\boldsymbol{\kappa}} = 1$ geometry. Most of the experiments carried out so far have been by Hicks and his co-workers (see Hicks 1979a,b), using the LONGPOL polarization analysis instrument at Lucas Heights. It is unfortunate that most of the data measured by this group have been with polycrystalline samples, since spherical averaging leads to a loss of potentially useful information. This is particularly serious when the defects do not have a cubic symmetry, which appears to be the case, for example, in γ-Mn alloys. Rainford (1982) has pointed out that these alloys probably have a type-I antiferromagnetic structure with ferromagnetic (001) sheets which are antiferromagnetically coupled.

Single-crystal polarization analysis measurements of the nuclear and magnetic scattering along three crystallographic directions in γ-$Mn_{73}Si_{27}$ were reported by Moze and Hicks (1982). The nuclear disorder scattering was isolated and shown to be consistent with a short-range atomic ordering which was only significant for the first few neighbours, with the nickel atoms showing a preference towards a manganese-rich environment. A major problem with the analysis of the magnetic scattering arose because a significant spin-flip intensity was observed for $\boldsymbol{\kappa}$ parallel to the [001] direction. The geometrical factor $[1-(\tilde{\boldsymbol{\kappa}}\cdot\tilde{\boldsymbol{\eta}})^2]$ is zero in this case, and if equns (5.3.9) and (5.3.11) are valid, no magnetic scattering is expected. The most likely explanation for this anomaly is that γ-Mn does not have a simple collinear magnetic structure.

The atomic-disorder scattering and magnetic-defect scattering from an antiferromagnetically commensurate **Cr**-6.5 at% Fe polycrystalline alloy was studied by Cywinski and Hicks (1980). The object of the experiment was to evaluate the role played by iron impurities in the breakdown of antiferromagnetism in the **Cr**Fe system, and to determine the effects of iron impurities on the chromium matrix. The polarization analysis measurements yielded the total, spin-flip (SF) and non-spin-flip (NSF)

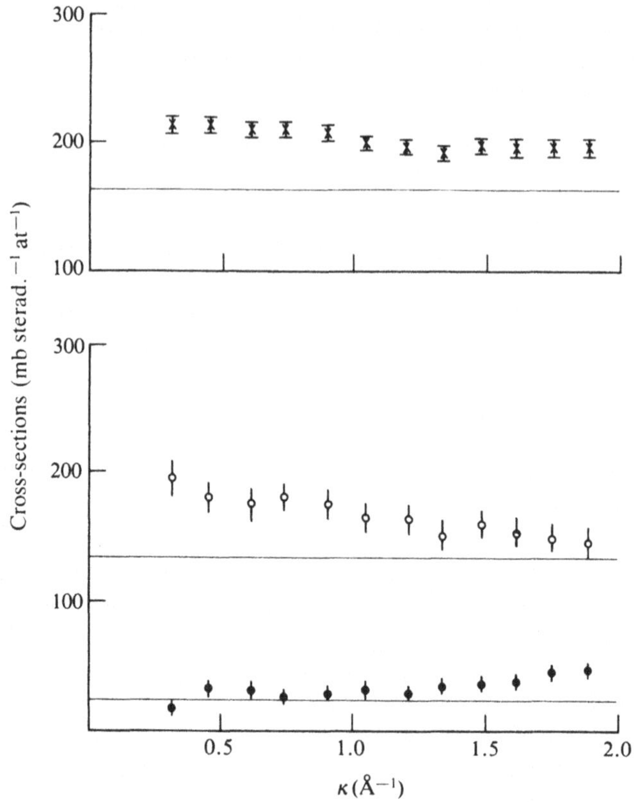

Fig. 5.12. The total (×), non-spin-flip (ϕ), and spin-flip (♦) cross-sections in Cr 6.5 at.% Fe measured by polarization analysis. The broken lines represent estimates of the total incoherent and multiple scattering from the Cr host (from Cywinski and Hicks 1980).

scattering cross-sections shown in Fig. 5.12. Careful estimates of the nuclear spin incoherent, isotopic incoherent, and multiple scattering from the Cr host were required, and these are shown as the broken lines in the figure. The nuclear disorder scattering was analysed to give a model in which the impurity Fe atoms are strongly clustered, each Fe atom having on average 16 further Fe atoms in its first-nearest-neighbour shell.

This alloy has the b.c.c. structure and therefore the magnetic scattering, for linearly superposing defects, can be described by equn (5.3.11) with an averaged orientation geometrical factor $[1-(\tilde{\boldsymbol{\kappa}}\cdot\tilde{\boldsymbol{\eta}})^2]$ equal to 2/3. The corrected SF cross-sections, together with spherically averaged values of $\overline{S(\boldsymbol{\kappa}-\boldsymbol{\tau})}$, were then used to give experimental values of $M^2(\boldsymbol{\kappa})$, and from this $(\mu_{Fe}-\mu_{Cr})=-0.6\pm0.07\beta$ was deduced. Finally, by using the mean ordered moment of the alloy, it was concluded that $\mu_{Fe}=$

$0.08 \pm 0.08\beta$ and $\mu_{Cr} = 0.68 \pm 0.08\beta$, i.e. that there is zero magnetic moment at the Fe impurity site. The Fe impurity was also found to perturb the first- and second-nearest-neighbour Cr atoms by reducing their moments to almost zero. Cywinski and Hicks (1980) suggested that, despite the observation $\mu_{Fe} \sim 0$ in their experiment, the Fe atoms do in fact possess localized moments which couple to form superparamagnetic clusters. The magnetic form factors of these clusters were assumed to be of significance only at wave vectors below those covered by the experiment, and that in fact only short-range magnetic defects associated with the immediate environment of the Fe impurity atoms were observed experimentally. The existence of a superparamagnetic behaviour in **Cr**Fe alloys is also supported by Mössbauer and small-angle neutron scattering studies.

5.3.6 Magnetic ordering in a dilute antiferromagnet

Steigenberger and Schärpf (1986) used the three-dimensional polarization analysis method on the D7 instrument at the Institut Laue–Langevin to study the formation of magnetic ordering in a single crystal of $^{114}Cd_{0.35}Mn_{0.65}Te$, which is a dilute magnetic semiconductor. The magnetic scattering from this material is extremely weak, but it was known to develop antiferromagnetic short-range order below about 30 K. By measuring the non-spin-flip (NSF) and spin-flip (SF) intensities in the three orthogonal directions X, Y and Z, equns (5.1.26) show

$$\sum |I_X + I_Y - 2I_Z|_{\text{NSF or SF}} \propto I_{\text{mag}}, \tag{5.3.12}$$

and the magnetic scattering I_{mag} can therefore be clearly isolated. We discuss the result obtained at a relatively high temperature $T = 100$ K, at which the sample was expected to be paramagnetic. The magnetic scattering should then possess the characteristics of the Mn paramagnetic form factor around $(0, 0, 0)$, and should not reveal any structure. Steigenberger and Schärpf in fact observed that the low-temperature (6 K) antiferromagnetic diffuse peaks with the same $|\boldsymbol{\kappa}|$ values merged together at the high temperature to form a ring of antiferromagnetic intensity around $(0, 0, 0)$. This intensity was independent of the orientation of the single-crystal sample, and is shown as a function of $|\boldsymbol{\kappa}|$ in Fig. 5.13. The regions between the pairs of broken lines correspond to those $|\boldsymbol{\kappa}|$ values which could not be accessed on the D7 instrument, owing to obscuration by the spin-turn coils. The arrows in the figure show the positions of the $(1\,\frac{1}{2}\,0)$ and $(1\,\frac{3}{2}\,0)$ antiferromagnetic diffraction peaks which were observed in the low-temperature phase. It is evident that, even at 100 K, the nearest- and next-nearest-neighbour interactions are still appreciable in this material.

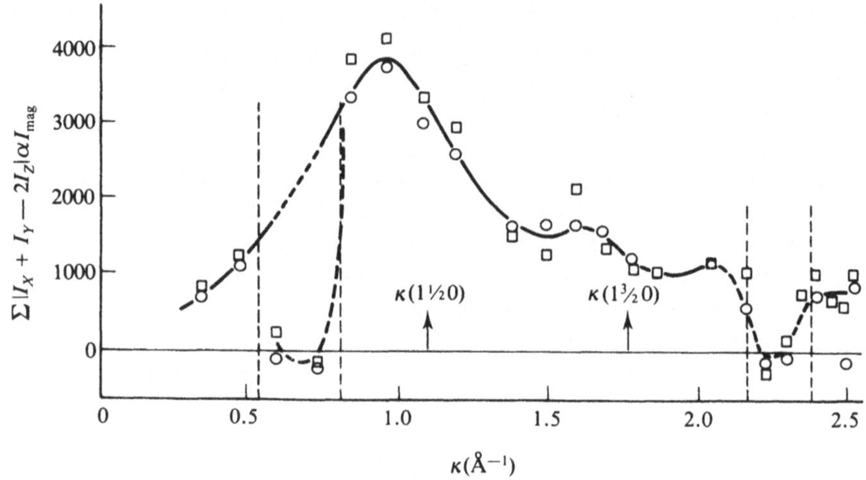

Fig. 5.13. The magnetic scattering function in $^{114}Cd_{0.35}Mn_{0.65}Te$ at 100 K determined by three-dimensional polarization analysis; □, SF scattering; ○, NSF scattering (from Steigenberger and Schärpf 1986).

5.4 POLARIZATION ANALYSIS OF INELASTIC MAGNETIC SCATTERING

In discussing the results of inelastic polarization analysis experiments it is useful to distinguish between two types of measurements: partial polarization analysis and full polarization analysis. The energy analysis in both of these experiments is done in a conventional way as with unpolarized beams, e.g. triple-axis spectrometry or time-of-flight methods. Full polarization analysis requires two instrument polarizers, as for example shown in Fig. 2.9, and the technique may be used to investigate a wide range of oriented or unoriented scatterers. Partial polarization analysis is only applicable for oriented samples which create neutron polarization on scattering. In these experiments the scatterer acts as its own spin analyser and only one instrument polarizer, which may be placed either before or after the scattering sample, is needed. The term partial polarization analysis is now in common usage, though only for inelastic measurements; it is the inelastic analogue of polarized neutron diffraction.

5.4.1 Dynamic structure factors in amorphous ferromagnets

One of the first polarized neutron inelastic experiments on an amorphous magnet was that by Mook and Tseui (1977) who used the time-of-flight method described in Section 3.3.1 to study $Fe_{75}P_{15}C_{10}$. The most remarkable result of their experiment was the observation of a minimum in the magnetic excitation energy near the first maximum in $S(\boldsymbol{\kappa})$, and these authors concluded that there was a large magnon energy gap, of magnitude ~20 meV, in this material at room temperature. The magnetic excitations in $Fe_{75}P_{15}C_{10}$ were later reexamined by Shirane *et al.* (1982), who also reported experiments on $(Fe_{93}Mo_7)_{80}B_{10}P_{10}$ and $Fe_{84}B_{16}$. These measurements were made on a triple-axis spectrometer using a pyrolytic graphite monochromator, and the Heusler alloy (111) reflection was used as a combined (−) spin and energy analyser. A horizontal magnetic field of 0.75 T was applied along the scattering vector **κ**, and a spin-flipper placed between the scatterer and the analyser allowed the following difference counts,

$$\Delta = I_{\text{OFF}} - I_{\text{ON}} \propto \left[\left(\frac{d^2\sigma}{d\Omega\, dE'}\right)_{+-} - \left(\frac{d^2\sigma}{d\Omega\, dE'}\right)_{-+}\right], \qquad (5.4.1)$$

to be determined as a function of the energy transfer $\hbar\omega$. Note that $\hbar\omega$ is defined to be positive for neutron energy loss, and corresponds to the (+ −) magnon creation cross-section. The measurements were limited to a final energy E' between 24 and 60 meV and to a maximum energy transfer of 15 meV.

The difference counts observed for the $(Fe_{93}Mo_7)_{80}B_{10}P_{10}$ sample are illustrated in Fig. 5.14, and show both energy loss and energy gain peaks at $\hbar\omega \sim \pm 2$ meV. These peak positions increased as the analyser energy was increased reaching $\hbar\omega \sim \pm 6$ meV for $E' = 60$ meV. When the sample

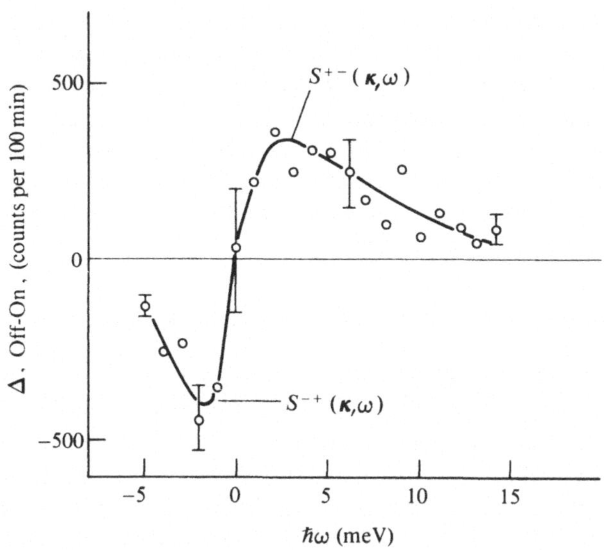

Fig. 5.14. Spin-flipper 'off' minus spin-flipper 'on' difference counts for the amorphous ferromagnet $(Fe_{93}Mo_7)_{80}B_{10}P_{10}$ observed by partial polarization analysis with a Heusler alloy analyser; $E' = 30$ meV, $\kappa = 3.05$ Å^{-1}, T = 290 K. Positive $\hbar\omega$ denotes neutron energy loss or magnon creation (from Shirane *et al.* 1982).

was studied by the polarized-beam cross-correlation method at an incident energy $E = 120\,\text{meV}$, it gave $\hbar\omega \sim \pm 11\,\text{meV}$, which was approximately equal to the instrumental resolution. It therefore became apparent that the peak separations observed in these experiments were closely related to the energy resolutions of the instruments, and it was concluded that if a magnon gap existed in $(Fe_{93}Mo_7)_{80}B_{10}P_{10}$, it must be less than 2 meV.

The results for $Fe_{75}P_{15}C_{10}$ and $Fe_{84}B_{16}$ were qualitatively the same, though the difference counts profiles were less symmetrical than that displayed in Fig. 5.14; this was shown to be a result of neutron beam depolarization within the samples. Shirane *et al.* (1982) concluded that all these measurements were consistent with a mechanism involving 'umklapp scattered' long-wavelength spin waves with no energy gap. In crystalline solids the difference between the momentum transfer of the scattering event $\hbar\boldsymbol{\kappa}$ and the momentum of the excitation $\hbar\mathbf{q}$ is taken up by the whole lattice, $\hbar(\boldsymbol{\kappa} - \boldsymbol{q}) = \hbar\boldsymbol{\tau}_{hkl}$, and Shirane *et al.* used an essentially identical model to describe the magnetic excitations in amorphous systems. When the results of their model, which predicts the form of $(S^{+-}(\boldsymbol{\kappa}, \omega) - S^{-+}(\boldsymbol{\kappa}, \omega))$ in equn (3.3.2), were convoluted with the instrumental resolution, they gave calculated curves which agreed well with the results of the partial polarization analysis experiments.

The importance of the effect of instrumental resolution was also acknowledged by Mook and Lynn (1984), and it is now clear that no experimental data exists to support the magnon gap hypothesis. An independent set of partial polarization analysis measurements on Co_4P and $Fe_{75}P_{15}Co_{10}$ by Paul *et al.* (1982) also concurs with this viewpoint. The precise nature of the excitations in amorphous magnets is still not completely resolved, since Mook and Lynn (1984) observed a clear maximum in the $S^{+-}(\boldsymbol{\kappa}, \omega)$ function of $Fe_{40}Ni_{40}P_{14}B_6$ at an energy well beyond their instrumental resolution width; this cannot be explained by the 'umklapp' scattering model.

5.4.2 Paramagnetic scattering from 3d electron systems above T_c

The paper by Mook *et al.* (1973) which announced that spin wave modes persist into the paramagnetic phase in Ni was responsible for initiating many theoretical developments in magnetism, and the observation has been explained on the basis of local band theories (see Moriya (1983)). Similar observations were made later in Fe by Lynn (1975). If modes which are associated with magnetic order are seen above T_c, then this can reasonably be interpreted as providing evidence for the existence of short-range magnetic order. At present the subject is highly controversial, both experimentally and theoretically. The early measurements were all obtained using constant-E scans on an unpolarized triple-axis spectrometer. Much more definitive experiments were performed later using polarized-beam triple-axis spectrometers.

Experiments using full polarization and energy analysis have been performed by two main groups: the Brookhaven group and the Oak Ridge group. Many of their results have been summarized in the reviews of Shirane (1984) and Mook and Lynn (1985) respectively. The success of these experiments again follows because this technique allows the weak magnetic scattering to be identified unequivocally, and this can be accomplished in one of three ways. Firstly, in the $\boldsymbol{P} \| \boldsymbol{\kappa}$ 'horizontal field' (HF) geometry, a measurement of the spin-flip scattering gives the magnetic cross-section directly, while the non-spin-flip scattering gives the nuclear cross-section. A more reliable determination of the magnetic scattering can be achieved by evaluating the difference scattering between this configuration and that observed in the $\boldsymbol{P} \perp \boldsymbol{\kappa}$ 'vertical field' (VF) geometry. In the latter, the spin-flip scattering corresponds to half the magnetic cross-section, and the non-spin-flip scattering contains the nuclear cross-section plus half the magnetic scattering cross-section. The magnetic scattering can then be isolated either by subtracting the VF counts from the HF counts in the flipper 'on' or spin-flip instrument setting, or by subtracting HF counts from the VF counts with the spin-flipper 'off'. The method favoured by the Brookhaven group is the differencing of the spin-flip scattering data (see for example Böni and Shirane (1985)). The Oak Ridge group have obtained consistent data using all three methods (see Mook and Lynn (1985)).

A similar technique, but which measures the energy integral of the dynamic scattering function $S(\boldsymbol{\kappa}, \omega)$, has been developed by the Grenoble group to determine the paramagnetic scattering in Cr, Mn, Fe, Ni and a number of intermetallic compounds at temperatures well beyond their respective T_c values (see Brown *et al.* (1984)). Their method has been discussed briefly in Section 5.1.2 and gives a paramagnetic cross-section expressed by equn (5.1.16). The essential difference between this technique and those used by the USA groups is that all the measurements are obtained with poor energy resolution using constant-ω scans at $\Delta\omega = 0$, this window being typically ~60 meV wide, and the data is collected over a range of $\boldsymbol{\kappa}$ values. Within this approximation, the measured paramagnetic scattering can be written in terms of an 'effective moment' $M(\boldsymbol{\kappa})$ such that

$$[M(\boldsymbol{\kappa})F(\boldsymbol{\kappa})]^2 = \int_0^E S(\boldsymbol{\kappa}, \omega)\, \delta E, \qquad (5.4.2)$$

where $F(\boldsymbol{\kappa})$ is the form factor of the paramagnetic carriers, and the integration is over the incident energy of the instrument, the D5 triple-axis polarization analysis spectrometer. If the energy resolution is sufficiently wide to include all the fluctuations which contribute to the susceptibility, then

$$M^2 = 12k_B T\chi(\boldsymbol{\kappa} = 0), \tag{5.4.3}$$

where $\chi(\boldsymbol{\kappa} = 0)$ is the static magnetic susceptibility.

We now present some typical results obtained by these three groups. The objective is to report their findings factually, rather than to make any critical evaluation of their significance. It will be evident, first, that some of the experimental results are inconsistent, and, secondly, that there is no consensus view of their interpretation. It is clear that further progress can only be achieved by obtaining additional and higher-quality polarization analysis data.

Nickel and iron

The magnetism of the itinerant ferromagnets Ni and Fe has received considerable attention owing to the complicated nature of their magnetic interactions. The study of the spin dynamics in these metals above T_c is well suited to the neutron polarization analysis method, and many experiments have been performed. The most significant measurements have been made at Brookhaven and Oak Ridge using both constant-ω and constant-$\boldsymbol{\kappa}$ scans on polarized triple-axis spectrometers. The appropriate scattering function $S(\boldsymbol{\kappa}, \omega)$ for an isotropic magnet takes on the double Lorentzian function (see for example Böni and Shirane (1985), or Mook and Lynn (1985)):

$$S(\boldsymbol{\kappa}, \omega) = M^2 F(\boldsymbol{\kappa})^2 \left(\frac{\kappa_1^2}{\kappa_1^2 + \kappa^2}\right) W(\boldsymbol{\kappa}, \omega) \left(\frac{\hbar\omega/k_B T}{1 - \exp(-\hbar\omega/k_B T)}\right), \tag{5.4.4}$$

where $W(\boldsymbol{\kappa}, \omega)$ is the spectral weight function which contains the information about the shape of the scattering function, and is the quantity of prime interest here. The simplest form of $W(\boldsymbol{\kappa}, \omega)$ pertains when $\hbar\omega \ll k_B T$, where spin diffusion theory is valid, and the response peaks at $\hbar\omega = 0$, i.e.

$$W(\boldsymbol{\kappa}, \omega) = \Gamma/[\hbar\pi(\Gamma^2 + \omega^2)]. \tag{5.4.5}$$

The other quantities in these expressions are the static susceptibility M as given in equn (5.4.3), the inverse correlation range κ_1, and the Lorentzian half-width at half maximum height $\hbar\Gamma$ (in energy units).

Calculated fits using these expressions have been shown to agree well with the experimental polarization analysis data on Ni above T_c for low values of κ. The diffusive model seems to be an excellent one in the low-κ region, and is supported by the constant-$\boldsymbol{\kappa}$ results of Böni and Shirane (1985), and Mook and Lynn (1985). Figure 5.15 shows the agreement between the calculated curves and polarization analysis results

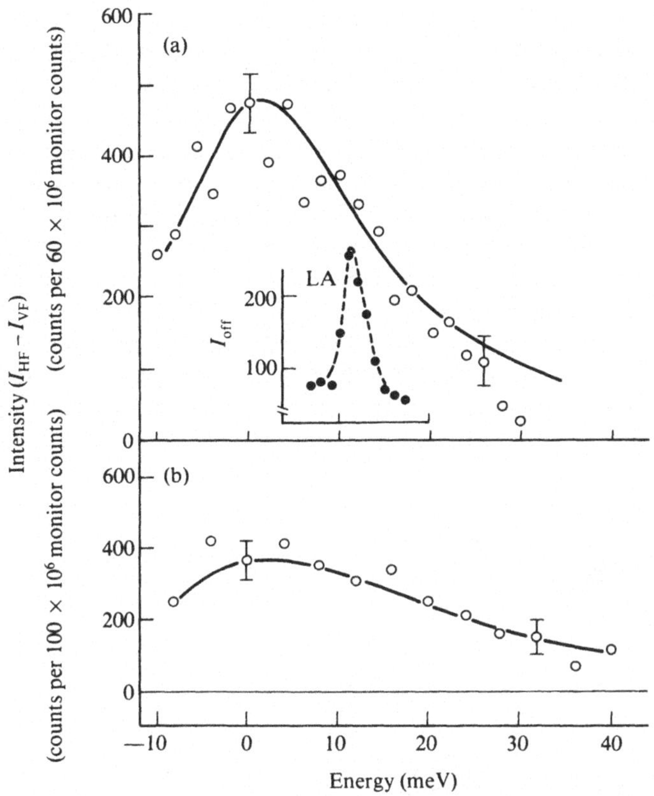

Fig. 5.15. Paramagnetic scattering from Ni at $1.06T_c$ observed by Martinez *et al.* (1985) in (a) $\zeta = 0.10$, $\kappa \sim 0.31$ Å^{-1} and (b) $\zeta = 0.12$, $\kappa \sim 0.37$ Å^{-1} constant-$\boldsymbol{\kappa}$ scans on a triple-axis polarization analysis spectrometer; full curves, calculated. The inset shows the longitudinal acoustic phonon measured with the spin-flipper 'off', and is indicative of the good resolution conditions in this experiment.

obtained by Martinez *et al.* (1985) for Ni at $1.06T_c$, and for two constant-$\boldsymbol{\kappa}$ scans.

The higher-$\boldsymbol{\kappa}$ data obtained by the Brookhaven group on both Ni and Fe above T_c were all shown to be consistent with this spin diffusion model, though a closer agreement between theory and measurements resulted from using a $W(\boldsymbol{\kappa}, \omega)$ factor

$$W(\boldsymbol{\kappa}, \omega) = (\hbar\pi)^{-1}[\Gamma/(\Gamma^2 + \omega^2)]^{\varepsilon} \tag{5.4.6}$$

rather than equn (5.4.5), with

$$\begin{aligned} \varepsilon(\omega) &= 1 && \text{for } \omega \leqslant \Gamma \\ &= 1 + \alpha[(|\omega| - \Gamma)/\Gamma] && \text{for } \omega \geqslant \Gamma. \end{aligned} \tag{5.4.7}$$

The improvement obtained by using a finite α factor is typified by the results of Wickstead *et al.* (1984) for b.c.c. Fe at $1.02T_c$. A further analysis of their Fe data led these authors to conclude that, in constant-ω scans, the peaks observed in the magnetic scattering are simply energy slices of the paramagnetic scattering function, and that they should not be interpreted as spin-wave peaks. Ni behaves in the same way (see Böni and Shirane (1985)), and the view of the Brookhaven group is that there is no experimental evidence to support the existence of propagating spin waves or giant short-range magnetic order in Fe and Ni above T_c.

The constant-$\boldsymbol{\kappa}$ measurements on Ni at $1.06T_c$ reported by Mook and Lynn (1985) are at variance with this conclusion. Although their low-$\boldsymbol{\kappa}$ results are readily explained by the above diffusive spin theory, they showed that with increasing $\boldsymbol{\kappa}$ the scattering develops structure and no longer peaks at $\hbar\omega = 0$. Figure 5.16 shows their results for three

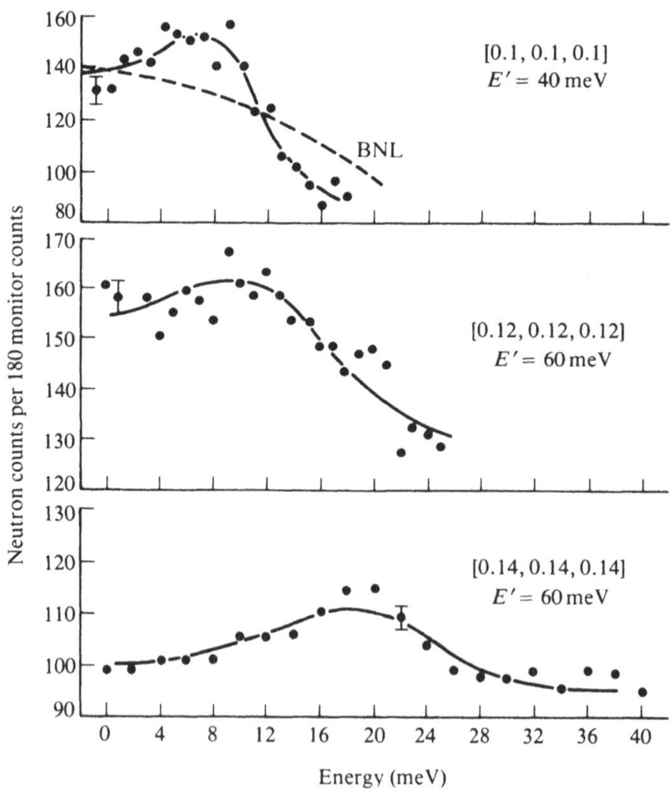

Fig. 5.16. Paramagnetic scattering in Ni at $1.06T_c$ observed by measuring the spin-flip scattering with a polarization analysis triple-axis spectrometer. The full curves were calculated using a damped harmonic oscillator form for the spectral weight function $W(\boldsymbol{\kappa}, \omega)$ (from Mook and Lynn 1985).

constant-$\boldsymbol{\kappa}$ scans, which they claim establishes the propagating nature of the scattering. The results of similar measurements on Fe were less convincing, though later constant-$\boldsymbol{\kappa}$ experiments on Ni (Lynn and Mook, 1986), also revealed a peaked magnetic cross-section.

The solid curves of Fig. 5.16 are fits to a damped harmonic oscillator model for $W(\boldsymbol{\kappa}, \omega)$ which was convoluted with an instrumental resolution function. Mook and Lynn (1985) simulated the effect of coarse resolution on this type of measurement, and suggested that the Brookhaven Ni results labelled BNL in Fig. 5.16, could be explained by this instrumental factor. We have seen, however, from Fig. 5.15 that the high-resolution polarization analysis investigation of Ni, by Martinez *et al.* (1985), does not reveal the propagating mode behaviour reported by Mook and Lynn (1985). A theoretical discussion of the implications of these experiments has been published by Korenman (1985). This author argues that constant-ω experiments are more easily interpreted, and that the measurements performed so far provide good evidence for magnetic ordering above T_c. The criterion of whether there is a peak at $\hbar\omega \neq 0$ in the dynamic correlation function at constant $\boldsymbol{\kappa}$ should not be regarded as a definitive one in deciding whether spin-wave modes exist above T_c in these transition metals.

We now discuss the Grenoble group's polarization analysis experiments of the paramagnetic scattering above T_c in Fe and Ni. Their technique ideally determines the energy integral of $S(\boldsymbol{\kappa}, \omega)$ for all energies, and the measured quantity is $(M(\boldsymbol{\kappa})F(\boldsymbol{\kappa}))^2$; see equn (5.4.2). The results for a single crystal of Fe containing 5% Si were published by Brown *et al.* (1982a). The most striking result in this series of measurements is that showing the variation of $\boldsymbol{\kappa}^2 M(\boldsymbol{\kappa})^2$ versus $\boldsymbol{\kappa}$; data taken at two temperatures above $T_c = 1000$ K are reproduced in Fig. 5.17. The plot gives a direct indication of the $\boldsymbol{\kappa}$ values which contribute substantially to the Fourier spectrum of the spin density–spin density correlation function. If there were a complete lack of magnetic order, $S(\boldsymbol{\kappa})$ would be constant, and the plot of $\boldsymbol{\kappa}^2 M(\boldsymbol{\kappa})^2$ would be expected to increase with $\boldsymbol{\kappa}$. We see that a peak is observed at $\boldsymbol{\kappa}_M \sim 0.4$ Å^{-1}, which gives a typical wavelength in the Fourier spectrum $\lambda_M = 2\pi/\boldsymbol{\kappa}_M$ of the order of 16 Å, and this result is only weakly dependent on temperature. The fact that very little scattering occurs at $\boldsymbol{\kappa} > \boldsymbol{\kappa}_M$ was interpreted as indicating nearly perfect ferromagnetic correlations up to distances $\sim\lambda_M$ in Fe. By integrating $\boldsymbol{\kappa}^2 M(\boldsymbol{\kappa})^2$ over all $\boldsymbol{\kappa}$ values out to the zone boundary, the magnetization per Fe atom was deduced to be $\sim 1.3\beta$, a value which is much smaller than that observed in the ordered phase. Measurements on a ^{60}Ni single crystal by Brown *et al.* (1983) gave qualitatively the same features, though the ferromagnetic correlations had a significantly longer characteristic length of ~20 Å.

The interpretation of these experiments has been strongly challenged

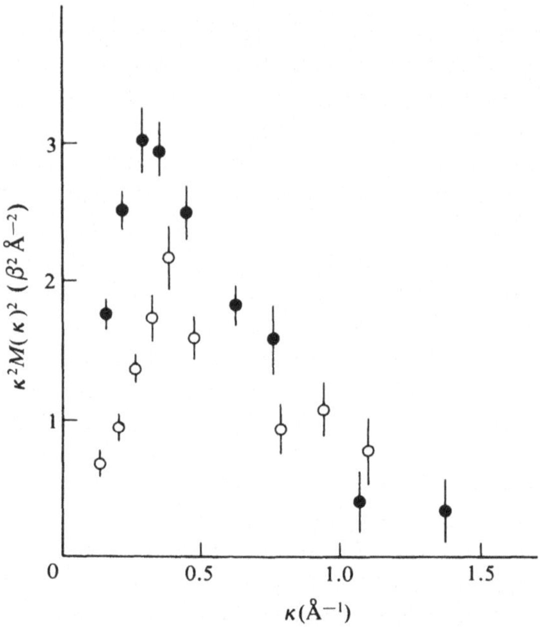

Fig. 5.17. $\boldsymbol{\kappa}^2 M(\boldsymbol{\kappa})^2$ versus $\boldsymbol{\kappa}$ for Fe 5 at% Si deduced from polarization analysis triple-axis spectrometer measurements taken in the zone centred on (110); ● 1273 K; ○ 1473 K (from Brown *et al.* 1982a).

by Edwards (1983). The main problem may be an experimental one, since it is not possible to determine the energy integral of $S(\boldsymbol{\kappa}, \omega)$ between infinite limits. The experiments could miss a substantial amount of weight, particularly at large $\boldsymbol{\kappa}$ values, and this would overestimate the degree of short-range order. However, the main conclusions following the experiments of Brown *et al.* (1982a) on Fe were later substantiated by the triple-axis polarization analysis measurements of Shirane *et al.* (1986). The agreement was seen as confirming the viability of the low-resolution quasielastic scattering method for studying the paramagnetic scattering in these systems.

MnSi

MnSi is generally regarded to be a prototype of weak itinerant systems, which are characterized as having low transition temperatures (<50 K) and small ordered moments (<0.5β). The paramagnetic spin fluctuations were first investigated by high-resolution triple-axis measurements, and the results of Ishikawa *et al.* (1982) showed these to have strong temperature and wave vector dependences with the Lorentzian-type energy spectrum $\Gamma/(\Gamma^2 + \omega^2)$ as given in equn (5.4.5); these are known as

Moriya–Kawabata fluctuations. Polarization analysis experiments were later carried out by Ishikawa *et al.* (1985), and the magnetic scattering was placed on an absolute scale by simultaneously measuring the phonon cross-sections. The magnetic excitation spectra, obtained over a wide energy range from 0 to 20 meV ($\hbar\omega/k_B T_c = 7.9$), and up to temperatures $\sim 10T_c$, were shown to be consistent with the Moriya–Kawabata model. On the basis of these results, Ishikawa *et al.* (1985) attributed the origin of the increase in the amplitude of the local magnetic moment $\langle M_L \rangle^2$, which occurs with increasing temperature above T_c in the system, to a thermal excitation of the high-energy spin fluctuations in the Stoner continuum.

A different conclusion was reached by Ziebeck *et al.* (1982) using the energy integral polarization analysis method. These authors found the magnetic correlation to be confined to the low-$\boldsymbol{\kappa}$ region even at 580 K, which is $\sim 20T_c$. Also the paramagnetic moment $M(\boldsymbol{\kappa})^2$, calculated from the diffuse scattering, was found to be appreciable ($\sim 0.83\beta$ at 11 K) in comparison with the spontaneous ferromagnetic moment of 0.4β. Their results for the $M(\boldsymbol{\kappa})^2\boldsymbol{\kappa}^2$ versus $\boldsymbol{\kappa}$ plot obtained at two temperatures are reproduced in Fig. 5.18, together with a calculated curve for the variation expected for an ideal paramagnetic having a localized moment of 0.4β. The main difference between the two experimental sets of results occurs at $\boldsymbol{\kappa} \sim 0.91\,\text{Å}^{-1}$. Most notable is the peak at $\boldsymbol{\kappa} = 0.5\,\text{Å}^{-1}$ in the 11 K measurements; this corresponds to a wavelength ~ 12.6 Å, and this was interpreted to be the spatial dimension of an exchange hole in a heavy

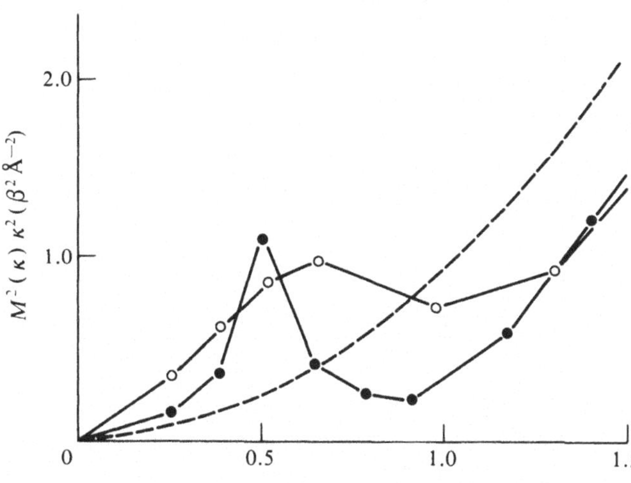

Fig. 5.18. $M(\boldsymbol{\kappa})^2\boldsymbol{\kappa}^2$ as a function of $\boldsymbol{\kappa}$ observed for MnSi at two temperatures: ●, 11 K; ○, 580 K. The broken curve corresponds to the behaviour predicted for an ideal paramagnet having a localized moment of 0.4β (from Ziebeck *et al.* 1982).

Fermi liquid. This peak becomes broader and shifts towards larger **κ** in the paramagnetic phase, but it clearly does not disappear.

The Heusler alloy Pd_2MnSn

Experiments analogous to those discussed above for itinerant magnets have also been performed on the localized metallic ferromagnet Pd_2MnSn, mainly with the aim of comparing their different properties. The excitation energies in these systems, up to $\hbar\omega \sim 50$ meV, are much lower than in Fe and Ni, and it is therefore possible to measure the whole magnetic scattering function $S(\boldsymbol{\kappa}, \omega)$ with thermal neutrons. The full polarization analysis triple-axis measurements of the Brookhaven group have been published by Shirane *et al.* (1985). Figure 5.19 shows the spin wave dispersion curves observed for Pd_2MnSn below T_c (190 K), and the 'constant-E ridges' observed for constant-ω triple-axis scans above T_c.

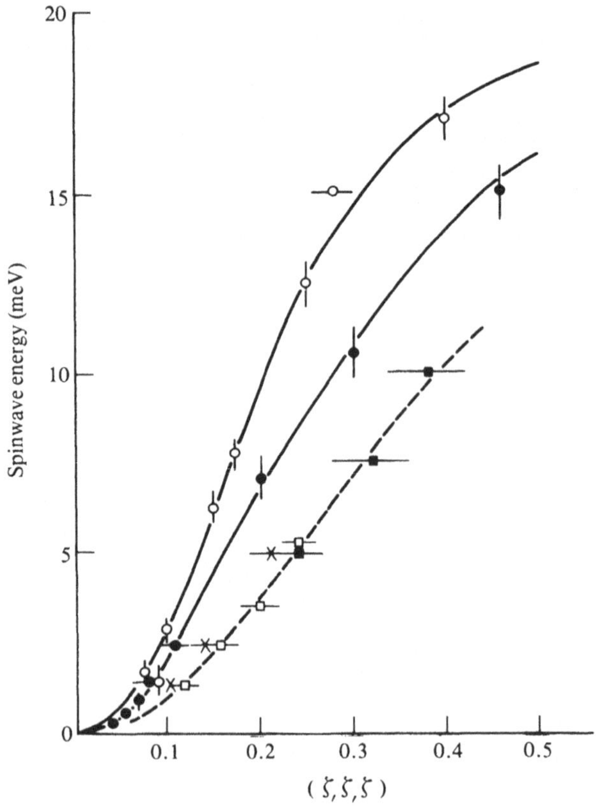

Fig. 5.19. Spin waves below T_c (○, 100 K; ●, 150 K) and 'constant-E' ridges above T_c (×, 193 K; □, 210 K; ■, 295 K) observed by polarization analysis of the scattering from Pd_2MnSn (from Shirane *et al.* 1985).

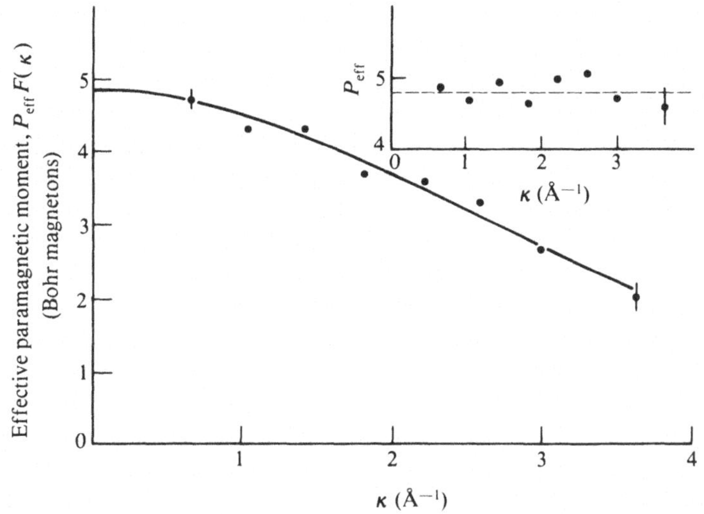

Fig. 5.20. The wave vector dependence of the paramagnetic scattering in Pd_2MnSn at $4T_c$. The solid curve is the effective moment calculated using $P_{eff} = g(S(S+1))^{1/2} = 4.81\beta$ at^{-1} and a Hartree–Fock Mn^{2+} form factor $F(\boldsymbol{\kappa})$. The inset shows that the effective moment is wave vector independent (from Ziebeck *et al.* 1981).

The scattering function above T_c can be described by a double Lorentzian function, similar to equn (5.4.4), containing a temperature-dependent 'normal' short-range order parameter κ_1. The Brookhaven group conjectured that the double Lorentzian scattering law is valid for all cubic ferromagnets above T_c, and that the relevant parameters which explain detailed differences are the value of T_c and the respective diffusion constants.

The experiments of the Grenoble group on Pd_2MnSn above T_c are described by Ziebeck *et al.* (1981). Figure 5.20 shows the $\boldsymbol{\kappa}$-dependence of the observed effective moment, including the magnetic form factor $F(\boldsymbol{\kappa})$, and the results agree well with the calculated curve using $P_{eff} = g[S(S+1)]^{1/2} = 4.81\beta$ per Mn atom, and a Hartree–Fock Mn^{2+} form factor. The inset shows the results divided by $F(\boldsymbol{\kappa})$, from which it was concluded that there are no long-range spatial correlations present above T_c in Pd_2MnSn.

Amorphous $Fe_{76}Cr_4P_8C_{12}$

The first observation of magnetic correlations in the paramagnetic phase of an amorphous magnetic alloy, $Fe_{76}Cr_4P_8C_{12}$, was reported by Mangin *et al.* (1986). Full polarization analysis was used on the D5 spectrometer at the Institut Laue–Langevin to isolate the magnetic scattering within an energy window ±20 meV centred at zero energy. The measurements were made at 553 K ($T/T_c = 1.11$) over a $\boldsymbol{\kappa}$ range between 0.5 and 3.6 Å^{-1}. The most striking feature of the results was that a strong magnetic intensity was observed both in the forward direction, and also at a $\boldsymbol{\kappa}$ value corresponding to the first maximum in the nuclear static structure factor $S(\boldsymbol{\kappa})$. This is strongly indicative of the continued existence of strong magnetic correlations above T_c in the paramagnetic phase; analogous observations had previously been made around the Bragg peaks in crystalline iron (4% Si); see Brown *et al.* (1982a). Analysis of the results showed that the spin–spin correlation function in the amorphous alloy was best described by the exponential form $\exp(-r/r_0)$ with $r_0 \simeq 4.5$ Å. This contrasts with the result in the crystalline alloy, where the correlation function is approximately Gaussian, i.e. $\exp[-(r/r_0)^2]$, though the correlation range r_0 was found to be approximately the same in both cases.

5.4.3 Spin wave scattering in ferromagnetic single crystals

We now give two illustrations of the use of partial polarization analysis to separate coherent magnetic excitations from phonon scattering. The scattering specimens must be magnetized and, since polarization is created in the scattering process, only one polarizing element is required in the neutron spectrometer.

Pd_3Fe

Partial polarization analysis was used by Holden and Stirling (1977) to provide an unambiguous separation of the broad magnetic scattering in the ordered alloy Pd_3Fe from the phonon scattering. A Heusler alloy monochromator on the first axis of the D5 triple-axis spectrometer at Grenoble provided a (−) spin incident beam, and the energy analysis on the third axis was performed using the (111) reflection planes of a Cu monochromator. For neutron energy loss, or magnon creation, we see from equn (2.3.53) that the neutron polarization dependence of the cross-section is given by

$$\left(\frac{d^2\sigma}{d\Omega\, dE'}\right) \propto [1 + (\tilde{\boldsymbol{\kappa}} \cdot \tilde{\boldsymbol{M}})^2 + 2(\boldsymbol{P} \cdot \tilde{\boldsymbol{\kappa}})(\tilde{\boldsymbol{\kappa}} \cdot \tilde{\boldsymbol{M}})]; \qquad (5.4.8)$$

hence, in the experiment under discussion, and with $\tilde{\boldsymbol{\kappa}} \cdot \tilde{\boldsymbol{M}} = 1$, this factor can be switched from zero (spin-flipper 'off') to 4 (spin-flipper 'on'). In practice this switching factor was less than 4 mainly because it was not possible to arrange for $\tilde{\boldsymbol{M}}$ to be exactly parallel to $\tilde{\boldsymbol{\kappa}}$, and a geometric correction factor was applied to the data.

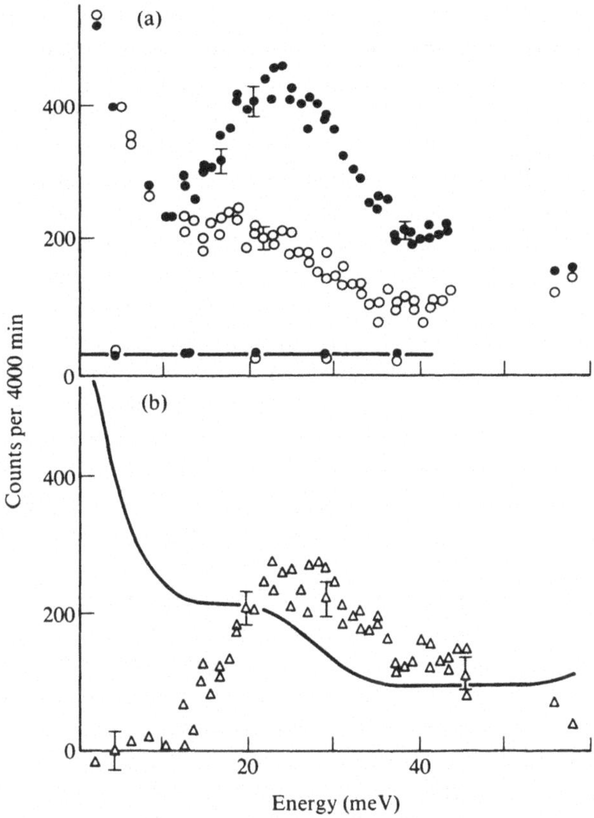

Fig. 5.21 Partial polarization analysis separation of magnon and phonon scattering in Pd_3Fe at 300 K by Holden and Stirling (1977). The solid and open circles are flipper 'on' and flipper 'off' data respectively, and the isolated magnetic ($\triangle$) and nuclear (full curve) scattered intensities are obtained by differencing.

The measurements of the [100] zone boundary magnon in Pd_3Fe at 300 K are shown in Fig. 5.21. The solid and open circles of Fig. 5.21(a) show the data collected with the spin flipper 'on' $((+)\rightarrow(-)$ magnon scattering) and 'off' $((-)\rightarrow(-)$ phonon scattering) respectively. After correcting and differencing this data, the broad magnetic scattering (shown as triangles) and the nuclear scattering (shown as the solid line) were separated; see Fig. 5.21(b). Holden and Stirling attributed the broad magnetic response observed in their experiment to the interaction between the collective excitations and the Stoner modes.

Nickel

Lowde *et al.* (1983) used the partial polarization analysis method to investigate deviations from Stoner-theory behaviour in nickel. In this theory of ferromagnetism the spin-up and spin-down electrons occupy two bands whose energy splitting diminishes with temperature. The overall moment is simply a function of the populations of the two levels. Stoner and non-Stoner behaviour can be distinguished by partial polarization analysis in a saturated ferromagnet by observing the spin-wave creation cross-sections. Let us consider a neutron energy loss experiment in which the neutron polarization, scattering vector and sample magnetization directions are all parallel. In the ideal Stoner magnet we may assign the net spin direction at all sites as always 'down', and these can only be flipped by (+) spin neutrons which flip in the converse sense. The 'Stoner' cross-section in the polarization analysis experiment is therefore $(\mathrm{d}^2\sigma/\mathrm{d}\Omega\,\mathrm{d}E')_{+-}$, and this is associated with a χ^{-+} magnetic susceptibility (the signs on χ refer to the electron spin magnetization). This picture is of course far too simplistic, and in practice finite $(\mathrm{d}^2\sigma/\mathrm{d}\Omega\,\mathrm{d}E')_{-+}$ cross-sections are also observed in Ni as the temperature is increased, until they become equal at T_c and above. The object of the measurements of Lowde *et al.* (1983) was to observe how the two cross-sections came together as T_c was approached from below.

The experiments were conducted at Oak Ridge with a ^{57}FeSi polarizing monochromator, and the down-scattered neutrons were energy-analysed using the (002) planes of a pyrolytic graphite crystal. Corrections for non-parallelism of $\boldsymbol{P}$ and $\boldsymbol{M}$ were insignificant since the maximum deviation angle was only 9.9°. The spin-wave scattering was always analysed by performing constant-ω scans through the $(2, 0, 0)$ point in reciprocal space, and this produced both κ and $-\kappa$ peaks. Figure 5.22 shows a set of pure magnon creation scattering data at $\hbar\omega = 16$ meV from ^{60}Ni in an applied field of 0.85T. Both background and phonon intensities have been subtracted, and a correction has been made for the incomplete polarization of the beam; the full lines are 'best fit' theoretical curves. The upper curves correspond to the 'Stoner' $(+-)$ neutron spin scattering, and the lower curves to the so-called 'forbidden' $(-+)$ scattering. The Curie temperature of the Ni single crystal, which contained a 0.8% molybdenum impurity, was $T_c = 584$ K.

The outstanding feature of the results of several constant-ω scans was the onset, with rising temperature, of the 'forbidden' magnon intensities. The integrated 'forbidden' intensity in the $\hbar\omega = 16$ meV measurements is shown as a fraction of the total scattering in Fig. 5.23. It is clear that the non-Stoner intensity begins to increase dramatically at about $0.9T_c$ and, even in the presence of a 0.85 T applied field, it reaches 48% of its final value at T_c. It is also probably not coincidental that it is at temperatures between $0.8T_c$ and $1.2T_c$ that Lynn and Mook (1981) observed large linewidth increases in the magnon peaks from Ni.

The experiments established that Stoner theory was able to give a satisfactory explanation for the spin wave behaviour in Ni up to about

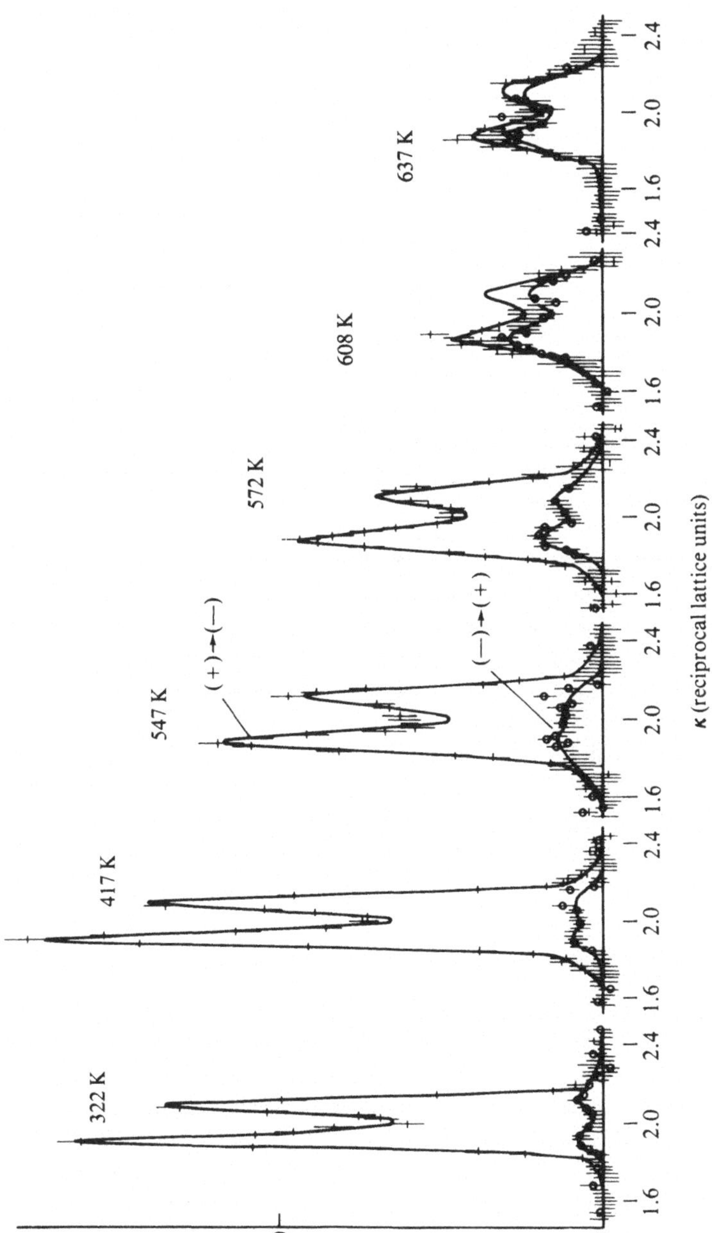

Fig. 5.22. Partial polarization analysis constant-ω scans at $\hbar\omega = 16$ meV for ^{60}Ni in an applied field of 0.85 T. The upper curves represent the 'Stoner' intensities and the lower curves give the 'non-Stoner' intensities (from Lowde *et al.* 1983).

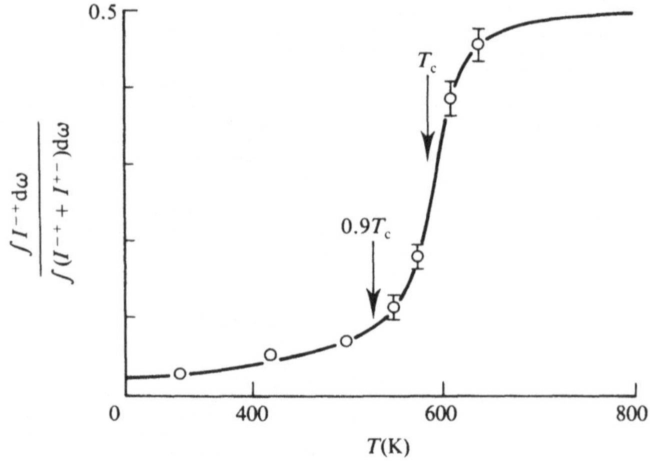

Fig. 5.23. The ω-integrated 'non-Stoner' spin wave intensity at a κ corresponding to 16 meV energy transfer expressed as a fraction of the total spin wave intensity at the appropriate κ (from Lowde *et al.* 1983).

half the Curie temperature. The observations above $\sim 0.5T_c$ can be accounted for by invoking the idea that regions of deviant magnetization direction are produced as the temperature is increased. Neutron spin wave scattering from this deviant magnetism becomes appreciable over the same temperature range as that at which large line-broadening effects are observed in the spin wave spectra. It is therefore tempting to suggest a physical picture in which large regions of spin distribution rotate away from the Z-direction at these temperatures. Lowde *et al.* discussed their results in the context of three different models which contained some form of spin deviation, but were not able to reach any definitive conclusions.

5.4.4 Spin dynamics in a one-dimensional ferromagnet

The in-plane dynamics of one-dimensional spin systems can be modelled by the classical sine-Gordon equation which allows both linear (spin wave) and non-linear (soliton) solutions. The latter give a zero-frequency peak in the dynamic structure factor, and unpolarized neutron experiments, for example those of Steiner *et al.* (1982), have provided data which were ascribed to soliton scattering. $CsNiF_3$ is a classical one-dimensional ferromagnet and Kakurai *et al.* (1984) carried out inelastic neutron polarization experiments to study its low-temperature spin dynamics. The aim of their experiment was to analyse the neutron polarization created by magnon scattering, and to examine the polarization dependence of the quasielastic scattering which had earlier been attributed to solitons.

The measurements were carried out on the IN12 triple-axis spectrometer at Grenoble. We restrict ourselves here to a discussion of the full polarization analysis results obtained with a (+) spin incident beam provided by a supermirror bender/pyrolytic graphite monochromator combination, and for which the scattered beam was analysed by a Heusler polarizing crystal in front of which was placed a Mezei flat coil spin-flipper. The arrangement allows the $(d\sigma/d\Omega\, dE')_{+-}$ and $(d\sigma/d\Omega\, dE')_{++}$ cross-sections to be observed. The $Cs^{58}NiF_3$ sample was

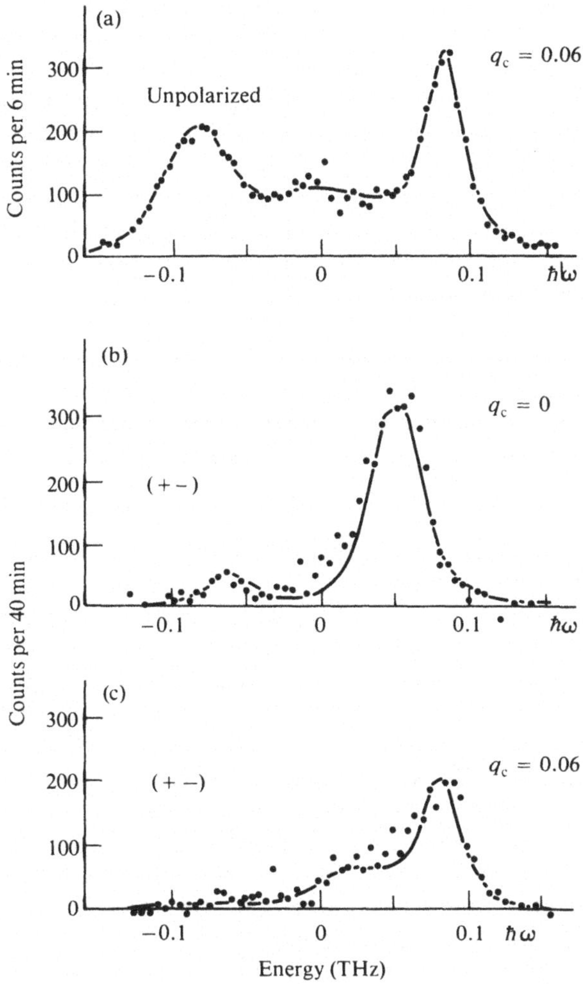

Fig. 5.24. Magnetic inelastic spectra for $CsNiF_3$ at 12 K, where the q_c values are in reciprocal lattice units. (a) An unpolarized triple-axis experiment; (b) and (c) full polarization analysis experiments with a (+) spin polarizer and a (−) spin analyser (from Kakurai *et al.* 1984).

oriented with the a^* and c^* reciprocal axes in the scattering plane, and a 1 T magnetic field was applied along a^* which is perpendicular to the chain direction c^*. We have seen that when an isotropic Heisenberg ferromagnet is subjected to a saturating magnetic field, a peak corresponding to magnon creation can only appear as $(+\,-)$ scattering.

Figure 5.24 shows the inelastic spectra observed by Kakurai *et al.* at a temperature of 12 K. The backgrounds and the zero-frequency (incoherent) peak measured at low temperature have been subtracted in each case. The spectra are therefore due entirely to magnetic scattering resulting from spin fluctuations which are perpendicular to the applied field or a^* direction, and these can only give spin-flip scattering. The $\hbar\omega = 0$ peak in Fig. 5.24(a), obtained using an unpolarized triple-axis spectrometer, has been interpreted as being due to soliton scattering; it reaches a maximum at $q_c \sim 0.06$ (in reciprocal lattice units) but is zero at $q_c = 0$. This latter result is confirmed by the $(+\,-)$ spin-flip scattering spectrum of Fig. 5.24(b), measured by full polarization analysis at $q_c = 0$, which can be described by a single resolution-broadened Lorentzian peak with positive $\hbar\omega$, and a weak negative $\hbar\omega$ peak. The $q_c = 0.06$ $(+\,-)$ spin-flip scattering shown in Fig. 5.24(c) appears only at positive $\hbar\omega$ values.

The peak at negative $\hbar\omega$ in Fig. 5.24(b) can easily be explained by invoking an anisotropic rather than an isotropic spin Hamiltonian for $CsNiF_3$. The neutron polarization induced by single magnon scattering in an anisotropic Heisenberg system may be expressed by

$$P(\boldsymbol{\kappa}, \omega) = [\sigma_{+-}(\boldsymbol{\kappa}, \omega) - \sigma_{-+}(\boldsymbol{\kappa}, \omega)]/[\sigma_{+-}(\boldsymbol{\kappa}, \omega) + \sigma_{-+}(\boldsymbol{\kappa}, \omega)], \quad (5.4.9)$$

where $(\mathrm{d}^2\sigma/\mathrm{d}\Omega\,\mathrm{d}E')$ has been abbreviated to σ. This function can be evaluated theoretically (see Hood (1983)) for various types of anisotropy. The full curves in Figs 5.24(b) and (c) were calculated using the data of the unpolarized neutron triple-axis experiment together with P values obtained using Hood's theory. The agreement between experiment and calculation provides strong evidence that the P values deduced for magnon scattering also pertain for the quasielastic 'soliton' scattering. Kakurai *et al.* (1984) also argued that if the $\hbar\omega = 0$ scattering in Fig. 5.24(a) was due exclusively to 'in-plane' fluctuations of the soliton, then the central peak shape would be the same for both the polarized and unpolarized beam measurements. It was therefore concluded from the data of Fig. 5.24(c) that the soliton scattering from $CsNiF_3$ at 12 K involves some 'out-of-plane' fluctuations.

6

Precessing neutron polarization

It was shown in Section 1.4 that the neutron polarization vector $\boldsymbol{P}$ rotates about its local magnetic field $\boldsymbol{B}$ at the Larmor frequency $|\gamma|\, B = 2\,|\mu_n B|/\hbar$. If the neutron experiences this field for a time t, then it precesses through an angle $\phi_L = |\gamma|\, Bt$, and a determination of ϕ_L can be used to provide information on either $\boldsymbol{B}$ or t. This chapter describes three applications of precessing neutron polarization. In the first, neutron spin echo spectroscopy, a careful control of $\boldsymbol{B}$ provides an accurate method for determining t, or more precisely changes in the neutron velocity v. The second type of experiment, which involves measuring the neutron depolarization in either one or three dimensions as the beam traverses a ferromagnetic specimen, gives magnetic domain structural information on a spatial dimension of approximately 1 μm. The third application measures the local field experienced in samples containing polarized nuclei, and these experiments provide a very accurate way of determining the difference between the neutron scattering amplitudes b_+ and b_- for the $(i \pm \frac{1}{2})$ spin compound nuclei; see Section 2.1.1.

6.1 NEUTRON SPIN ECHO SPECTROSCOPY

The term 'spin echo' was first coined by Hahn (1950) in relation to various types of transient responses observed in radiofrequency (r.f.) nuclear magnetic resonance experiments. The technique relies on the use of either $\pi/2$ or π r.f. 'pulses' which are applied to the nuclear spin system at well-defined times, and which rotate one of the nuclear spin components either through $\pi/2$ or π radians. Let us consider an assembly of nuclear spins in an (X, Y, Z) coordinate system, in which the spins are initially oriented in the Z direction by applying a steady magnetic field. At a time $\tau = 0$ the spins are subjected to a $\pi/2$ pulse which rotates them into the Y direction, and they then start to precess in the XY plane; see for example Fig. 1.9(a) with the nuclear spin vector $\boldsymbol{i}$ replacing $\boldsymbol{P}$. If the effective fields at the different nuclei are inhomogeneous, there will be a range of spin components which precess at different rates with the result that the precessing nuclear spin components 'fan out'. After some time τ, a π pulse is applied which reverses the X component of the nuclear spins, and the result is that the precessing spin components start to coalesce, and they appear in phase again along the Y direction after a time 2τ. This produces an induced 'spin echo' signal at a time 2τ. For an elementary introduction to nuclear magnetic resonance spin echo methods, see White (1983).

An analogous phenomenon occurs with polarized neutron beams; it is termed 'neutron spin echo' and was devised by Mezei (1972). The counterparts of the r.f. pulses in this technique are $\pi/2$ and π spin-turn coils, and steady precession magnetic fields are provided along the neutron beam trajectories. The important variable in all 'echo' experiments is time, and for the fixed flight-paths of the neutron spin echo experiment this essentially means the neutron velocity. Neutron spin echo is, in fact, primarily used to measure very small neutron velocity changes which occur on scattering, and we now present a detailed description of the underlying principles of the method.

6.1.1 Principle of neutron spin echo

The essential features of the neutron spin echo (NSE) method are illustrated in Fig. 6.1. The polarization vector P_Z in the applied field or $\boldsymbol{B}_0$ direction is rotated into the X-direction by means of a $\pi/2$ spin turner at position A, and the P_X polarization vector then precesses in the XY plane around $\boldsymbol{B}_0$ over the flight path l_0, where the precession field is effective. P_X provides a convenient parameter with which to characterize the Larmor precession, and its value at A is unity. The polarization

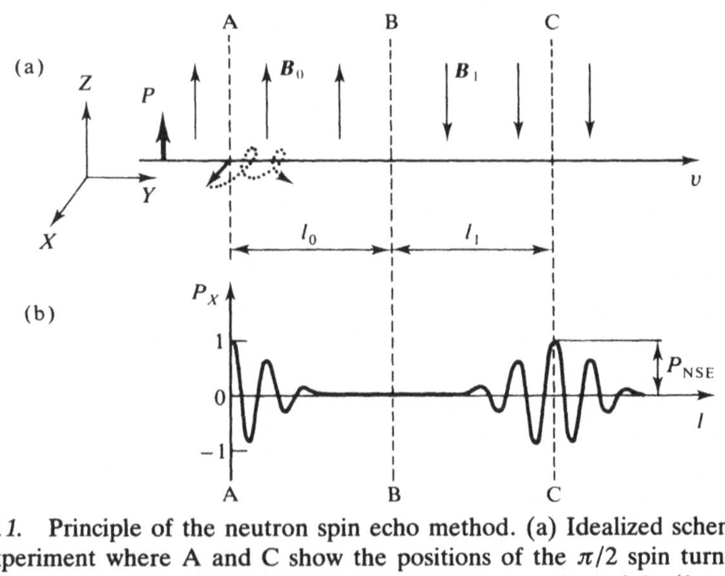

Fig. 6.1. Principle of the neutron spin echo method. (a) Idealized schematic of the experiment where A and C show the positions of the $\pi/2$ spin turners. (b) Illustration of the NSE groups formed due to the precession of P_X (from Mezei 1980a).

component P_X follows the same classical equation of motion as given by equn (1.4.13) for the neutron spin vector $\langle \hat{s}' \rangle$, and the number of radians precessed by a neutron with velocity v at a distance l from A is

$$\phi_L = \gamma l B_0 / v, \tag{6.1.1}$$

where γ, which is numerically equal to 1.833×10^8, is expressed in the units rad s^{-1} T^{-1}. The polarization component P_X for a neutron beam with a velocity distribution $f(v)$ at a distance l from A is then simply given by

$$P_X = \langle \cos \phi_L \rangle = \int f(v) \cos(\gamma l B_0 / v) \, dv. \tag{6.1.2}$$

As l increases, the differences between the ϕ_L for the different velocity neutrons in a real beam become larger, i.e. the Larmor precessions of P_X become dephased, and P_X approaches zero at point B of Fig. 6.1(b). The period of the damped oscillation along AB relates to the average beam velocity. In order to fully exploit the Larmor precession method to measure velocity changes in a scattering experiment, the dephasing effect due to the velocity distribution $f(v)$ must be eliminated, and this can be realized by making the neutron spins precess in the opposite direction after a pre-set time. This occurs in the region BC of Fig. 6.1, where the magnetic induction field $\boldsymbol{B}_1$ is oppositely directed to $\boldsymbol{B}_0$. The accumulated precession angle in radians at point C is

$$\phi_L = \phi_{L(AB)} - \phi_{L(BC)} = (\gamma / v)(l_0 B_0 - l_1 B_1) \tag{6.1.3}$$

and if the field integrals $l_0 B_0$ and $l_1 B_1$ are equal, ϕ_L is zero for all neutron velocities v, and $P_X = \langle \cos \phi_L \rangle = 1$; this is called the 'symmetric' field configuration. The signal shown at C in Fig. 6.1(b) is known as a 'spin echo group' and the essential condition for its appearance is that

$$(\partial \phi_L / \partial v) = 0 \tag{6.1.4}$$

for all the neutrons in the beam; this is satisfied for $\phi_L = 0$, but is also possible for finite ϕ_L. The precessing polarization P_X appears as a maximum when all the neutrons in the beam have the same direction at the point at which P_X is measured, and this does not depend on the way in which this common direction is established. The shape of the spin echo group is a measure of the local dephasing of the spin precessions around the echo maximum, and in the scheme of Fig. 6.1, where there is no scatterer, this is determined by the velocity distribution $f(v)$ of the neutron beam. More precisely, P_X follows the relation given by equn (6.1.2), i.e. it is the Fourier transform of the distribution function for $1/v$. This is equal to $v^2 f(v)$, and is in fact identical to the neutron wavelength spectrum of the beam.

In practice a spin echo group is measured by making observations as a

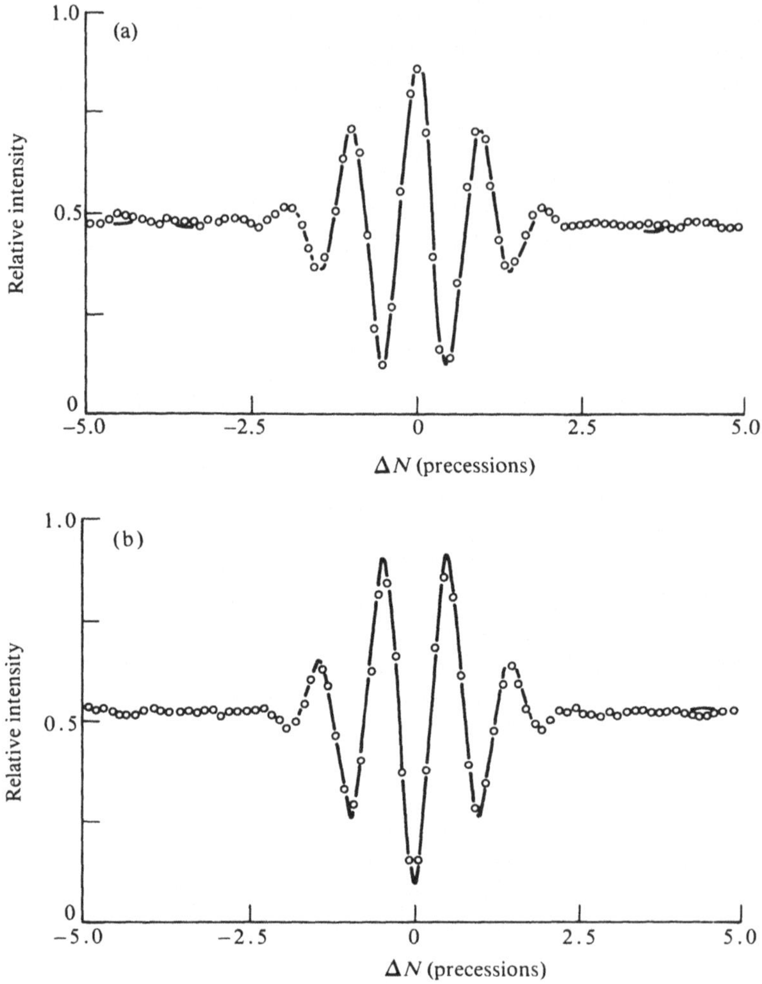

Fig. 6.2. Asymmetric neutron spin echo scans in which there is a 180° phase difference in the P_X component between (a) and (b). The open circles are measurements and the full curves are calculated responses using a matrix analysis simulation method (from Hayter and Penfold 1979).

function of B_1 rather than l_1, and the type of scan in which $\Delta\phi_L$ (or $2\pi\Delta N$ in terms of the number of precessions N) changes is called an asymmetric scan; Fig. 6.2 shows examples of such scans. Hayter and Penfold (1979) have described a fast and efficient method of analysing the wavelength spectra of neutron beams which uses an asymmetric spin echo scan. Another way in which actual experiments differ from the idealized scheme of Fig. 6.1 is that $\boldsymbol{B}_0$ and $\boldsymbol{B}_1$ are actually kept parallel, and the change in the precession direction is achieved by flipping the neutron spin at position B. Finally, since the polarization vector can only be analysed in the direction of applied field at the analyser, P_X is rotated by a $\pi/2$ spin turner to give P_Z, and the measured echo signal is actually the oscillation of P_Z, though its form is precisely the same as that of P_X. The essential components in a real NSE spectrometer are shown, for example, in Fig. 6.3.

Before completing this general description of the basic NSE experiment we define, following the method given by Hayter (1980), what is meant by the spin echo polarization term $\langle P_Z \rangle$ or P_{NSE}. If $I^{(E)}$ and $I^{(-E)}$ are complementary pairs of spin echo intensities obtained by reversing the current in the second $\pi/2$ spin turner of a NSE spectrometer (see Fig. 6.3), and as shown for example in the asymmetric scans of Fig. 6.2, then these intensities relate to an effective polarization

$$\begin{aligned} P_{\mathrm{eff}} &= (I^{(E)} - I^{(-E)})/(I^{(E)} + I^{(-E)}) \\ &= \langle P_Z \rangle P P_{\mathrm{A}}, \end{aligned} \tag{6.1.5}$$

where P is the polarizing efficiency of the polarizer, and P_{A} is that for the analyser. The average intensity $\bar{I}$, or the intensity outside the NSE group, is given by

$$\bar{I} = (I^{(E)} + I^{(-E)})/2 \tag{6.1.6}$$

In order to evaluate $\langle P_Z \rangle$ the flipper 'off' (I_+) and flipper 'on' (I_-) intensities must be determined with the $\pi/2$ spin-turn coils 'off'. If the flipper or π-turner has an efficiency f, we then have

$$\begin{aligned} I_+ - I_- &= \bar{I}\{[1 + PP_{\mathrm{A}}] - [1 + (1 - 2f)PP_{\mathrm{A}}]\} \\ &= \bar{I}(2fPP_{\mathrm{A}}) = fPP_{\mathrm{A}}(I^{(E)} + I^{(-E)}). \end{aligned} \tag{6.1.7}$$

From equns (6.1.5) and (6.1.7) we obtain

$$\langle P_Z \rangle = f(I^{(E)} - I^{(-E)})/(I_+ - I_-), \tag{6.1.8}$$

which is the general result for the NSE polarization. Mezei (1980a) gives a similar expression for this quantity which assumes that the π-turner or spin flipper is 100% efficient, and defines

$$P_{\mathrm{NSE}} = (I_{\mathrm{Max}} - I_{\mathrm{Min}})/2\bar{I}, \tag{6.1.9}$$

where I_{Max} and I_{Min} are the highest maximum and lowest minimum intensities in the NSE group.

We shall see in Section 6.1.3 that in real inelastic scattering experiments only relative values of this polarization are important, and there is no need to make absolute determinations of the instrument component efficiencies. For example, the true $\langle P_Z \rangle$ is obtained from the measured values $\langle P_Z \rangle_{\mathrm{M}}$ by a simple normalization using a perfectly elastic scattering standard, e.g.

$$\langle P_Z \rangle = \langle P_Z \rangle_{\mathrm{M}} / \langle P_Z \rangle_M^S. \tag{6.1.10}$$

where $\langle P_Z \rangle_M^S$ is the NSE polarization measured with the standard. One consequence of this normalization procedure is that $\langle P_Z \rangle$ and P_{NSE} become identical, since the f term in equn (6.1.8) cancels when the ratio is taken.

6.1.2 Neutron spin echo with scattered beams

In the NSE scattering experiment, the scatterer is placed at position B of Fig. 6.1; see also Fig. 6.3. If the scattering is purely elastic then the accumulated precession angle at the end of the B_1 field is still given by equn (6.1.3), and the echo condition is defined by $\phi_{\mathrm{L}} = 0$. In general we shall be concerned with inelastic scattering processes in which the accumulated precession angle (in radians) is

$$\phi_{\mathrm{L}} = \phi_{\mathrm{L}(0)} - \phi_{\mathrm{L}(1)} = \gamma[(l_0 B_0/v_0) - (l_1 B_1/v_1)], \tag{6.1.11}$$

and where we have used the subscripts 0 and 1 to denote the appropriate quantities in the incident and scattered beams.

The basic idea of the experiment is to use ϕ_{L} as given by this equation to measure the energy transfer

$$\hbar\omega = E_1 - E_0 = (m/2)(v_1^2 - v_0^2). \tag{6.1.12}$$

This is done by measuring ϕ_{L} with respect to its average value $\bar{\phi}_{\mathrm{L}}$ in a restricted region around an arbitrarily chosen value of $\hbar\omega_0$. This condition may be summarized by

$$\phi_{\mathrm{L}} - \bar{\phi}_{\mathrm{L}} = t(\omega - \omega_0), \tag{6.1.13}$$

where t may be regarded for the present as a proportionality constant; the equation clearly describes an echo since it implies that the ϕ_{L} values are all the same for those neutrons which undergo the same energy transfer $\hbar\omega$. In the vicinities of $\bar{\phi}_{\mathrm{L}}$ and ω_0, equns (6.1.11) and (6.1.12) can be expressed in their differential forms

$$\phi_{\mathrm{L}} - \bar{\phi}_{\mathrm{L}} = -\gamma\left(\frac{l_0 B_0}{\bar{v}_0^2}\right)\delta v_0 + \gamma\left(\frac{l_1 B_1}{\bar{v}_1^2}\right)\delta v_1 \tag{6.1.14}$$

and

$$\hbar(\omega - \omega_0) = -m\bar{v}_0\,\delta v_0 + m\bar{v}_1\,\delta v_1, \tag{6.1.15}$$

where, for the average velocities $\bar{v}_0$ and $\bar{v}_1$ we have $\phi_{\mathrm{L}} = \bar{\phi}_{\mathrm{L}}$, $\omega = \omega_0$, and $\delta v_{\mathrm{i}} = v_{\mathrm{i}} - \bar{v}_{\mathrm{i}}$. The NSE condition as expressed by equn (6.1.13) can therefore only be satisfied when the coefficients of δv_i are proportional, hence we have the alternative NSE condition

$$l_0 B_0/\bar{v}_0^3 = l_1 B_1/\bar{v}_1^3 \tag{6.1.16}$$

and the proportionality parameter t, with dimension of time, becomes

$$t = \gamma\hbar l_0 B_0/(m\bar{v}_0^3) = \gamma\hbar l_1 B_1/(m\bar{v}_1^3). \tag{6.1.17}$$

It is obvious that for either elastic scattering or the simple direct beam NSE experiment of Fig. 6.1, where $\bar{v}_0 = \bar{v}_1 = v$, we obtain $l_0 B_0 = l_1 B_1$ for the echo condition $\phi_L = 0$. Equation (6.1.13) shows that ϕ_L depends only on $\hbar\omega$, and not explicitly on v_0 and v_1 individually, and this may be restated as

$$(\partial\phi_L/\partial v_0)_{\omega=\omega_0} = (\partial\phi_L/\partial v_1)_{\omega=\omega_0} = 0. \tag{6.1.18}$$

It follows therefore that neutrons which undergo an energy change $\hbar\omega_0$ on scattering produce essentially the same type of NSE signal as shown in Fig. 6.1, but centred around a position at which $(l_1 B_1/l_0 B_0) \neq 1$.

We now express the NSE signal P_{NSE} for a scatterer with a scattering law $S(\boldsymbol{\kappa}, \omega)$, where ω does not have a strong $\boldsymbol{\kappa}$ dependence. $S(\boldsymbol{\kappa}, \omega)$ describes the probability that neutrons are scattered with energy transfer $\hbar\omega$, and this gives, via equn (6.1.13), the distribution of ϕ_L in the scattered beam. We therefore obtain

$$P_{NSE} = P_S\langle\cos(\phi_L - \phi_L')\rangle = P_S \int S(\boldsymbol{\kappa}, \omega)\cos[t(\omega - \omega_0)]\,d\omega \Big/ \int S(\boldsymbol{\kappa}, \omega)\,d\omega, \tag{6.1.19}$$

where an additional polarization factor P_S has been introduced to account for any polarization-dependence of the scattering process itself. We see that P_{NSE} is the Fourier transform of the scattering function $S(\boldsymbol{\kappa}, \omega)$ with respect to $(\omega - \omega_0)$, and that the proportionality constant t in equn (6.1.17) is the Fourier time. In the inelastic NSE scattering experiment the P_{NSE} signal is determined for a range of t values by scanning B_0 and B_1 while keeping their ratio constant. The foregoing is only strictly valid at the centre of the spin echo group, and this is the 'symmetric' scan as defined by Hayter and Penfold (1979), where the difference between a number of precessions before and after the scatterer is maintained identical to zero.

For quasielastic scattering studies of diffusion, spin relaxation, etc., where $\omega_0 = 0$, the Fourier parameter t which is conjugated to ω becomes the real physical time. The NSE polarization then gives a direct measure of the normalized intermediate scattering law $S_N(\boldsymbol{\kappa}, t)$ since

$$P_{NSE}(t)/P_S = S_N(\boldsymbol{\kappa}, t) = \int_{-\infty}^{+\infty} S(\boldsymbol{\kappa}, \omega)\exp(i\omega t)\,d\omega \Big/ \int_{-\infty}^{+\infty} S(\boldsymbol{\kappa}, \omega)\,d\omega. \tag{6.1.20}$$

The case of the Lorentzian scattering law is particularly instructive.

This is centred at ω_0 with a half-width at half maximum Γ such that

$$S(\kappa, \omega) \propto \Gamma/[\Gamma^2 + (\omega - \omega_0)^2] \tag{6.1.21}$$

and the integration between $-\infty$ and $+\infty$ gives (see Mezei (1980a)),

$$P_{NSE}(t) = P_S \int_{-\infty}^{+\infty} [\Gamma^2 + (\omega - \omega_0)^2]^{-1} \times \cos(\omega - \omega_0)t \, d\omega \bigg/ \int_{-\infty}^{+\infty} [\Gamma^2 + (\omega - \omega_0)^2]^{-1} \, d\omega$$

$$= P_S S(\kappa, t)/S(\kappa) \tag{6.1.22}$$

$$= P_S \exp(-\Gamma t). \tag{6.1.23}$$

Thus a measurement of the t-dependence of P_{NSE} provides an excellent means for establishing whether a line broadening is truely Lorentzian, and it can also supply the value of the linewidth parameter Γ.

We now give typical values of the sample polarization factors P_S in different NSE experiments. This parameter depends first on the type of scattering process being investigated, and any spin-dependence of this scattering must be accounted for by using the same rules as described in Section 2.3 for full polarization analysis experiments. Secondly, the action of any π-turners, or in the case of ferromagnetic scatterers $\pi/2$-turners, placed near the scatterer must be included, since the quantity which determines the amplitude of the NSE signal is the resultant spin vector change in the scattering sample region. A detailed discussion of these effects has been given by Mezei (1980a), and his main conclusions are summarized in Table 6.1.

The essential principles of the neutron spin echo scattering experiment have now been explained. Further specific details will be discussed later in the chapter with the presentation of typical experimental results. Before this, it is appropriate to give a more comprehensive description of NSE instrumentation as well as some indication of how the echo data are handled.

TABLE 6.1 Polarization factors P_S in neutron spin echo experiments with different types of scattering samples and spin-turn coil configurations

Types of scatterer	Sample (S) region: Spin-turn coils	Sample (S) region: B_S field	Polarization factor, P_S
Coherent nuclear	π	small	1
Nuclear spin-coherent	π	small	$-1/3$
Paramagnets	none	small	1/2
Ferromagnets	$\pi/2 - S - \pi/2$	high (∥ Z)	1/2
Antiferromagnets	none	small	$1/2 \leq P_S \leq 1$

6.1.3 Neutron spin echo instruments and methods

The Institut Laue–Langevin IN11 spectrometer

The only neutron scattering instrument so far constructed exclusively for spin echo spectroscopy experiments is the IN11 instrument at the Institut Laue–Langevin, Grenoble; it has been used largely for quasielastic scattering experiments. Figure 6.3 shows a schematic layout of the spectrometer as described by Dagleish *et al.* (1980).

A mechanical velocity selector defines the wavelength and wavelength resolution of the neutrons taken from a cold neutron guide. The wavelength can be varied between 4 Å and 9 Å with $\delta\lambda_0/\lambda_0$ between 10% and 25%. If better monochromatization is required, a graphite crystal analyser can be inserted after the polarization analyser to give $\delta\lambda_0/\lambda_0 \sim 3\%$. The 'background' spectrometer is therefore a 2-axis diffractometer and the scattering vector resolution is determined by $\delta\lambda_0/\lambda_0$ together with the usual $(\cot\theta)\cdot\Delta\theta$ term, where 2θ is the scattering angle. In its original form the spectrometer had only one detector and 2θ was varied between 2° and 30°; this was later replaced by a multidetector array.

The polarizer and analyser are Co–Ti polarizing supermirrors with a polarization product PP_A which varies between 0.90 and 0.97 depending

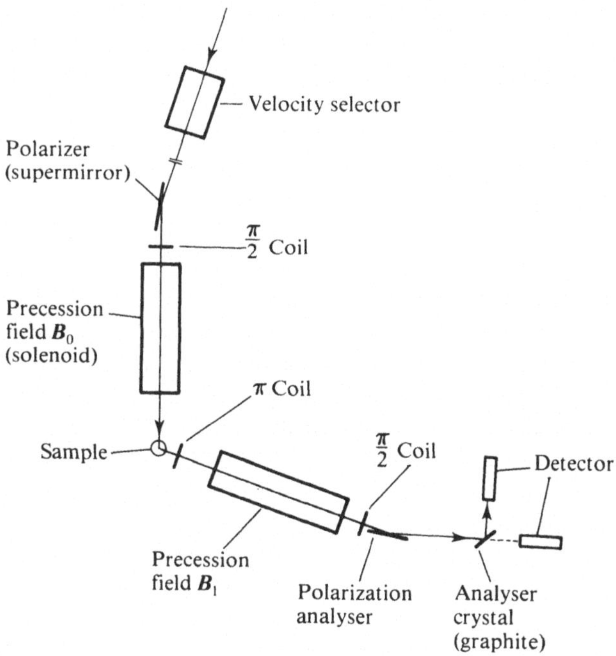

Fig. 6.3. Schematic layout of the IN11 neutron spin echo spectrometer (from Dagleish *et al.* 1980).

on the instrument configuration used. The Larmor precession fields are provided by solenoids, each of length 2 m, which produce maximum fields ~0.05 T. Neutrons of wavelength 6 Å make about 2000 precessions around the beam trajectory in one of these coils, and this corresponds to a Fourier time $t \sim 10^{-8}$ s. The $\pi/2$ spin turners are standard Mezei coils which are described in Section 3.2.5, as is the π-turner or spin flipper. This flipper, which is close to the sample position, reverses the spin direction so that the precessions in the second solenoid are in the opposite sense to those in the first. The off-axis field integral inhomogeneities of the main solenoids are corrected using spiral coils positioned at the entrance to the B_0 field and the exit of the B_1 field; these are not shown in the diagram. Also omitted is the tuning coil placed around the B_1 main coil, which is used to identify the maximum echo condition in asymmetric NSE scans. A tuning coil current of ~0.6 amperes adds sufficient field that 6 Å neutrons make one extra precession in the B_0 field. The 'Lorentzian' energy resolution of the spectrometer is of the order 10 neV.

Matrix analysis methods for calculating the response of neutron spin echo spectrometers and their components have been developed by Hayter (1978b), and Hayter and Penfold (1979). Briefly, their method depends on calculating operator products of the form $\hat{C}\hat{B}_1\hat{F}\hat{B}_0\hat{C}$, where $\hat{C}$ and $\hat{F}$ are operators corresponding to $\pi/2$ and π spin turn coils, and $\hat{B}_0$ and $\hat{B}_1$ are Larmor precession operators for the B_0 and B_1 precession fields. A detailed discussion of their technique is beyond the scope of this book. However, since the method allows for the inclusion of the effect of scatterers, it is illustrative to reproduce one of their results. The example chosen in the scattering from a system undergoing self-diffusion with a Lorentzian scattering law

$$S(\boldsymbol{\kappa}, \omega) = D\kappa^2/\{\pi\hbar[(D\kappa^2)^2 + \omega^2]\}, \tag{6.1.24}$$

where D is the diffusion coefficient. The spin echo intensity calculated by Hayter and Penfold (1979) for a diffusive scatterer with $D = 2 \times 10^{-5}\,\text{cm}^2\,\text{s}^{-1}$ in an ideal spectrometer with $PP_A = 1$, $\lambda_0 = 6$ Å, $\delta\lambda_0/\lambda_0 \simeq 0.30$, and $\kappa = 0.18$ Å^{-1}, is shown in Fig. 6.4. This result provides an excellent illustration of the concepts of the 'asymmetric' scan obtained by varying the difference between the number of Larmor precessions (ΔN) in the incident and scattered beams, and the 'symmetric' scan when ΔN is kept constant. Inelastic scattering information is contained in the symmetric scan when $\Delta N = 0$, and we have seen that this is accomplished by varying both the B_0 and B_1 fields such that the ratio B_0/B_1 is kept constant.

The response shown in Fig. 6.4 is that due to the scatterer itself, and we now return to discuss how this can be isolated from the measured NSE signal. The most serious instrumental effect on the sample's NSE

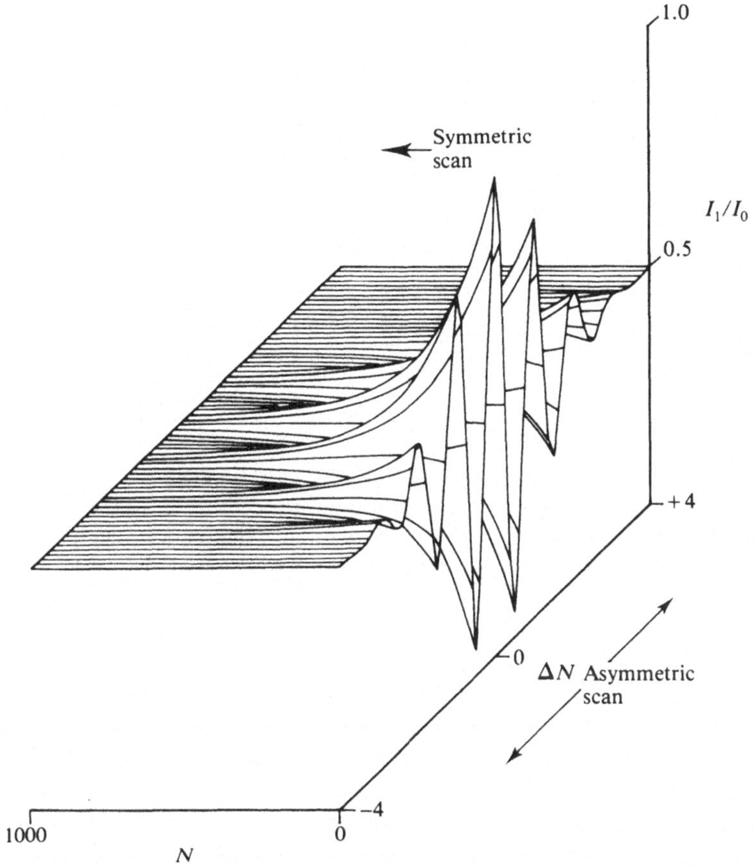

Fig. 6.4. Spin echo response from a diffusive scatterer calculated using a matrix analysis procedure (from Hayter and Penfold 1979). (I_1/I_0) is the ratio of the intensities observed with and without the $\pi/2$ coils in operation.

signal $P_{\mathrm{NSE}}(t)$ is due to variations in the magnetic field line integrals (lB) in the neutron beam paths, since these can lead to a partial dephasing of the precessions. We may write the effective or measured accumulated precession angle $\phi_{\mathrm{L}}^{\mathrm{M}}$ as follows,

$$\phi_{\mathrm{L}}^{\mathrm{M}} = t(\omega - \omega_0) + \phi_{\mathrm{L}}', \tag{6.1.25}$$

where ϕ_{L}' is the additional instrumental phase shift due to field integral variations whose magnitude depends on the neutron trajectory within the beam. By substituting this relation into the argument of the cos function in equn (6.1.19) we see that, following a Fourier transform, the measured echo signal $P_{\mathrm{NSE}}^{\mathrm{M}}(t)$ becomes

$$P_{\mathrm{NSE}}^{\mathrm{M}}(t) = P_{\mathrm{NSE}}^{\mathrm{S}}(t) P_{\mathrm{NSE}}(t), \tag{6.1.26}$$

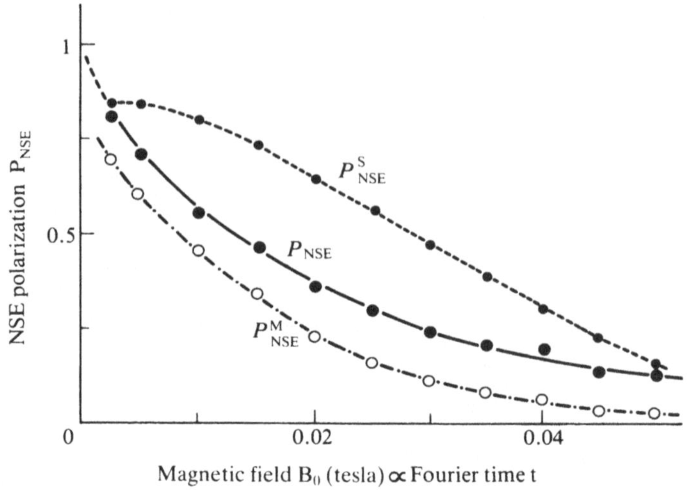

Fig. 6.5. A typical example of the calibration procedure in a neutron spin echo experiment in which P^S_{NSE} is the instrumental polarization dephasing factor measured with a standard elastic scatterer (from Mezei 1980a).

where $P^S_{NSE}(t)$ is the Fourier transform of the instrumental dephasing factor ϕ'_L. The instrument calibration therefore involves measuring the NSE signal for a standard elastic scatterer with $P_S = 1$, and this yields $P^S_{NSE}(t)$ directly. Note the equivalence between equns (6.1.26) and (6.1.10). Thus, in contrast to conventional neutron spectrometers, the instrument resolution effects in a NSE spectrometer can be accounted for by a simple division rather than by a deconvolution, a not insignificant advantage. Figure 6.5 shows a typical example of the magnitude of this correction for a symmetric NSE scan on the IN11 spectrometer. It should be noted that the absolute values of the field integral inhomogeneities increase with increasing B_0 and B_1, thus $P^S_{NSE}(t)$ is itself strongly dependent on t.

Neutron spin echo triple-axis spectroscopy

The combination of NSE with a classical triple-axis spectrometer was initially proposed by Mezei (1977). Its main areas of application were recognized to be the separation of elastic and inelastic scattering, and the study of phonon life-times. The basic experimental arrangement is illustrated in Fig. 6.6, and Zeyen (1982) has used the D10 triple-axis spectrometer at the Institut Laue–Langevin, Grenoble, to carry out initial tests to assess the potential of the method. His spectrometer used focusing Heusler alloy polarizing monochromators on the first and third axes. The value of the NSE addition is, of course, that it provides a much

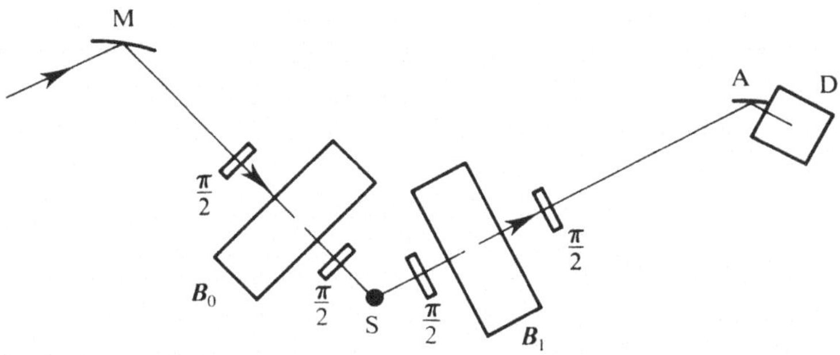

Fig. 6.6. Schematic diagram of a neutron spin echo triple-axis spectrometer. M and A are focusing Heusler alloy polarizing crystals; $\boldsymbol{B}_0$ and $\boldsymbol{B}_1$ are transverse precession fields; the two $\pi/2$ coils around the sample S can be replaced by a single π spin-flipper; D is the detector (from Zeyen 1982).

superior energy resolution to that of the background spectrometer, and the resolution improvement in elastic–inelastic separation experiments was approximately two orders of magnitude. Experiments on phonon linewidths were really only successful for non-dispersive scattering processes, where linewidths down to ~1 μeV were measured.

Some form of NSE focusing is required to measure dispersive coherent excitations, and one of the methods suggested to achieve this was the 'tilted-magnet' technique (see Mezei (1977) and Pynn (1978)). The fundamental idea, see Fig. 6.7, is to use the Larmor precession (around the field which is perpendicular to rectangular precession region shown) to determine a selected component of the neutron velocity or wave vector. The total Larmor precession angle between the $\pi/2$ spin turner and the reference surface F in the diagram, is

$$\phi_L = 2\pi N = A/(\tilde{\boldsymbol{n}} \cdot \boldsymbol{k}), \tag{6.1.27}$$

where $\tilde{\boldsymbol{n}}$ is the unit vector perpendicular to the magnet faces, A is a

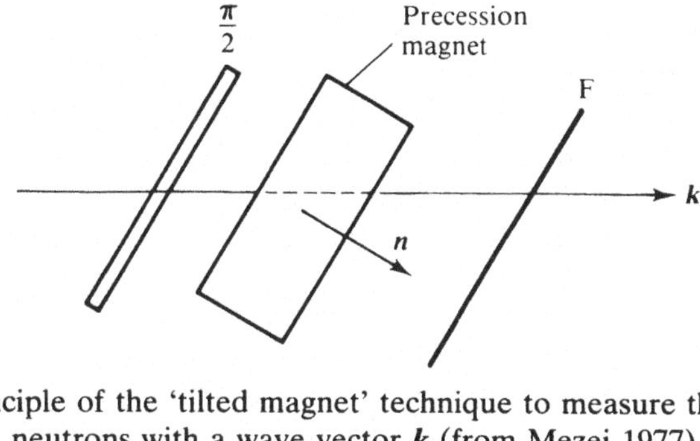

Fig. 6.7. Principle of the 'tilted magnet' technique to measure the precession of neutrons with a wave vector $\boldsymbol{k}$ (from Mezei 1977).

constant proportional to the field integral across the magnet, and we have placed this arrangement in the incident arm of the spectrometer to select neutrons with wave vector $\boldsymbol{k}$. A similar tilted magnet precession field can be placed in the scattered beam, which is tuned to select neutrons with wave vector $\boldsymbol{k}'$, hence

$$\phi'_{\mathrm{L}} = 2\pi N' = A'/(\tilde{\boldsymbol{n}}' \cdot \boldsymbol{k}'). \tag{6.1.28}$$

A detailed discussion of the theory of spin echo focusing in triple-axis spectrometers has been presented by Pynn (1978). His calculation showed that, by using the 'tilted magnet' method, the instrument can be focused on any phonon branch, and the Fourier transform of the spectral density can be measured. The resolution improvement compared with a non-NSE measurement was shown to be substantial, particularly in the case of longitudinal phonon modes.

Following experimental tests, however, Zeyen (1982) concludes that the 'tilted magnet' scheme, though possible in principle, is not compatible with a compact instrument design. The main difficulty arises because of the complicated behaviour of the fringe fields following tilting, since this requires many other spectrometer parameters to be reset. It now appears that, owing to recent advances in calculating and designing superconducting coil magnetic field profiles, it may be advantageous to perform the phonon focusing 'electrically' rather than geometrically. Currently available magnet design techniques should allow the magnetic fields to be shaped in such a way that the curvature of the NSE constant precession surface follows the local curvature of the dispersion surface for the excitation under investigation, and this development is being actively pursued at the Institut Laue–Langevin.

The pulsed source neutron spin echo method

Mezei (1979) was the first to point out the special advantages of combining NSE with the neutron time-of-flight method. Such a spectrometer, which is shown in Fig. 6.8, becomes particularly powerful on a pulsed neutron source if it is equipped with a large analyser–multidetector system. The instrument shown is suitable for quasielastic scattering measurements, in which the precession field regions before and after the scatterer have identical field integrals $l_0 B_0 = l_1 B_1 = lB$. The total time-of-flight T over the combined distance $(L_0 + L'_0 + L_1)$ measures a neutron velocity $\bar{v}$, which for quasielastic scattering gives a κ value

$$\kappa \simeq (2m\bar{v}/\hbar)\sin(\phi/2), \tag{6.1.29}$$

where ϕ is the scattering angle. As shown in Section 6.1.2, quasielastic scattering information is obtained by scanning the NSE polarization,

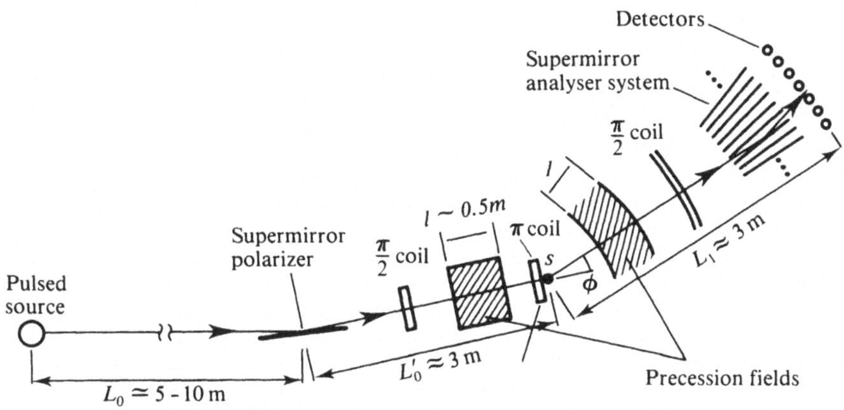

Fig. 6.8. Schematic diagram of a neutron spin echo spectrometer for quasielastic scattering on a pulsed source (from Mezei 1979).

which takes the form, for $\omega_0 = 0$,

$$P_{NSE} = P_0 \int S(\boldsymbol{\kappa}, \omega) \cos(\omega t) \, d\omega, \tag{6.1.30}$$

where the Fourier time t is

$$t = (\hbar \gamma B l)/(m \bar{v}^3). \tag{6.1.31}$$

By using equn (6.1.29) to eliminate $\bar{v}$ from this expression we may also write the Fourier time as

$$t = [2 \sin(\phi/2)]^3 \gamma B l m^2/(\hbar^2 \kappa^3) \tag{6.1.32}$$

or

$$t = (\hbar \gamma B l m^2 T^3)/(L_0 + L_0' + L_1)^3 \tag{6.1.33}$$

Thus, dynamical information can be obtained not only by scanning B, as in the continuous beam instrument, but also by scanning the neutron time-of-flight at constant B. The $(\boldsymbol{\kappa}, t)$ ranges probed in a wide wavelength band $(4 < \lambda(\text{Å}) < 15)$ pulsed-source NSE spectrometer which has a large multidetector is illustrated in Fig. 6.9. As in all NSE instruments the κ measurements are provided by the 'background' spectrometer. By defining the dynamic range R of the spectrometer as the ratio of the maximum (t_{max}) and minimum (t_{min}) Fourier times which can be accessed during one scan, we obtain from equn (6.1.33),

$$R = (t_{max}/t_{min}) = (B_{max}\lambda^3_{max})/(B_{min}\lambda^3_{min}). \tag{6.1.34}$$

A wide wavelength-band pulsed source instrument is therefore potentially much more powerful than its steady-state source counterpart,

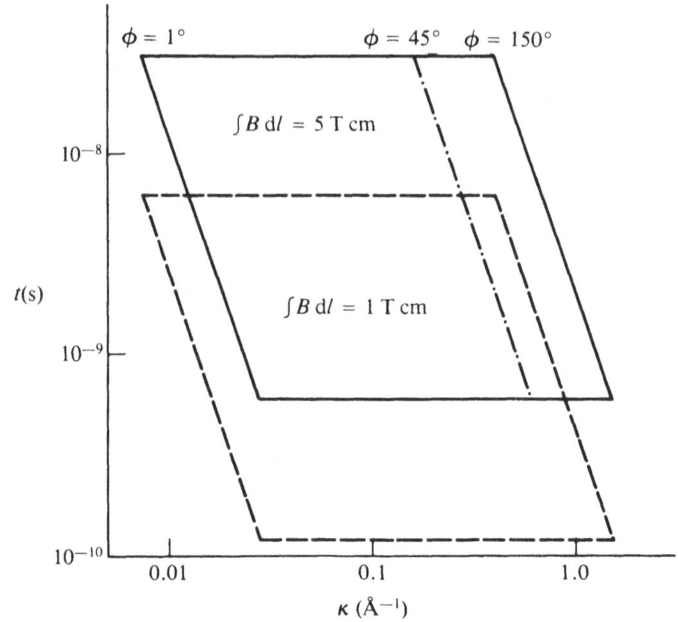

Fig. 6.9. κ–t range covered by a time-of-flight spin echo spectrometer using wavelengths between 4 Å and 15 Å and a large range of detector angles ϕ (from Mezei and Richter 1984).

whose R range is limited to B_{max}/B_{min}. The main drawback of the pulsed instrument lies in the technical difficulty of designing large scattered beam precessional fields, with suitably uniform magnetic field line integrals. There have not been any pulsed-source NSE spectrometers built so far.

6.1.4 Scientific applications of neutron spin echo spectroscopy

Neutron spin echo (NSE) spectroscopy has now been applied to the investigation of a wide range of scientific problems. In the main, the experiments concern either measuring the motions of atoms and molecules, or spin relaxation studies in magnetic systems. It is not appropriate to review all the experiments, but in order to give the reader an impression of the diversity of the applications, we now describe one typical example from each of six different scientific areas. These are the dynamics of polymer solutions, the dynamics of micelle solutions, the dynamics of a spin-glass system, the critical dynamics in a transition metal above T_c, soliton diffusion in a one-dimensional antiferromagnet, and finally the study of elementary excitations in superfluid ^{4}He.

Dynamics of polymer solutions

The first application of NSE to the study of polymers was that by Richter *et al.* (1978), who used the IN11 spectrometer to examine dilute solutions of atactic poly(methyl methacrylate) (PMMA) and poly(dimethysiloxane) (PDMS) in deuterated benzene (B). The aim of the experiment was to verify the scaling predictions of the Zimm model for dilute polymer solutions; in the single-chain regime this states that the half-width of the relaxation spectrum ($\Delta\omega$) scales with $(T/\eta)\kappa^3$, where T is the temperature and η is the viscosity. Deviations from this behaviour are expected in the many-chain regime at higher concentrations, where the influence of entanglements becomes important. A calculation by Dubois-Violette and de Gennes (1967) predicts the proportionality constant in the single-chain limit, and these authors gave

$$\Delta\omega(\mu\text{eV}) = 0.28(T/\eta)\kappa^3 \quad \text{K cP}^{-1}\,\text{Å}^3. \tag{6.135}$$

NSE measurements of the quasielastic linewidths were made over a range of concentrations in the case of PDMS/B, and also over a κ range between 0.027 and 0.137 Å^{-1}. The dominant scattering was that of the coherent cross-section of the polymer, and the $\Delta\omega$ widths were obtained from $\ln(P_{\text{NSE}})$ versus precession field results, as described earlier. The fit parameter used in the data analysis was $T/\eta z$, where z allows for a deviation in the nominal 0.28 pre-factor. Figure 6.10(a) shows a test of the applicability of a κ^3 power law, where $\Delta\omega/\kappa^3(\propto T/\eta z)$ is plotted as a function of κ for PMMA and PDMS solutions of concentrations $c = 0.05\,\text{g cm}^{-3}$. The solid lines show the values obtained by a common fit, and the results provide excellent evidence for the validity of the κ^3 law. A fit of the exponent x in κ^x by a common variation of the parameters gave $x = 3.02 \pm 0.03$ for PDMS and $x = 2.93 \pm 0.03$ for PMMA.

The PDMS results at higher concentrations are shown in Fig. 6.10(b), where $\Delta\omega/\kappa^2(\propto \kappa T/\eta z)$ is plotted as a function of κ. The results for $c = 0.05\,\text{g cm}^{-3}$ follow a straight line, as expected for a κ^3 law. However, the data obtained at higher concentrations show marked deviations from this behaviour, especially at low κ. These measurements in fact support a cross-over from a κ^3 to a κ^2 behaviour, an effect which had been predicted by de Gennes (1976) for entangled systems. The influence of concentration can be treated by introducing a correlation length ξ, which roughly represents the distance between entanglement points of different polymer molecules. In the de Gennes theory the position of the cross-over point κ^* allows a dynamical estimate of ξ to be made, and the values shown in Fig. 6.10(b) were obtained assuming $\kappa^*\xi = 1$. The $\kappa^*\xi$ product may in reality take on larger values; however, the order of

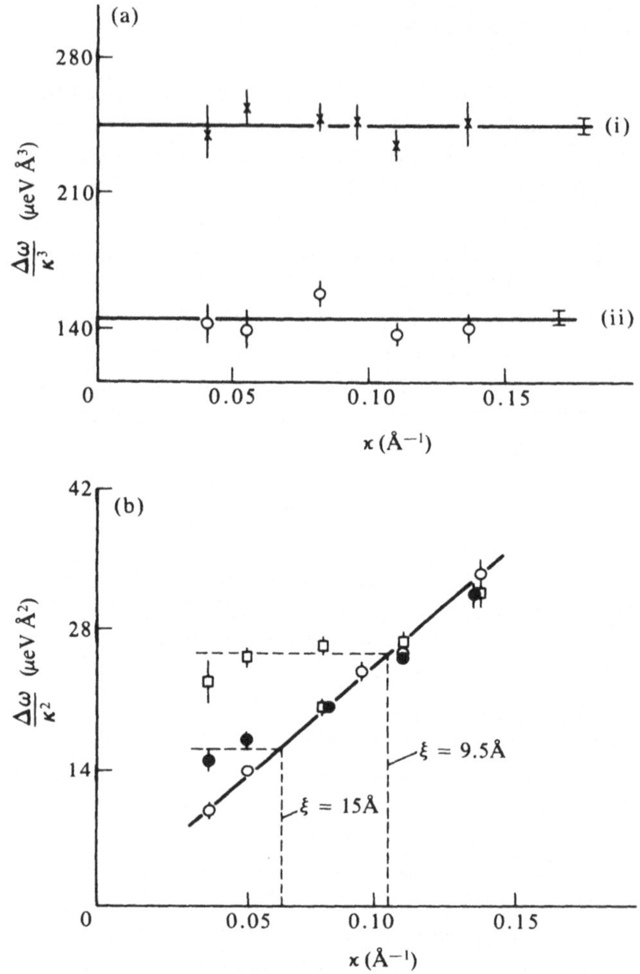

Fig. 6.10. (a) NSE linewidths $\Delta\omega$ divided by κ^3 as a function of κ; (i) PDMS, 0.05 g cm^{-3}; (ii) PMMA, 0.05 g cm^{-3}. (b) $\Delta\omega$ divided by κ^2 as a function of κ for the polymer solutions described in the text; ○, PDMS/B, 0.05 g cm^{-3}; ●, PDMS/B, 0.15 g cm^{-3}; □, PDMS/B, 0.30 g cm^{-3} (from Richter *et al.* 1978).

magnitude of the ξ values obtained in this NSE experiment agrees well with those found by other techniques.

Dynamics of concentrated micelle solutions

Micelles are clusters of charged surface-active molecules with characteristic dimensions of order 50–100 Å. Neutron small-angle scattering, using contrast variation, can be used to study their structures, though in concentrated solutions the scattering cross-section is complicated by the fact that it contains both a single-particle structure factor $S_0(\boldsymbol{\kappa})$ and a structure factor $S(\boldsymbol{\kappa})$ due to intermicellar correlations, i.e.

$$(\mathrm{d}\sigma/\mathrm{d}\Omega) = KS_0(\boldsymbol{\kappa})S(\boldsymbol{\kappa}), \tag{6.1.36}$$

where K is a contrast factor. $S_0(\boldsymbol{\kappa})$ is usually derived by extrapolation to infinite dilution, though the procedure is not appropriate for micelles which dissociate below a critical micellar concentration (c.m.c.). A self-consistent analysis of the scattering data from concentrated solutions of charged spherical micelles has, however, been developed by Hayter and Penfold (1981a), which allows $S(\boldsymbol{\kappa})$ to be derived explicitly.

Neutron spin echo spectroscopy can be used to probe the interparticle dynamics, and since these correlations only involve $S(\boldsymbol{\kappa})$, such measurements are very useful in substantiating the results obtained by small-angle diffraction. We recall from equn (6.1.20) that the quasielastic NSE experiment measures the normalized intermediate scattering law $S_N(\boldsymbol{\kappa}, t)$, and for $P_S = 1$,

$$\begin{aligned} P_{NSE} &= \int_{-\infty}^{+\infty} S(\boldsymbol{\kappa}, \omega)\cos(\omega t)\,\mathrm{d}\omega \Big/ \int_{-\infty}^{+\infty} S(\boldsymbol{\kappa}, \omega)\,\mathrm{d}\omega \\ &= S(\boldsymbol{\kappa}, t)/S(\boldsymbol{\kappa}) \\ &= S_N(\boldsymbol{\kappa}, t). \end{aligned} \tag{6.1.37}$$

The general form of the normalized intermediate scattering law may be written as

$$S_N(\boldsymbol{\kappa}, t) = \exp\{-c_1(\kappa)t + c_2(\kappa)t^2/2! + \cdots\} \tag{6.1.38}$$

and for simple diffusion in a non-interacting system we have the result

$$c_1(\kappa) = D_0\kappa^2, \tag{6.1.39}$$

where D_0 is the diffusion constant, and the higher-order terms $c_2(\kappa)$, etc., are zero. Concentrated micellar solutions are characterized by two time-scales: $\sim 10^{-12}$ s due to rapidly fluctuating Brownian motions, and an interaction time $\sim 10^{-7}$ s which is approximately equal to the time that a particle takes to travel the distance of the average interparticle separation. If a measurement is made in a time regime which is intermediate between these two values, then Pusey (1975) has shown that, for the interacting system,

$$D_0/D_{\mathrm{eff}} = S(\boldsymbol{\kappa}), \tag{6.1.40}$$

where

$$c_1(\kappa) = D_{\mathrm{eff}}\kappa^2. \tag{6.1.41}$$

Hydrodynamic effects may also be included into this theory, but are outside the scope of this book.

Hayter and Penfold (1981b) carried out NSE small-angle scattering measurements on sodium dodecyl sulphate (SDS) solutions with various quantities of added NaCl, and used the above equations to compare the dynamic (D_0/D_{eff}) and static structure factors $S(\boldsymbol{\kappa})$. The spin echo data were obtained on the IN11 spectrometer using a 5% solid solution of perdeuteropolystyrene in polystyrene as the scattering standard. Normalized polarizations ($P_{\text{NSE}}/P_{\text{S}}$) were calculated by correcting for the spin incoherent scattering. The equivalent 'Lorentzian' spectral energy resolution of the spectrometer was 14 neV.

The decay of the NSE polarization P_{NSE} was found to be non-linear in accordance with

$$\ln P_{\text{NSE}} = -c_1(\kappa)t - c_2(\kappa)t^2 \tag{6.1.42}$$

and the short-t behaviour of data taken at various κ values was used to give $c_1(\kappa)$, and hence, from equn (6.1.41), D_{eff}. D_0 was calculated from the Stokes–Einstein relation

$$D_0 = (1-\phi)k_{\text{B}}T/(6\pi\eta a), \tag{6.1.43}$$

where $k_{\text{B}}T$ is the thermal energy, and η is the solvent viscosity. The external radius of the micelle is denoted by a, and ϕ is the micelle volume fraction. The factor $(1-\phi)$ accounts for higher-order terms due to hydrodynamic interactions, though further detailed calculations showed that a correction term $(1-0.65\phi)$ is more precise.

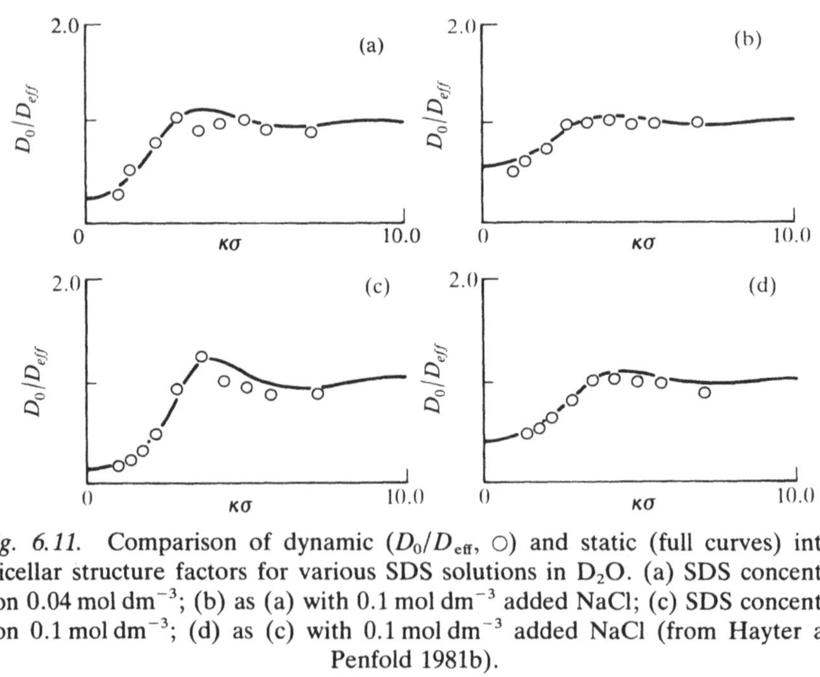

Fig. 6.11. Comparison of dynamic (D_0/D_{eff}, ○) and static (full curves) intermicellar structure factors for various SDS solutions in D_2O. (a) SDS concentration 0.04 mol dm^{-3}; (b) as (a) with 0.1 mol dm^{-3} added NaCl; (c) SDS concentration 0.1 mol dm^{-3}; (d) as (c) with 0.1 mol dm^{-3} added NaCl (from Hayter and Penfold 1981b).

The D_0/D_{eff} values obtained in this way for a range of SDS solutions are shown in Fig. 6.11, and compared with $S(\boldsymbol{\kappa})$ results derived from fitting static small-angle scattering data. The scattering vectors κ were multiplied by the micelle diameter σ to provide a dimensionless scale. The agreement between these two independent methods of determining the intermicellar structure factor is excellent, and the NSE measurements can therefore provide supportive evidence for the validity of the single-particle structure factors measured in a small-angle diffraction experiment.

Dynamics in a spin glass system

The time-dependent spin correlation function $S(\boldsymbol{\kappa}, t)$ for a Cu 5at% Mn spin glass alloy has been determined by Mezei and Murani (1979) over a wide range of times t using both neutron spin echo and three-dimensional polarization analysis.

We consider first the NSE experiments and define the scattering triangle geometry as shown in Fig. 6.12. The Larmor precession plane is (X, Y) and X is along the scattering vector direction for elastic scattering. For paramagnetic and isotropic magnetic scatterers the scattered beam polarization P'_X is equal to $-P_X$, and this pure spin-flip scattering allows spin echo focusing to be achieved without the use of a π spin-flipper at the sample position. From equn (6.1.22) we see that, for perfect instrument polarizers, the experiment measures the polarization component

$$\langle P_X \rangle = P_{\text{NSE}} = S(\boldsymbol{\kappa}, t)/S(\boldsymbol{\kappa}, 0) = \exp(-\Gamma t). \tag{6.1.44}$$

The measurements were made at the IN11 spectrometer with an

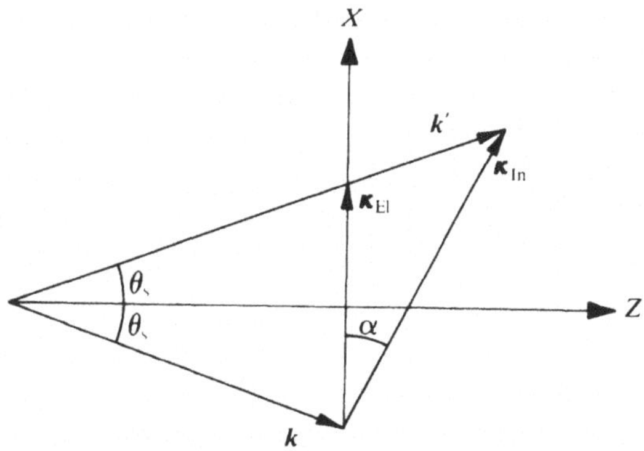

Fig. 6.12. Scattering triangle geometry used by Mezei and Murani (1979) to describe their NSE and three-dimensional polarization analysis experiments for an isotropic magnetic scatterer.

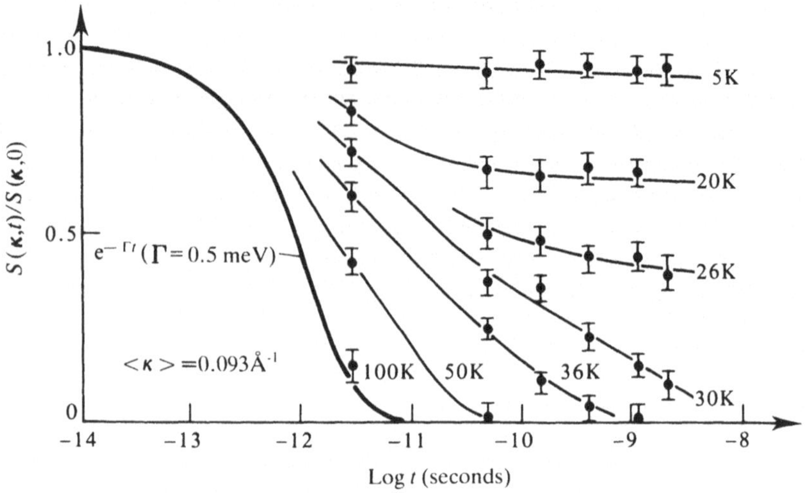

Fig. 6.13. Time-dependence of the normalized spin correlation function for Cu 5 at% Mn measured by the combined neutron spin echo/polarization analysis method at various temperatures. The curves on the left shows the simple exponential decay behaviour (from Mezei and Murani 1979).

incident wavelength $\lambda = 5.9$ Å and $\delta\lambda/\lambda \sim 0.35$. $S(\boldsymbol{\kappa}, t)$ was therefore averaged over a wide $\boldsymbol{\kappa}$ range, though in fact it was found that $S(\boldsymbol{\kappa}, t)/S(\boldsymbol{\kappa}, 0)$ was $\boldsymbol{\kappa}$-independent within the accuracy of the measurements. The results for the intermediate scattering law $S(\boldsymbol{\kappa}, t)$ obtained at a scattering angle $2\theta_s = 5°$, $\langle\kappa\rangle = 0.092$ Å^{-1}, are shown in Fig. 6.13. The data are normalized to the spectrometer polarizing efficiency product PP_A, and give the ratio $S(\boldsymbol{\kappa}, t)/S(\boldsymbol{\kappa}, 0)$, where $t = 0$ corresponds to times $t < 10^{-14}$ s, a limit already known from earlier neutron scattering experiments. A simple relaxation process described by $\exp(-\Gamma t)$ with $\Gamma =$ 0.5 meV has been included in Fig. 6.13 for illustrative purposes. The Fourier times accessed by this NSE experiment were in the range $5 \times 10^{-11} \leqslant t \leqslant 5 \times 10^{-9}$ s.

The values of the spin correlation functions shown in Fig. 6.13 at the time $t = 3 \times 10^{-12}$ s were determined using the three-dimensional polarization analysis method. The scattered beam polarizations P'_X and P'_Z were measured first with the polarization vector P parallel to X, and then with P parallel to Z. Mezei and Murani (1979) showed that

$$\langle P'_Z\rangle - \langle P'_X\rangle = -P(\cos^2\alpha - \sin^2\alpha) \tag{6.1.45a}$$

$$= PS(\boldsymbol{\kappa}, t)/S(\boldsymbol{\kappa}, 0), \tag{6.1.45b}$$

where α is the angle between the elastic scattering vector $\boldsymbol{\kappa}_{\text{El}}$ and the inelastic scattering vector $\boldsymbol{\kappa}_{\text{In}}$; see Fig. 6.12. The second expression is valid only for small energy transfers and for a specific time $t =$

$\cot\theta_s/(2E)$, where E is the incident neutron energy. A measurement of α by determining the scattered beam polarizations in two directions therefore gives the inelasticity of paramagnetic scattering processes.

NSE and three-dimensional polarization analysis measurements of $S(\boldsymbol{\kappa}, t)$ at sample temperatures <45 K were obtained over the $\boldsymbol{\kappa}$ range 0.047–0.37 Å^{-1}, and were observed to be $\boldsymbol{\kappa}$-independent. Previous unpolarized neutron time-of-flight experiments had also shown that the width of $S(\boldsymbol{\kappa}, \omega)$ became increasingly more $\boldsymbol{\kappa}$-independent as the sample temperature was lowered, in agreement with the polarized neutron observations. In this dynamic range it is therefore valid to express $S(\boldsymbol{\kappa}, t)$ as

$$S(\boldsymbol{\kappa}, t) = S(\boldsymbol{\kappa})S(t), \tag{6.1.46}$$

where $S(\boldsymbol{\kappa})$ describes the spatial spin correlations, and $S(t)$ is the time-dependent correlation function, which in this example is the same as the self-correlation function $\langle S_i(0)S_i(t)\rangle$. The latter functions can be evaluated by Monte Carlo simulation, and have the form

$$S(t) = \text{const} - \ln(t). \tag{6.1.47}$$

This relation is in qualitative agreement with the behaviour shown in Fig. 6.13 at times greater than 10^{-10} s. The simulation model only contains solute–solute exchange interactions, and the distinct curvature observed at times $t \leqslant 10^{-11}$ s was explained by invoking the onset of a rapid Korringa relaxation mechanism.

Critical dynamics in Fe *above* T_c

The critical dynamics of iron above its Curie temperature $T_c \simeq 1040$ K has been investigated by Mezei (1982) using both neutron spin echo and high-resolution time-of-flight (TOF) spectroscopy. The improved resolution of the techniques used showed that certain conclusions derived from earlier low-resolution neutron measurements were erroneous. The lifetime of the critical fluctuations is contained in the Fourier-transformed spin–spin correlation function $S(\boldsymbol{\kappa}, \omega)$, which for this experiment may be written as

$$S(\boldsymbol{\kappa}, \omega) = S(\boldsymbol{\kappa})\Gamma_\kappa/[\pi\hbar(\Gamma_\kappa^2 + \omega^2)]. \tag{6.1.48}$$

The object of the investigation was to use the IN5 time-of-flight (resolution ~18 μeV) and IN11 spin echo spectrometers at the Institut Laue–Langevin to determine the energy width $\hbar\Gamma_\kappa$ as a function of κ. The scattering of interest is paramagnetic and the NSE measurements were performed in the same way as those for the **Cu**Mn experiment just described. The similarity between equn (6.1.48) and equn (6.1.21) is self-evident, and the NSE experiment provides the value of Γ_κ directly.

One of the fundamental predictions of dynamical scaling theory (see Halperin and Hohenberg (1969)), is that at $T = T_c$,

$$\Gamma_\kappa = A\kappa^z, \quad \text{with } z = (5 - \eta)/2, \tag{6.1.49}$$

where A is a constant, and the factor η in the exponent is expected to be small for three-dimensional systems. Previous neutron scattering data gave $z = 2.7 \pm 0.3$ in the range $0.05 < \kappa < 0.3$ Å^{-1}. Mezei's results added a further two decades to this earlier data, and showed unequivocally that $z = 2.48 \pm 0.05$ down to $\kappa \sim 0.01$ Å^{-1}. A second prediction of the theory is that at temperatures $T > T_c$, and for the so-called hydrodynamic limit where $\kappa \rightarrow 0$,

$$\Gamma_\kappa = D\kappa^2 \tag{6.1.50}$$

and the spin-diffusion constant D decreases as T approaches T_c according to

$$D = D_0 T^{*y}, \quad \text{with} \quad T^* = (T - T_c)/T_c. \tag{6.1.51}$$

The energy widths $\hbar\Gamma_\kappa$ measured at various temperatures by both the NSE and TOF techniques are shown as a function of κ in Fig. 6.14. The

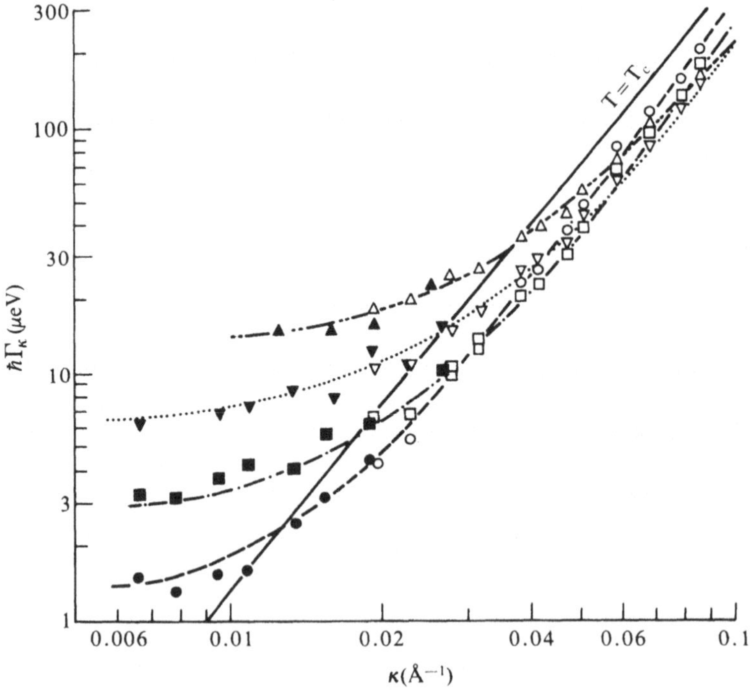

Fig. 6.14. Relaxation rates or energy widths above T_c for iron shown as a function of the scattering vector. The discontinuous curves result from an empirical fit to the data, and the straight line shows the $T = T_c$ results where $\Gamma_\kappa \propto \kappa^{2.5}$. NSE: ●, $T_c + 1.4$ K; ■, $T_c + 5.8$ K; ▼, $T_c + 20.1$ K; ▲, $T_c + 51$ K. TOF: ○, $T_c + 1.1$ K; □, $T_c + 5.5$ K; ▽, $T_c + 21.6$ K; △, $T_c + 51$ K (from Mezei 1982).

straight line corresponds to the $T = T_c$ results where $\Gamma_\kappa \propto \kappa^{2.5}$. It is clear that the data do not follow the $D\kappa^2$ law over any sizeable κ range, and that dynamical scaling breaks down at small κ values. The data were fitted to an empirical formula, and the low-κ behaviour was shown to be due to random spin-non-conserving forces, which were tentatively identified to have a thermodynamic origin.

Soliton diffusion

The nature of the soliton dynamics in the one-dimensional antiferromagnet $(CD_3)_4NMnCl_3$ (TMMC) has been investigated using NSE by Boucher *et al.* (1985). The experiment was able to distinguish between two models: the 'coherent' model in which the solitons behave ballistically, and the 'incoherent' model in which the solitons diffuse randomly along the chains. The main purpose of the measurements was to make a frequency analysis of the antiferromagnetic mode at high resolution, and the IN11 spectrometer was used with the graphite analyser crystal; see Fig. 6.3. The crystal was oriented with the chain axis in the horizontal plane and with the scattering vector $\boldsymbol{\kappa}$ fixed at the position $\boldsymbol{\kappa}_0 = (0.15, 0, 1)$ in reciprocal lattice units. The magnetic fluctuations observed were therefore almost perpendicular to the chain direction [0 0 1] and were denoted by $S_\perp$. A vertical field $B \sim 0.38\,\mathrm{T}$ was applied to the sample, and because of this ferromagnetic NSE, with two $\pi/2$ coils at each side of the sample, was used so as to minimize any dephasing of the precessing neutron spins due to the large applied field. This meant that the time regime accessible in the experiment was more limited than in other NSE experiments, being between 2×10^{-11} s and 1.7×10^{-9} s.

In antiferromagnetic (AF) chains a soliton corresponds to a π rotation of the spins, and as the solitons move along the chains they induce spin-flips which reverse the spin direction in the magnetic sublattices. It is this flipping process which gives the AF short-range order its time-dependence, and the experiment of Boucher *et al.* (1985) measured the frequency width of the AF mode; these widths then gave a direct indication of the flipping rate. The results of the experiment are shown in Fig. 6.15. Reference should be made to the points at $t \sim 3.5 \times 10^{-12}$ s, which have a large experimental error. The decrease in the AF intensity with increasing temperature at this time is indicative of inelastic scattering outside the energy window of the spectrometer. For the ballistic soliton model Boucher *et al.* predicted a behaviour of the type

$$S_\perp(t) \simeq \exp(-At), \tag{6.1.52}$$

whereas for the diffusive model,

$$S_\perp(t) \simeq \exp[-(Bt)^{1/2}]. \tag{6.1.53}$$

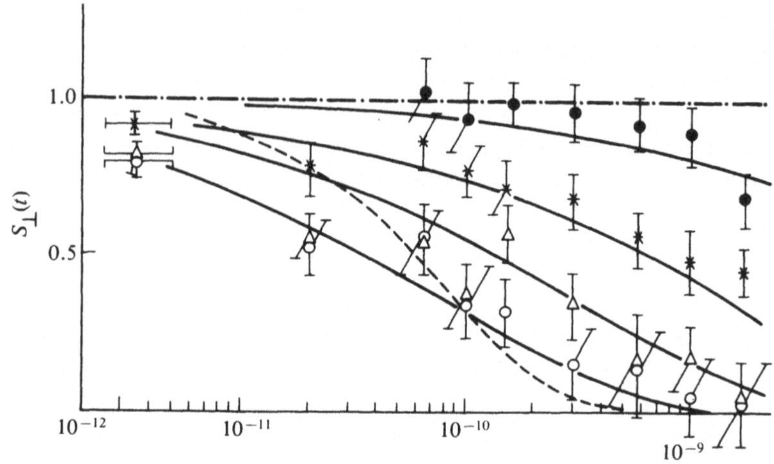

Fig. 6.15. The decay of the local antiferromagnetic order in TMMC at different temperatures measured by NSE ($B = 3.87$ T). The full curves correspond to the diffusive soliton model and the broken curve to the ballistic soliton model: ●, $T = 1.30$ K; $*$, $T = 1.60$ K; △, $T = 1.84$ K; ○, $T = 2.10$ K; from Boucher *et al.* (1985).

A and B in these expressions are both constants. The broken line in Fig. 6.15 shows the shape of the $S_\perp(t)$ versus t curve expected if equn (6.1.52) is valid, thus the results do not support the ballistic soliton model. However, the shape of the full curves in the figure, which were calculated using equn (6.1.53), agree well with the data, and it was concluded that the solitons diffuse randomly in this sample.

Elementary excitations in superfluid ^{4}He

This was the first NSE study of a dispersive excitation and was carried out by Mezei (1980b) using the IN11 spectrometer with a graphite analyser, as shown in Fig. 6.3. NSE focusing was achieved by tuning both the ratio of the precession fields B_0/B_1 and their tilt angles to give a resolution $\sim 0.4\,\mu$eV. The graphite crystal was set to analyse $\lambda = 4.75 \pm 0.7$ Å scattered neutrons, and this provided the κ resolution (± 0.02 Å^{-1}) needed for the measurements. The fundamental relationship in the NSE investigation of dispersive excitations is given by equn (6.1.13):

$$\phi - \bar{\phi}_L = t(\omega - \omega_d(\boldsymbol{\kappa})), \qquad (6.1.54)$$

where $\omega_d(\boldsymbol{\kappa})$ is the excitation energy. Most of the results reported by Mezei (1980b), and reproduced here, were obtained near the roton minimum in ^{4}He where $\omega_d(\boldsymbol{\kappa})$ represents the roton energy $\Delta(T)$; the corresponding κ value is ~ 1.925 Å^{-1}. The experiment entailed measuring the decay of the normalized polarization P_{NSE} as a function of B_0 (or t) at constant B_0/B_1, over a range of ^{4}He temperatures. Inelastic NSE allows both the roton energy $\Delta(T)$ and its linewidth $\Gamma(T)$ to be determined.

The NSE results for the temperature dependence of the linewidth $\Gamma(T)$, and of the energy $\Delta(T)$ of the roton excitation, are shown in Fig. 6.16, together with previous results obtained using classical neutron scattering methods. The importance of the enhanced resolution offered by the NSE method is self-evident. The roton energy data were fitted to the expression

$$\Delta(0) - \Delta(T) = 19T^{1/2} \exp[-\Delta(T)/T], \qquad (6.1.55)$$

with all quantities expressed in kelvin units, and where $\Delta(0)$ was given the accepted value of 8.62 K. The $\Delta(T)$ values from this best-fit curve

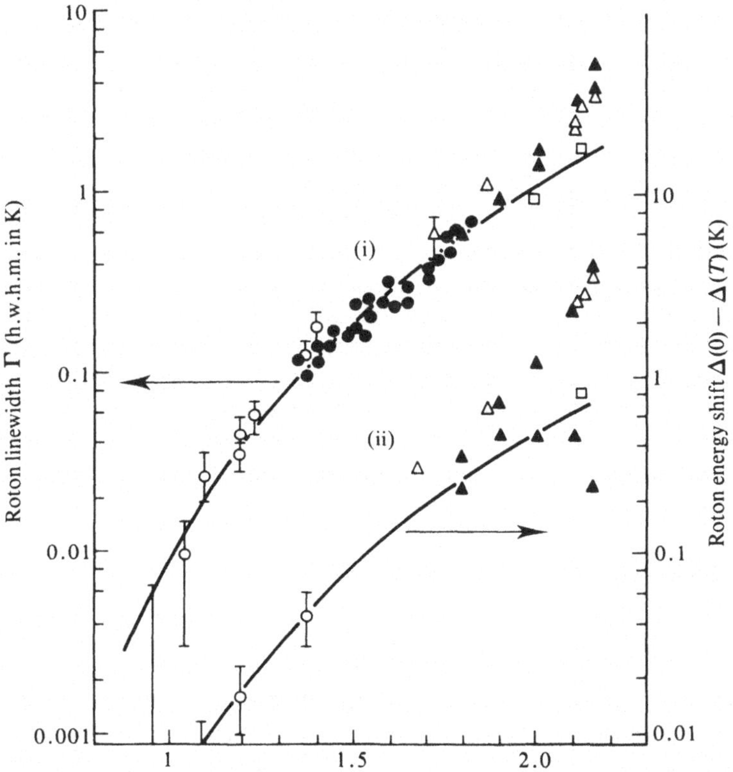

Fig. 6.16. Comparison of NSE results for the temperature dependence of the roton energy and linewidth in superfluid ^{4}He with conventional neutron scattering data. The continuous lines show theoretical predictions for these parameters: (i) $\Gamma(T) = 47T^{1/2} \exp[-\Delta(T)/T]$; (ii) $\Delta(0) - \Delta(T) = 19T^{1/2} \exp[-\Delta(T)/T]$ (from Mezei 1980b).

were then used to calculate the roton linewidths $\Gamma(T)$ using an early theory of the roton interaction developed by Landau and Khalatnikov (1948). These authors predicted that

$$\Gamma(T) = 47T^{1/2}\exp[-\Delta(T)/T], \tag{6.1.56}$$

where the constant coefficient was calculated from normal-fluid viscosity data. The excellent agreement between the theoretical and measured $\Gamma(T)$ values over two orders of magnitude in Γ provides excellent evidence that the roton life-time is indeed dominated by roton–roton interactions in the temperature range of these measurements.

6.2 NEUTRON DEPOLARIZATION IN MAGNETIC MATERIALS

When polarized beams are transmitted through magnetic materials they frequently suffer a large amount of depolarization. The theory of this process was formulated by Halpern and Holstein (1941) for ferromagnets, and first experiments were reported by Burgy *et al.* (1950). Major improvements in the experimental methods took place in the 1970s, largely due to the work of the Delft group, see Rekveldt (1973), and the Leningrad group, see Okorokov *et al.* (1976), who extended the simpler one-dimensional experiment into three dimensions. The experiments give information on the micrometre distance scale, and most of the early one-dimensional measurements were only able to reveal mean domain sizes. The three-dimensional depolarization method provides directional information of the static domain structure, and it was further developed (see for example van Schaik and Rekveldt (1978)) to provide dynamical information on the movement of domain walls. Although the neutron depolarization technique has been mainly used in studying ferromagnets, some experiments have also been performed on antiferromagnets, metamagnets, dilute alloys and superconductors.

6.2.1 Depolarization in one dimension

The basic theory which describes the depolarization of neutron beams as they are transmitted through ferromagnetic domains was set out by Halpern and Holstein (1941), and this remains the most authoritative work on this subject. Their treatment was a classical one in which they followed the precession of the average neutron spin component $\langle \boldsymbol{s}' \rangle$, that is,

$$\mathrm{d}\langle \boldsymbol{s}' \rangle/\mathrm{d}t = |\gamma|\,\langle \boldsymbol{s}' \rangle \times \boldsymbol{B}_i, \tag{6.2.1}$$

through a series of regions or domains i, in each of which the magnetic induction field $\boldsymbol{B}_i$ was constant in direction. It is apparent from equn (6.2.1) that in passing through the ith domain the spin component parallel to $\boldsymbol{B}_i$ is unchanged, and the component perpendicular to $\boldsymbol{B}_i$

precesses about $\boldsymbol{B}_i$ at the Larmor frequency $\omega_L = |\gamma| B_i$. We now describe the results of Halpern and Holstein's calculations using two models to describe the precessions within the domains.

In the first model it is assumed that the magnetic field $\boldsymbol{B}_i$ in the ith domain varies randomly in direction from domain to domain, but that it is constant in both magnitude and direction within each domain. The precessional behaviour of $\langle \boldsymbol{s}' \rangle$ is identical to that of the polarization $\boldsymbol{P}$, and the general result which relates the transmitted neutron polarization P' to the incident beam polarization P is

$$P'/P = \prod_i [1 - \tfrac{4}{3}\sin^2(\tfrac{1}{2}\gamma \boldsymbol{B}_i \tau_i)], \tag{6.2.2}$$

where τ_i is the time spent by the neutron within the domain i. We may distinguish between two limiting cases. The first of these is the fast neutron or low-B_i limit, $\frac{1}{2}\gamma B_i \tau_i \ll 1$, when

$$P'/P = \exp(-\tfrac{1}{3}\gamma^2 \langle \overline{B^2} \rangle T\bar{\tau}) = \exp(-\tfrac{1}{3}\gamma^2 \langle \overline{B^2} \rangle d\delta/v^2). \tag{6.2.3}$$

Here, T is the time taken by a neutron of velocity v to traverse a magnetic sample of thickness d; δ and $\bar{\tau}$ are the average domain length and the average time spent by the neutron in this domain, respectively. In the slow neutron or high-field limit $\frac{1}{2}\gamma B_i \tau_i \gg 1$, we have

$$P'/P \simeq (1/3)^{T/\bar{\tau}} \simeq \exp(-d/\delta). \tag{6.2.4}$$

The physical distinction between these two limits arises because when $\frac{1}{2}\gamma B_i \tau_i \ll 1$, depolarization is caused by a series of small rotations, each of which is given by $\gamma B_i \tau_i$. On the other hand, when $\frac{1}{2}\gamma B_i \tau_i \gg 1$, the rotation within each domain is so large that the component of $\langle \boldsymbol{s}' \rangle$ (or $\boldsymbol{P}$) which is initially perpendicular to $\boldsymbol{B}_i$ averages out to zero, and only the parallel component which is indepenent of $\frac{1}{2}\gamma B_i T_i$ remains. Since the average of $\langle \boldsymbol{s}_i' \rangle (\| \boldsymbol{B}_i)$ over all directions of $\boldsymbol{B}_i$ is $\frac{1}{3}\langle \boldsymbol{s}_i' \rangle$, the depolarization factor per domain is 1/3, or $3^{-n} = 3^{-d/\delta}$ for a specimen in which n domains are traversed.

Burgy *et al.* (1950) were the first to apply this model, and were able to make estimates of the mean magnetic domain sizes in a large number of unmagnetized samples of iron. They used the double transmission effect with two magnetized iron polarizing filters to measure the depolarization in their samples. Subsequent one-dimensional depolarization experiments have all been analysed using a second model, also introduced by Halpern and Holstein (1941), in which the total $\boldsymbol{B}$ field in each domain is resolved into its components: $\boldsymbol{B}_0$ parallel to the neutron polarization, and $\boldsymbol{B}_\perp$ perpendicular to this direction. The general result for the depolarization in this case is

$$P'/P \simeq \exp[-2 \sum_i (\boldsymbol{B}_\perp^2/\boldsymbol{B}_0^2) \sin^2(\tfrac{1}{2}\gamma \boldsymbol{B}_0 \tau_i)], \tag{6.2.5}$$

where the summation is over all the domains i in the specimen. We again have the two limiting conditions, $\frac{1}{2}\gamma B_0 \tau_i \ll 1$ when

$$P'/P = \exp(-\tfrac{1}{2}\gamma^2 \langle \overline{\boldsymbol{B}_\perp^2} \rangle T\tau) = \exp(-\tfrac{1}{2}\gamma^2 \langle \overline{\boldsymbol{B}_\perp^2} \rangle d\delta/v) \qquad (6.2.6)$$

and for $\frac{1}{2}\gamma B_0 \tau_i \gg 1$,

$$P'/P = \exp[-(\langle \overline{\boldsymbol{B}_\perp^2} \rangle / \boldsymbol{B}_0^2)(T/\bar{\tau})] = \exp[-(\langle \overline{\boldsymbol{B}_\perp^2} \rangle / \boldsymbol{B}_0^2)(d/\delta)]. \qquad (6.2.7)$$

The experimental determination of the domain dimension δ involves transmitting a neutron beam of known polarization P through the magnetic specimen, and then analysing the ongoing polarization P' with a second polarizer. The depolarization is neutron-velocity-dependent in the $\frac{1}{2}\gamma B_0 \tau_i \ll 1$ limit, see equn (6.2.6), thus it is advantageous to make measurements over a range of neutron velocities. In discussing the results of one-dimensional depolarization experiments, it is useful to introduce a depolarization parameter D_m which characterises the neutron spin-flip rate in the sample, and where D_m^{-1} is the mean free path for the spin reversal. By using ϕ_+ and ϕ_- to represent the fractions of (+) and (−) spin neutrons in the beam, the depolarization factor D_m may be written as

$$\frac{-\partial\phi_+}{\partial d} = D_\mathrm{m}\phi_+ - D_\mathrm{m}(1-\phi_+) = D_\mathrm{m}(2\phi_+ - 1), \qquad (6.2.8)$$

and integrating over the thickness d of the specimen gives

$$-\int_{\phi_+}^{\phi'_+} \frac{\partial\phi_+}{(\phi_+ - \frac{1}{2})} = 2D_\mathrm{m}\int_0^d \partial d \qquad (6.2.9)$$

or

$$(\phi'_+ - \tfrac{1}{2})/(\phi_+ - \tfrac{1}{2}) = \exp(-2D_\mathrm{m} d). \qquad (6.2.10)$$

Using

$$\begin{aligned} \phi_+ + \phi_- &= \phi'_+ + \phi'_- = 1 \\ P = \phi_+ - \phi_-; &\quad P' = \phi'_+ - \phi'_- \end{aligned} \qquad (6.2.11)$$

we obtain

$$P'/P = \exp(-2D_\mathrm{m} d) \qquad (6.2.12)$$

It is apparent from equns (6.2.6) and (6.2.7) that D_m is proportional to δ in the $\frac{1}{2}\gamma B_0 \tau_i \ll 1$ limit, but is inversely proportional to δ in the $\frac{1}{2}\gamma B_0 \tau_i \gg 1$ limit.

We now derive the D_m–δ relationship for the $\frac{1}{2}\gamma B_0 \tau_i \ll 1$ limit using first principles. Figure 6.17 shows the average neutron spin component $\langle \boldsymbol{s}' \rangle$ initially in the direction of the neutron polarization $P = 1$, and after a short interaction time Δt within a domain it changes by

$$\Delta\langle \boldsymbol{s}' \rangle = \gamma \langle \boldsymbol{s}' \rangle \boldsymbol{B} \sin\theta \, \Delta t \, \tilde{\boldsymbol{n}}, \qquad (6.2.13)$$

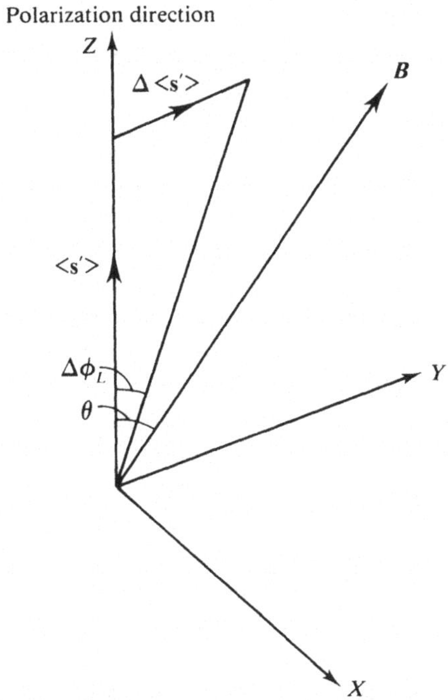

Fig. 6.17. Geometric construction showing the change in the average neutron spin component $\Delta\langle \boldsymbol{s}'\rangle$ due to a precession when traversing a magnetic domain in the limit $\frac{1}{2}\gamma B\tau \ll 1$. $\Delta\langle \boldsymbol{s}'\rangle$ is perpendicular to the plane containing $\langle \boldsymbol{s}'\rangle$ and $\boldsymbol{B}$.

where $\tilde{\boldsymbol{n}}$ is a unit vector perpendicular to the plane of $\langle \boldsymbol{s}'\rangle$ and $\boldsymbol{B}$. The neutron polarization after this time is

$$P'(\Delta t) = \cos \Delta\phi_{\mathrm{L}} \simeq 1 - \tfrac{1}{2}\Delta\phi_{\mathrm{L}}^2 \simeq \exp(-\tfrac{1}{2}\gamma^2\boldsymbol{B}^2 \sin^2\theta\, \Delta t^2) \quad (6.2.14)$$

and, since $P = 1$,

$$P'/P = \exp(-2D_{\mathrm{m}}\delta) = \exp(-\tfrac{1}{2}\gamma^2\boldsymbol{B}^2 \sin^2\theta\, \Delta t^2) \quad (6.2.15)$$

for the single domain. The depolarization produced by n domains each of dimension δ is

$$\begin{aligned}[P'(\Delta t)]^n &= \exp(-\tfrac{1}{2}\gamma^2\boldsymbol{B}^2(\sin^2\theta)\,.\,n\,.\,\Delta t^2)\\ &= \exp(-\tfrac{1}{2}\gamma^2\langle\overline{\boldsymbol{B}_\perp^2\delta^2}\rangle n/v^2)\\ &= \exp(-2D_{\mathrm{m}}d). \end{aligned} \quad (6.2.16)$$

By substituting $n\delta$ for d, this relation is seen to reduce the Halpern–Holstein equn (6.2.6). Furthermore, by writing the number of domains per unit length as $N = n/d$, we see that

$$D_{\mathrm{m}} = \tfrac{1}{4}\gamma^2\langle\overline{\boldsymbol{B}_\perp^2\delta^2}\rangle N/v^2. \quad (6.2.17)$$

In the $\frac{1}{2}\gamma B_0 \tau_i \gg 1$ limit, where the neutron makes more than one precession in each domain, the D_m–δ relationship becomes

$$D_m = \tfrac{1}{2}(\langle \overline{\boldsymbol{B}_\perp^2/\boldsymbol{B}_0^2} \rangle)N. \qquad (6.2.18)$$

As an example of the application of these equations we describe the depolarization results of Postma *et al.* (1962) for foils of holmium metal. P and P' were determined using two Co–Fe polarizing crystals at neutron energies between the thermal range and 0.5 eV. Two foils of thicknesses 0.32 mm and 0.62 mm were cooled to around 1 K and magnetized in a 1.79 T field. The D_m values obtained at a neutron energy $E = 0.062$ eV were the same, within experimental error, for both of the foils, that is $D_m = 17.3 \pm 0.9\ \text{cm}^{-1}$. The thicker foil was examined over a range of neutron energies and the results are shown in Fig. 6.18. At the highest energies, P'/P tends to be inversely proportional to E, in agreement with equn (6.2.6) for the $\frac{1}{2}\gamma B_0 \tau_i \ll 1$ limit, and a linear extrapolation of the measurements gave $D_m = 1.67/E\ \text{cm}^{-1}$ at these energies. The average $|\boldsymbol{B}_\perp|$ in these polycrystalline samples was estimated to be $0.61 B_i^2$, where $B_i \sim 3.8$ T is the domain induction in a holmium single crystal, and from this the domain size in the specimens measured was estimated to be ~0.4 μm.

A knowledge of the depolarization parameter D_m is important both in

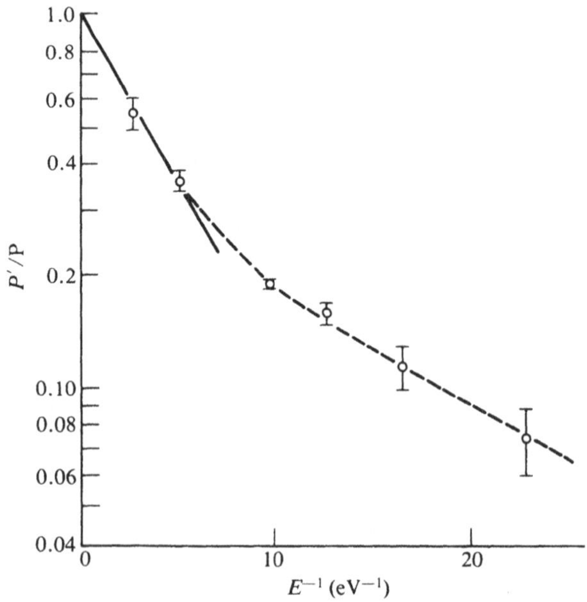

Fig. 6.18. The depolarization ratios P'/P measured by Postma *et al.* (1962) for a 0.62 mm thick polycrystalline holmium specimen as a function of the inverse neutron energy.

single-crystal diffraction experiments (see Section 4.1.3), and in evaluating the effectiveness of magnetic polarizing filter materials. In the latter, depolarization competes with the principal polarization selection process, and their combined effect can only be evaluated by using coupled differential equations to describe the changes in ϕ_+ and ϕ_- as a function of the filter thickness (see Williams (1974)). The effect of spontaneous depolarization on the properties of the permanent magnet polarizing filter material $SmCo_5$ has been studied by Mayers *et al.* (1985). As part of this development programme Mayers (1982a) carried out a detailed examination of the depolarization caused by magnetic domains, not only in the limiting regions where the analytic solutions of Halpern and Holstein are valid, but also in the intermediate region where $\frac{1}{2}\gamma B_0 \tau_i \sim 1$. A Monte Carlo program DEPOL was written to calculate the average depolarization per single-domain particle $P_1(\langle m \rangle)$ for $SmCo_5$ particles, where $\langle m \rangle = \gamma B_0 \tau / 2\pi$ is the average number of precessions with γ expressed in rad T^{-1} s^{-1}. The particle depolarization factors evaluated by this method for two values of the applied field, $\boldsymbol{B}_a = 0$ and $\boldsymbol{B}_a = 2$ T, are drawn as the circles in Fig. 6.19. The solid lines show the analytical values of $P_1(\langle m \rangle)$ as given by the high-$\langle m \rangle$ and low-$\langle m \rangle$ Halpern and Holstein (1941) expressions. This study showed that there is an oscillatory behaviour of the depolarization parameter $P_1(\langle m \rangle) = P'/P(\langle m \rangle) = \exp(-2D_m\,\delta)$ in the intermediate region where $\langle m \rangle \sim 1$.

The occurrence of oscillations in white-beam neutron depolarization experiments with ferromagnets was also predicted in a further theoretical analysis by Mayers (1982b). This author showed that the measurement of the wavelength variation of the depolarization parameter D_m should provide information on the domain size distribution function, the domain magnetic induction $\boldsymbol{B}_0$, and the mean value of $\overline{\cos^2 \theta}$, where θ is the domain orientation. The oscillatory nature of this wavelength dependence was later observed experimentally by Mitsuda and Endöh (1985) using a broad-band pulsed polarized neutron beam with time-of-flight analysis. In this technique it is the polychromaticity of the beam which provides the variation in τ and hence of $\langle m \rangle$.

6.2.2 Three-dimensional neutron depolarization

One-dimensional depolarization measurements can only provide information over correlation lengths in which the atomic magnetic moments are parallel; they thus allow only the mean domain size δ to be estimated. A major advance took place in the early 1970s with the introduction by the Delft group (see Rekveldt (1973)) and the Leningrad group (see Okorokov *et al.* (1976)), of the three-dimensional neutron depolarization technique. The 3-D method gives information on the mean domain size, the mean magnetization, the local direction cosines of

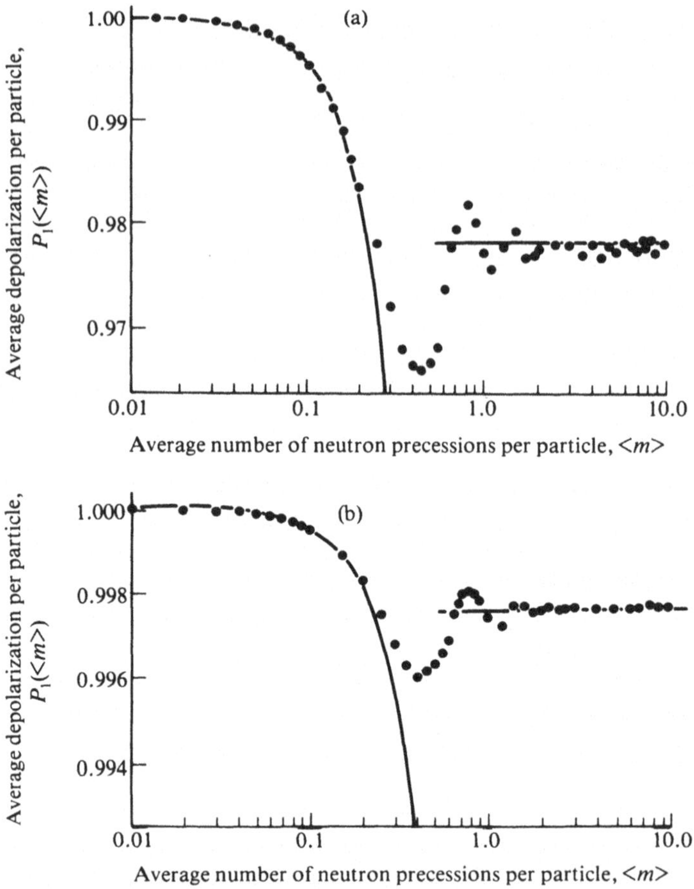

Fig. 6.19. Average depolarization per monodomain particle calculated by Mayers (1982a) for $SmCo_5$ in applied fields (a) $\boldsymbol{B}_a = 0$, and (b) $\boldsymbol{B}_a = 2\,T$, using a Monte Carlo program DEPOL (solid circles). The full curves show the analytically calculated depolarization factors using the Halpern–Holstein expressions in the limits $\langle m \rangle \ll 1$ and $\langle m \rangle \gg 1$.

the domain magnetizations, and in some cases on the correlations between the magnetization directions of neighbouring domains. The Russian group have formulated the depolarization process quantum mechanically, that is they ascribe it to small-angle neutron scattering, and this is exemplified for example in the work of Maleyev (1982). In this book we shall restrict ourselves to the classical 3-D depolarization formalism used by the Delft group. Early measurements were all made in a static mode, but in a later development (see van Schaik and Rekveldt (1978)), dynamic depolarization experiments were introduced. This extension allowed the time-dependence of the domain parameters to be followed to a resolution $\sim 10\,\mu s$, and was achieved by periodically changing the magnetic field at the sample while synchronously detecting the neutrons.

Experimental method and theory

Figure 6.20 shows a typical 3-D neutron depolarization instrument as described by Rekveldt and van Schaik (1979a). The polarizer (M_p) and analyzer (M_a) are Fe_3Si single-crystal polarizing monochromators oriented to reflect a neutron wavelength $\lambda = 1.6$ Å. A second instrument is also available at Delft which uses a magnetized mirror as the polarizer, and can operate over a wavelength range 2–6.5 Å. D_1 is a polarization turner which allows the polarization incident on the sample S to be directed along any one of the laboratory coordinate directions (X, Y, Z). D_2 is a similar 'coil' spin-turner which accepts the transmitted beam polarizations in any of these three directions. If I_{ji} is the intensity detected with the sample in the beam, and I_d and I_m are the intensities of the fully depolarized and maximum-polarized beams respectively, then the quantities

$$D_{1ji} = (I_d - I_{ji})/(I_d - I_m) \tag{6.2.19}$$

define the elements of a (3×3) depolarization matrix, with j and i as variable indices ($j, i = X, Y, Z$) which characterize the polarization changes due to the combined effect of the sample and the applied magnetic fields in the instrument.

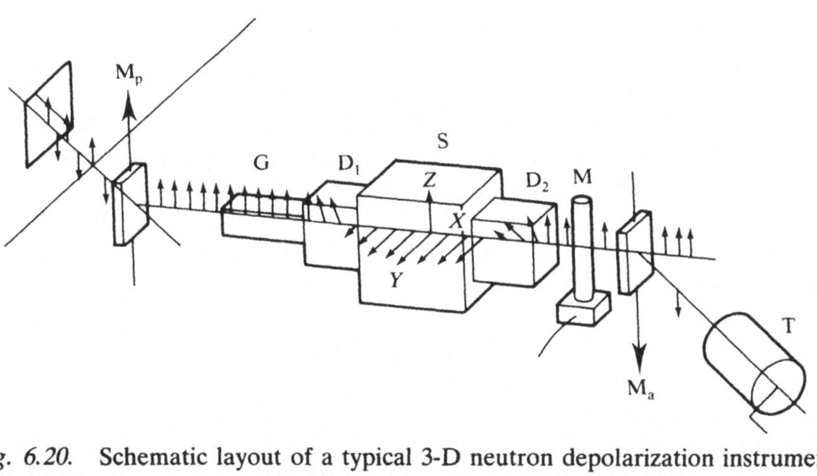

Fig. 6.20. Schematic layout of a typical 3-D neutron depolarization instrument. M_p and M_a are polarizing monochromators; D_1 and D_2 are spin turn coils which rotate the incident (or transmitted) beam polarization vectors into (or from) one of the three (X, Y, Z) coordinate directions; G is a magnetic guide field; M is a neutron monitor; and T is a neutron detector (from Rekveldt and van Schaik 1979a).

The depolarization matrix for a single domain is a pure rotation matrix, $D(\tilde{\boldsymbol{n}}, \omega_L t)$, where $\tilde{\boldsymbol{n}}$ is a unit vector in the direction of the magnetic induction, and t is the time that the neutron spends within the domain. Rekveldt (1977) showed that the change in polarization of a beam that is transmitted through N domains can be expressed as the product of N rotation matrices, that is

$$\boldsymbol{P}(t) = \int_0^\infty \cdots \int_0^\infty \int \cdots \int \mathrm{d}t_N \cdots \mathrm{d}t_1 \, \mathrm{d}\tilde{\boldsymbol{n}}_N \cdots \mathrm{d}\tilde{\boldsymbol{n}}_1$$

$$\cdot p(\tilde{\boldsymbol{n}}_N \cdots \tilde{\boldsymbol{n}}_1, t_N \cdots t_1) D(\tilde{\boldsymbol{n}}_N, \omega_L t_N) \cdots D(\tilde{\boldsymbol{n}}_1, \omega_L t_1) \boldsymbol{P}(0). \quad (6.2.20)$$

The integral $\int \mathrm{d}\tilde{\boldsymbol{n}}$ denotes an average over the orientations of the unit vector $\tilde{\boldsymbol{n}}$, and $p(\tilde{\boldsymbol{n}}_N \cdots \tilde{\boldsymbol{n}}_1, t_N \cdots t_1)$ is a normalized weighting function which gives the probability a matrix row with configuration $(\tilde{\boldsymbol{n}}_N \cdots \tilde{\boldsymbol{n}}_1, t_N \cdots t_1)$.

If we assume first that there is no correlation between the different domain orientations and the different interaction times, then the probability function can be factorized into a product of probability functions for each separate domain. The result is that a theoretical depolarization matrix can be derived containing the parameters

$$B_s^2(1 - m^2)\delta, \quad \text{and} \quad \gamma_v = \langle B_v^2 \rangle / B_s^2, \quad \text{with} \quad v = X, Y \text{ or } Z. \quad (6.2.21)$$

The brackets $\langle \quad \rangle$ represent an average over the part of the sample exposed to the polarized neutron beam, B_s is the spontaneous magnetic induction within the domains, B_v is the vth component of the local magnetic induction, m is the reduced magnetization B/B_s, γ_v are the mean-square direction cosines of the local magnetization, and δ is the mean domain size in the neutron beam direction.

This theory was also extended by Rekveldt (1977) to account for correlations between the orientations of adjacent domains. For this it is necessary to introduce a correlation parameter K given by

$$K = \langle (\tilde{\boldsymbol{n}}_i \cdot \tilde{\boldsymbol{n}}_{i+1}) \rangle - (\langle \tilde{\boldsymbol{n}}_i \rangle \cdot \langle \tilde{\boldsymbol{n}}_i \rangle) \quad (6.2.22)$$

and the mean square cosine of the orientations in which the correlation K occurs is

$$\gamma'_v = (\langle n_{iv} n_{jv} \rangle - \langle n_{iv} \rangle^2) / K^{|i-j|}. \quad (6.2.23)$$

At zero magnetization, K is the mean direction cosine between adjacent domains sampled in the neutron beam direction. The effect of a finite correlation is that the probability function $p(\tilde{\boldsymbol{n}}_N \cdots \tilde{\boldsymbol{n}}_1, t_N \cdots t_1)$ cannot be factorized. However, in the approximation, that (i) $\omega_L t \ll 1$ for each domain, (ii) γ'_v is independent on the sequence number i and j for $|i - j| \geqslant 1$, and (iii) the correlation parameter K is not close to unity, a theoretical depolarization matrix can be derived in terms of

$$B_s^2(1-m^2)\delta(1+K)/(1-K)$$
$$\gamma_v(1-K)/(1+K)+\gamma_v' 2K/(1-K)$$
$$\text{and } B. \qquad (6.2.24)$$

A measurement of the depolarization matrix allows all these quantities to be measured, as well as their dependences on the applied magnetic field, temperature, mechanical stress, etc. It should be emphasized that the correlation theory is only strictly applicable for small domains, where each domain in the neutron path obeys the same statistics. This is the case for example with some ferrites, but there are many instances in which a special depolarization theory has to be derived to match a particular domain model.

Static three-dimensional depolarization experiments

We start with an example in which the foregoing theory is valid, and the correlation parameter K is zero. This was the case for a single-domain NiZn ferrite sample investigated by Rekveldt (1977). The mean domain size δ and the mean-square Y-component of the internal magnetization derived from the neutron depolarization measurements, are shown as a function of the reduced magnetization $m = B/B_s$ in Fig. 6.21. The results were analysed in terms of a model involving an isotropic distribution of domain orientations, and the broken line in the γ_Y plot gives the predictions of this model; the agreement with experimental points is excellent. Figure 6.21 also indicates that the domain size δ is essentially independent of the magnetization, and of the order of the grain size, which is 2 μm. This shows that the correlation parameter K is constant for all m, and is probably zero on the basis that its definition makes a correlation very unlikely at higher values of m. The apparent increase of the domain size δ for $|m| > 0.7$ may be an artefact of a systematic error in m, since δ is determined from the quantity $B_s^2(1-m^2)\delta$ and will be overestimated if m is chosen to be too large.

The remaining examples described in this section involve specimens with large domains, for which the foregoing theory is inapplicable. First we discuss the experiments of Kraan and Rekveldt (1977) on electro-deposited nickel sheet. The 3-D depolarization method was used to measure the diagonal elements D_{XX}, D_{YY} and D_{ZZ} of the depolarization matrix. The coordinate system is defined as in Fig. 6.22(a), where the neutron beam travels in the X-direction and is inclined at an angle θ to the nickel sheet. The domain structure shown contains slabs of thickness d separated by 180° walls, and experimental results obtained at various θ

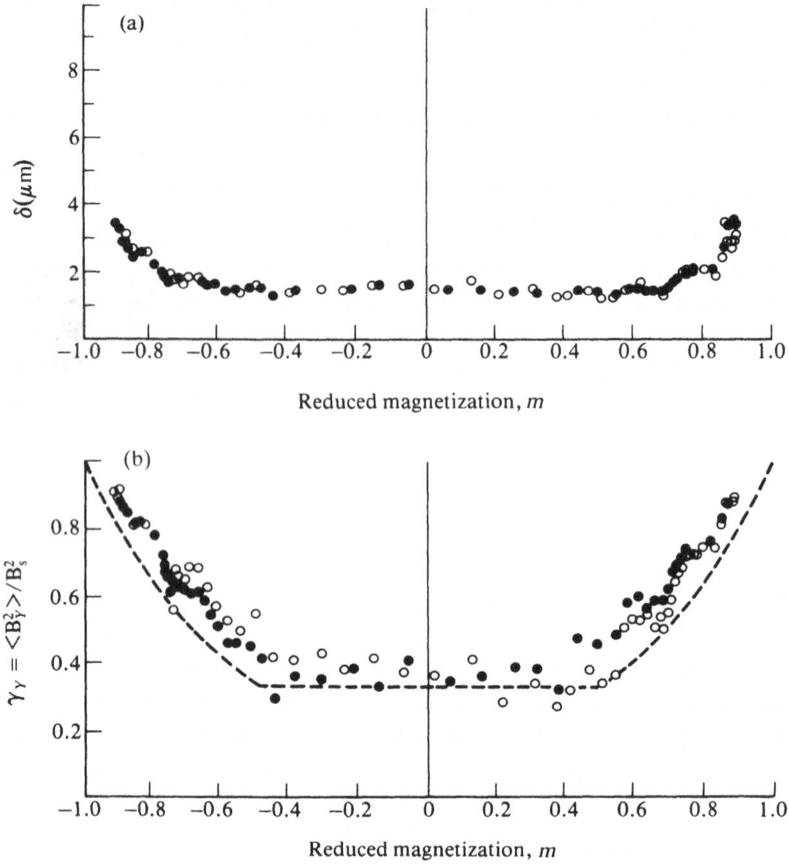

Fig. 6.21. (a) The mean domain size δ, and (b) the mean-square Y-component, of the internal magnetization of a NiZn ferrite derived by Rekveldt (1977) from three-dimensional neutron depolarization measurements, and plotted as a function of the reduced magnetization $m = B/B_s$. The open and solid circles are corresponding experimental points, and the broken curve is calculated from a theoretical model.

values, as well as at different wavelengths λ, were compared with the depolarization parameters predicted for several slab models. Figure 6.22(b) shows the measurements of D_{XX}, D_{YY} and D_{ZZ} as a function of λ when $\theta = 0°$. The observation that D_{XX} is nearly equal to unity at all wavelengths is indicative of a domain structure in which the major part of the volume of the slab is occupied by domains whose magnetizations are either parallel or antiparallel to the X direction. For a single domain structure (P'/P) and D are equivalent quantities, and in this example there is little precession about the Y and Z directions to cause a depolarization of P_X. The simplest description of the behaviour of P_Y and

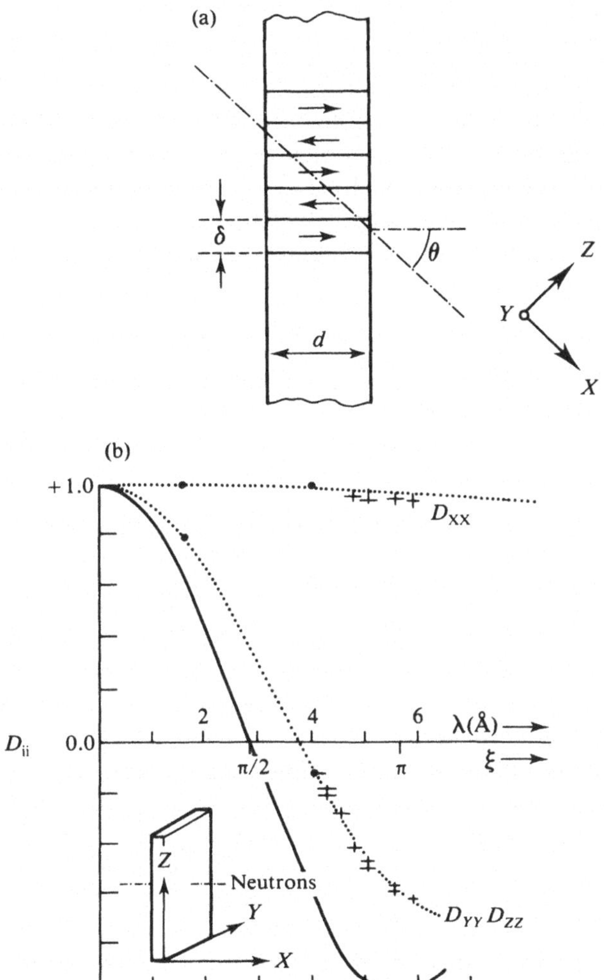

Fig. 6.22. Three-dimensional neutron depolarization measurements on electrodeposited Ni from Kraan and Rekveldt (1977). (a) The laboratory coordinate system and the slab domain structure; (b) the D_{XX}, D_{YY}, D_{ZZ} values measured as a function of the neutron wavelength λ, etc.

P_Z is that they precess perpendicular to the sheet at the Larmor frequency $\omega_L = 2\,|\mu_n|\,B_s/\hbar$, where B_s is the magnetic induction inside the domain, i.e.

$$(P'_{YY(ZZ)}/P_{YY(ZZ)}) = D_{YY(ZZ)} = \cos\xi = \cos(\omega_L d/v). \qquad (6.2.25)$$

The full curve in Fig. 6.22(b) shows this cos function for a domain of length $d = 20\ \mu m$ and $B_s = 0.61\ Wb\ m^{-2}$, which are the known parameters for the Ni sheet investigated. D_{YY} and D_{ZZ} decrease more slowly than anticipated by this simple model, and Kraan and Rekveldt (1977) suggested that this could be due to the effect of closure domains. They developed a sandwich model in which the closure domains on each side of the sheet were represented by two layers of constant thickness δ_c, which effectively reduce the sheet thickness to $(d - 2\delta_c)$. These authors also analysed the effects of a misalignment of the main 'slab' domains. By introducing the parameters

$$\eta = \delta/d; \qquad \eta_c = \delta_c/d; \qquad \phi_c = \xi\eta_c \tag{6.2.26}$$

Kraan and Rekveldt showed that the results of Fig. 6.22(b) could be explained using the relationships

$$D_{XX} = \cos^2 \phi_c \tag{6.2.27}$$

and

$$D_{YY} \sim D_{ZZ} \sim \tfrac{2}{3}(\Delta\theta/\eta)[(\sin \xi/\xi) - \cos \xi] + \cos \xi, \tag{6.2.28}$$

where $\Delta\theta$ is a domain misorientation factor; the best fit was obtained with $\eta_c = 0.06$ and $\Delta\theta/\eta = 0.33$. It was concluded that the most satisfactory model of the domain structure in Ni was one that contained both a misalignment of the principal domains, where the magnetic induction directions are approximately perpendicular to the face of the sheet, and closure domains.

Next we discuss the use of the 3-dimensional depolarization method to obtain precise values of the magnetic induction in iron and nickel. By carrying out measurements in the vicinity of their Curie temperatures, Stüsser *et al.* (1986) were able to determine the critical exponents β and γ which characterize these phase transitions, as well as the critical amplitudes b and their correction terms. The experiments were performed on annular rings (outside diameters 16 mm, inside diameters 10 mm) of 3 mm thick polycrystalline Fe and Ni, and the incident neutron polarization vector was directed along the X-axis, see Fig. 6.23(a). The ring shape of the specimens was chosen to minimize demagnetizing field corrections. The Larmor precession angle ϕ_L of the polarization vector around $\boldsymbol{B}$ during the time that it traverses the sample was determined by measuring the $Z(D_{ZX})$ and $X(D_{XX})$ components of the transmitted polarization. At the neutron wavelength ($\lambda = 1.6$ Å) of the experiments,

$$B(\text{tesla}) = \{\phi_L(\text{radian}) + 2\pi N\}/73d, \tag{6.2.29}$$

where N is the number of complete 2π rotations about B, and d is the sample thickness in millimetres. This technique gives B to an accuracy $\Delta B \simeq 0.0137\Delta\phi_L/d$, and this is essentially determined by the counting statistics in measuring D_{ZX} and D_{XX} (or ϕ_L). In the experiments of Stüsser *et al.* (1986), $\Delta\phi_L < 0.02$ rad, and hence the resolution in B was

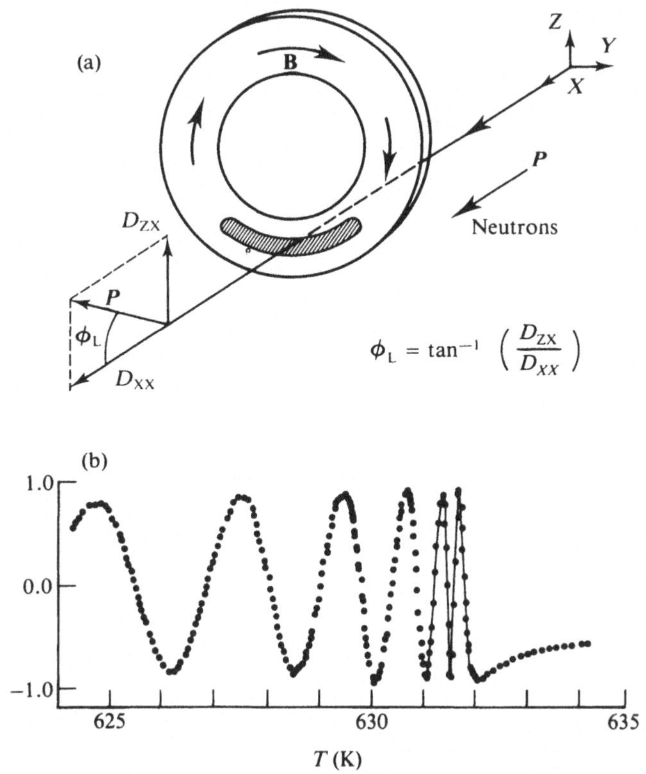

Fig. 6.23. (a) Geometry of neutron depolarization experiments in Ni and Fe; the shaded area of the ring denotes the region sampled by neutrons. (b) The depolarization element D_{ZX} observed in Ni around T_c (from Stüsser *et al.* 1986).

better than 2×10^{-4} T. Applied fields were generated by a toroidal coil wound around the specimen rings, and the temperature control for the measurements was within ±10 mK. Figure 6.23(b) shows the depolarization parameter D_{ZX} obtained for Ni near its Curie temperature.

Under the conditions of these experiments, in which the reduced temperatures $T^* = (T - T_c)/T_c$ were in the range $\sim 5 \times 10^{-4}$ to 5×10^{-3}, the critical behaviour just below the phase transition behaves as follows:

$$m = \frac{B(T)}{B_s(0)} \simeq \frac{B_s(T)}{B_s(0)} = bT^{*\beta}[1 + aT^{*\Delta}], \tag{6.2.30}$$

where the subscript s denotes the spontaneous magnetic induction, and β and b are the critical exponent and the critical amplitude respectively. The parameters a and Δ are amplitude and exponent correction terms. Figure 6.24 shows the $B(T)$ data taken for Ni, and this was analysed as a

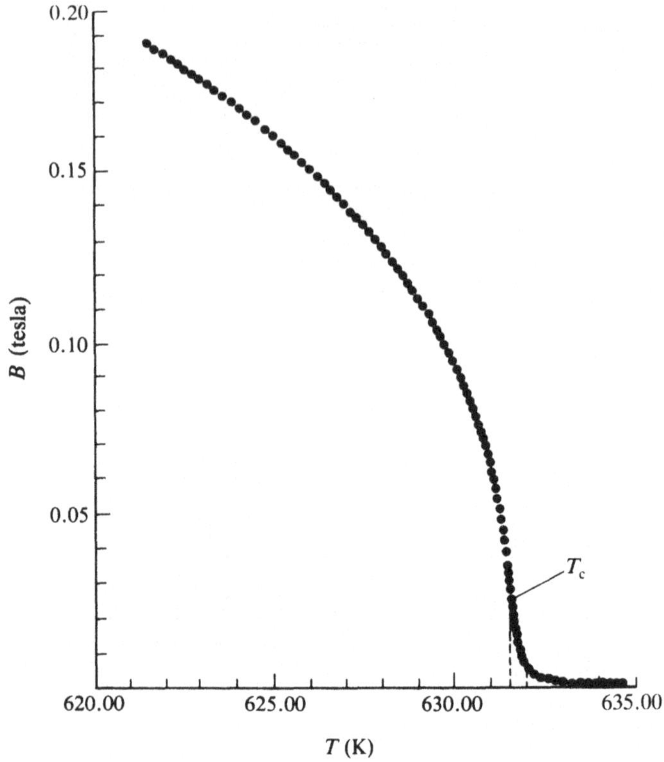

Fig. 6.24. The magnetic induction B observed in Ni ($H = 546$ A m^{-1}) around the Curie temperature as derived from neutron depolarization measurements (from Stüsser *et al.* 1986).

two-parameter fit in b_{eff} and β_{eff} using

$$m = b_{\text{eff}} T^{*\beta_{\text{eff}}}. \tag{6.2.31}$$

The relationship between β_{eff} and β is

$$\beta_{\text{eff}} = \beta + a\Delta \overline{T^{*\Delta}}, \tag{6.2.32}$$

where $\overline{T^*}$ is an average over the temperature range of the measurements. After fitting the experimental results obtained over several $\overline{T^*}$ intervals, and then using $\Delta = 0.55$, as appropriate for a three-dimensional Heisenberg system, it was possible to derive values for the universal critical exponent β as well as the correction to the scaling amplitude a. By combining the above equations we see that

$$b_{\text{eff}} = b\langle (1 + aT^{*\Delta})/(T^{*a\Delta\overline{T^{*\Delta}}}) \rangle. \tag{6.2.33}$$

Thus this analysis also provides the value of the critical amplitude b.

TABLE 6.2 The critical exponents β and γ, the critical amplitude b, and the correction to the scaling amplitude a derived for Fe and Ni using the neutron depolarization technique. The numbers in parentheses show the uncertainty in the last digit. The b and a values are based on a spontaneous magnetic induction at $T=0$ of 2.15 T in Fe, and 0.65 T in Ni (from Stüsser *et al.* 1986)

Ferromagnet	β	b	a	γ
Fe	0.373(4)	1.63	−0.53(8)	1.34(2)
Ni	0.390(4)	1.52(2)	−0.41(4)	1.32(2)

The neutron depolarization estimates of $B(T)$ obtained in the paramagnetic phase above T_c were fitted to the following expression for the normalized magnetic induction m,

$$m = B(T)/B(0) = GT^{*-\gamma}H, \tag{6.2.34}$$

and for the low applied field H used in the experiments m/H was assumed to be identical to the initial susceptibility. The upper bound to the temperature range over which the fit was attempted was typically a few degrees kelvin above T_c. The values of the critical exponents γ in Fe and Ni obtained by this procedure of fitting the B versus T^* results agreed well with those measured using other techniques.

Table 6.2 summarizes the measurements made by Stüsser *et al.* (1986) of the critical exponents β and γ, the critical amplitudes b, and the amplitude correction terms a, for iron and nickel.

Dynamic neutron depolarization experiments

The time-dependent or dynamic neutron depolarization technique, see for example van Schaik and Rekveldt (1978), can be applied to investigate a variety of ferromagnetic materials ranging from polycrystalline multidomain structures to single-crystal domain structures. We now describe the principal results of two experiments: a low-field study of the domain wall dynamics in an FeSi single crystal, and an examination of the magnetization dynamics in nickel near T_c.

The geometry of the FeSi (3.5 wt% Si) single-crystal experiment performed by Rekveldt and van Schaik (1979b) is shown in Fig. 6.25. The crystal was in the form of a picture frame, and polarized neutrons with velocity $v = 2500\ \mathrm{m\,s^{-1}}$ were transmitted in the X direction through one of its longer sides. The beam was defined by a rectangular $(5 \times 0.3)\mathrm{mm}^2$ Cd–B_4C aperture with the longer direction parallel to the Z-axis of the laboratory coordinate system. This side of the crystal was also magnetized by activating the coil winding shown in the diagram, and the object of the experiment was to follow the time evolution of the depolarization matrix elements by using a multichannel analyser to detect

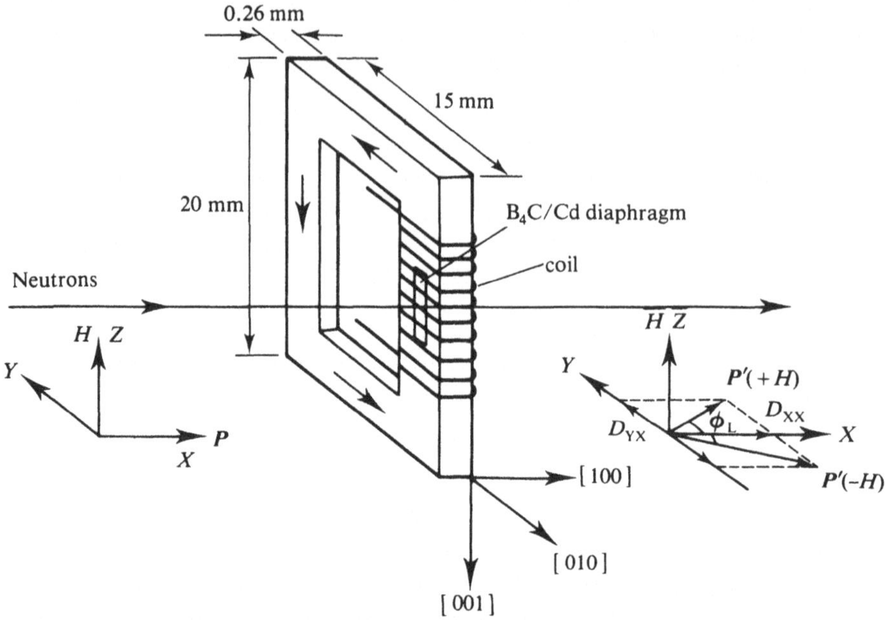

Fig. 6.25. Geometry of the dynamic three-dimensional depolarization experiment with the 'picture frame' FeSi (3.5 wt%) single crystal (from Rekveldt and van Schaik 1979b).

intensity changes as a function of time following the magnetization reversal pulse.

We shall assume that the crystal is homogeneously magnetized in the Z direction, and the net rotation angle in the XY plane is $\phi_L = \tan^{-1}(D_{YX}/D_{XX})$. The sign of D_{YX} changes with the sign of the mean magnetization M sampled in the crystal, and is in fact very sensitive to small changes in M. A rotation angle difference of 1° is easily detectable, and this corresponds to a change in the mean magnetization δM of the order 4×10^{-4}. D_{XX} is, in contrast, independent of the sign of M since, as is apparent from Fig. 6.25, $\cos(-\phi_L) = \cos(+\phi_L)$. Similarly, P_Z does not depend on M, provided that M remains parallel to Z; therefore D_{ZZ} is expected to remain at its maximum when the magnetization direction is reversed. The depolarization parameters D_{ZZ}, D_{YX} and D_{XX} measured by Rekveldt and van Schaik (1979b) are shown in Fig. 6.26, where the applied field was $7\ \mathrm{A\ m^{-1}}$ and the discontinuities show the times of the field reversals. The fact that D_{XX} does not change is consistent with the hypothesis that the domain wall in this example is a straight 180° wall perpendicular to the crystal surface. The change in D_{YX} takes place when this straight wall crosses the neutron diaphragm. These authors also discussed the behaviour expected in the case of a 'bowed' wall, and

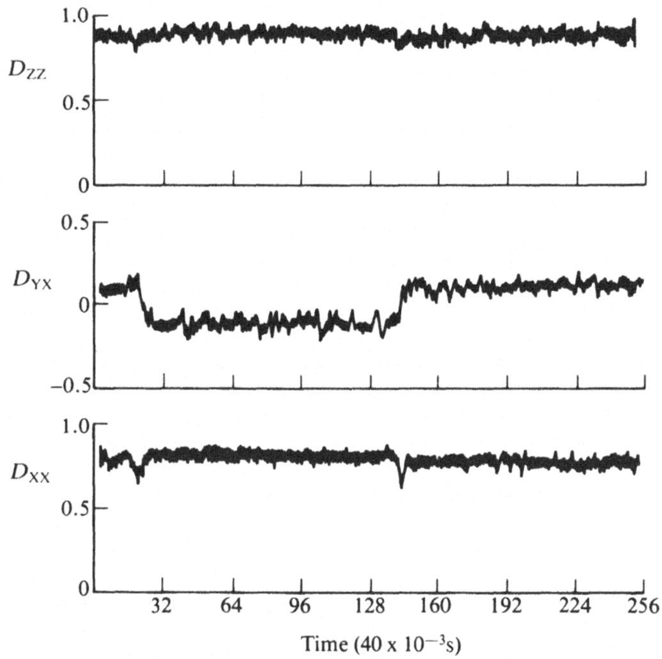

Fig. 6.26. The time dependences of D_{ZZ}, D_{YX} and D_{XX} in a FeSi single crystal measured by the dynamic three-dimensional depolarization method with a low magnetizing field (from Rekveldt and van Schaik 1979b).

predicted that large changes in D_{XX} would occur. If the domain wall was pronouncedly bowed, then D_{XX} would approach zero. The magnetization reversal in FeSi in high applied fields takes place by a completely different mechanism; for example, both D_{XX} and D_{YX} show a distinct oscillatory behaviour with time following the field reversal. The results were interpreted by Rekveldt and van Schaik (1979b) in terms of a sandwich domain model in which the magnetization reversal takes place through two domain walls which are parallel to the crystal surfaces, and which travel from the surfaces towards the interior of the specimen.

As a second example of a dynamical three-dimensional depolarization measurement, we now describe the experiment of Rekveldt, Stüsser and Spruijt (1986) who studied the magnetization reversal in a polycrystalline nickel ring over a temperature range from 20 K below T_c up to $T_c = 630$ K. The geometrical arrangement was identical to that for the static experiment described earlier, and is illustrated in Fig. 6.23(a). A coil was wound around the Ni ring, and by applying a periodic magnetic field to the sample it was possible to determine the time-dependences of the depolarization matrix elements D_{ji} with $j, i = X, Y$ or Z. Figure 6.27 shows the results for Ni at one temperature and field strength. D_{XX} and

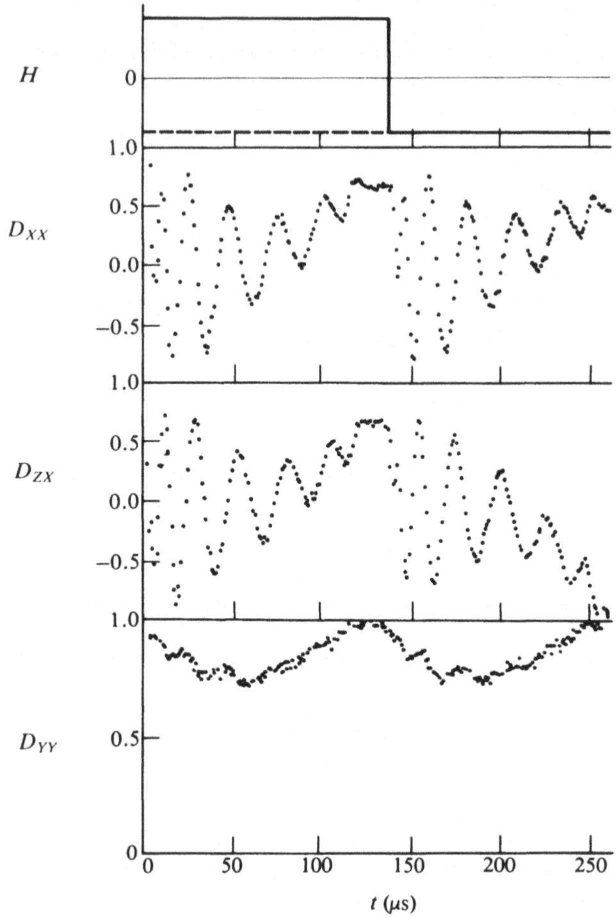

Fig. 6.27. Time dependences of the neutron depolarization elements D_{XX}, D_{ZX} and D_{YY} in Ni at $T = (T_c - 6)$ K following a reversal of the magnetic field H. The time resolution in these experiments was $\Delta t \sim 6\ \mu s$ (from Rekveldt *et al.* 1986).

D_{ZX} together determine the rotation angle ϕ_L of the polarization vector around the Y axis, whereas D_{YY} measures the mean-square of the angle between the crystallite magnetizations and the direction of the applied field.

The dynamics of the magnetization process can be deduced from the oscillatory behaviour of D_{XX} and D_{ZX}. The level around which these oscillations occur changes from zero, during the initial stages of the magnetization reversal, to approach the saturation values of D_{XX} and D_{ZX} during the final phase of the cycle. Rekveldt *et al.* (1986) showed that this behaviour was consistent with that expected for a cylindrical domain wall running from the surface of the specimen towards its centre, where it then disappears. The behaviour of D_{YY} immediately after the field reversal indicates that the sample remains essentially fully magnetized antiparallel to the applied field. The reduction in D_{YY} which then follows is a result of the relaxation of the individual crystallite magnetizations within a cylindrical domain as they are oriented towards their easy directions. D_{YY} then increases again to its maximum value and by the end of the reversal cycle all the crystallite magnetizations are parallel to the applied field.

6.3 SPIN PRECESSION IN SAMPLES CONTAINING POLARIZED NUCLEI

The experiments described in the previous section all rely on the appreciable interaction between the neutron magnetic moment and the magnetic fields generated by magnetic atoms. The equivalent purely magnetic interaction with nuclear moments is negligibly small by comparison, and is too weak to be useful in either scattering or spin precession experiments. We have, however, seen that the neutron spin does interact strongly with nuclear spins by a purely nuclear force, and this is manifested by the difference between the scattering amplitudes b_+ and b_- for the scattering through $(i+\frac{1}{2})$ and $(i-\frac{1}{2})$ compound nuclear spin states. This interaction has been formulated by Abragam (see for example Abragam and Goldman (1982)), in terms of a nuclear pseudomagnetism in which the nucleus can be assigned a pseudomagnetic moment. We now show that these moments can have values of the same order of magnitude as atomic moments, and that in specimens containing polarized nuclei, neutron spin precessional methods can be used to determine their magnitudes. Such experiments provide a very sensitive method for determining the nuclear scattering amplitude parameter $\mathscr{B}$, which is proportional to $(b_+ - b_-)$.

6.3.1 Nuclear pseudomagnetic moments

The concept of pseudomagnetic moments μ_N^* in nuclei which have a large spin incoherent scattering parameter $\mathscr{B}$, was introduced by Abragam and coworkers (see for example Abragam and Goldman (1982)). The easiest method of obtaining an expression for this moment is to compare the nuclear spin scattering length b_{NS} given by

$$b_{NS} = \tfrac{1}{2}\mathscr{B}\hat{\boldsymbol{\sigma}} \cdot \hat{\boldsymbol{i}} = \mathscr{B}\hat{\boldsymbol{s}}' \cdot \hat{\boldsymbol{i}} = \{\mathscr{B}[i(i+1)]^{1/2}\}/2, \tag{6.3.1}$$

where $\mathscr{B} = (b_+ - b_-)/(i+\frac{1}{2})$, with the usual electronic magnetic scattering amplitude p. In the case in which the atomic moment is perpendicular to the scattering vector, we see from equn (2.1.73) that p is given by

$$p = 2\,|r_0|\,\hat{\boldsymbol{s}}' \cdot \hat{\boldsymbol{s}}, \tag{6.3.2}$$

where $r_0 = -1.913$ is the classical electron radius (e^2/m_ec^2) multiplied by γ_n, the neutron magnetic moment expressed in nuclear magneton units. Both b_{NS} and p can therefore be formulated in terms of the coupling of the neutron spin $\hat{\boldsymbol{s}}'$ with either a nuclear ($\hat{\boldsymbol{\imath}}$) or atomic electron ($\hat{\boldsymbol{s}}$) spin, or alternatively in terms of the respective magnetic moments.

Let us first consider the 'conventional' magnetic scattering from unpaired electron spins. The neutron magnetic dipole moment $\boldsymbol{\mu}_n$ is

$$\boldsymbol{\mu}_n = \gamma_n(e\hbar/2mc)\hat{\boldsymbol{\sigma}} = \gamma_n(e\hbar/mc)\hat{\boldsymbol{s}}', \tag{6.3.3}$$

and the electron magnetic dipole moment $\boldsymbol{\mu}_e$ is

$$\boldsymbol{\mu}_e = -2\beta\hat{\boldsymbol{s}} = -2(e\hbar/2m_ec)\hat{\boldsymbol{s}}. \tag{6.3.4}$$

By combining these last three equations, we obtain a magnetic scattering amplitude:

$$p = (-2m/\hbar^2)\boldsymbol{\mu}_n \cdot \boldsymbol{\mu}_e. \tag{6.3.5}$$

We now introduce a nuclear pseudomagnetic moment μ_N^* which we write as

$$\boldsymbol{\mu}_N^* = (\mu_N^*/i)\hat{\boldsymbol{\imath}}, \tag{6.3.6}$$

and which we wish to equate with the nuclear spin scattering amplitude b_{NS}. By making the analogy with equn (6.3.5), we obtain

$$b_{NS} = -\frac{2m}{\hbar^2}\boldsymbol{\mu}_n\boldsymbol{\mu}_N^* = -\frac{2m}{\hbar^2}\gamma_n\left(\frac{e\hbar}{mc}\right)\frac{\boldsymbol{\mu}_N^*}{i}\hat{\boldsymbol{s}}' \cdot \hat{\boldsymbol{\imath}} = \mathscr{B}\hat{\boldsymbol{s}}' \cdot \hat{\boldsymbol{\imath}}, \tag{6.3.7}$$

and therefore

$$\boldsymbol{\mu}_N^* = -i\mathscr{B}\hbar c/(2e\gamma_n) = i\mathscr{B}hc/(2e\,|\gamma_n|). \tag{6.3.8}$$

A more instructive way of writing μ_N^* is to relate it to the Bohr magneton,

$$\beta = e\hbar/(2m_ec), \tag{6.3.9}$$

that is,

$$\frac{\mu_N^*}{\beta} = \frac{-i\mathscr{B}}{\gamma_n}\frac{\hbar c}{2e}\frac{2m_ec}{e\hbar} = \frac{-i\mathscr{B}}{\gamma_n}\frac{m_ec^2}{e^2} = \frac{-i\mathscr{B}}{r_0}, \tag{6.3.10}$$

with r_0 defined as

$$r_0 = \gamma_ne^2/(m_ec^2). \tag{6.3.11}$$

The spin incoherent scattering parameters $\mathscr{B}$ for many nuclei are comparable to and sometimes greater in magnitude than r_0. In these cases the pseudomagnetic moment μ_N^* can be of the order of a Bohr magneton, and the nuclear spin dependent scattering will be of a similar magnitude to electronic magnetic scattering. For the proton, $\mathscr{B} = 5.8 \times 10^{-12}$ cm and $\mu_P^* = \mu_N^*(^1H) = 5.4\beta = 3600\mu_P$, where μ_P is the 'true' magnetic moment of the proton. Thus, if polarizations comparable with those of electron spin systems can be produced with nuclear spins, then, as we have seen in Chapter 4, they can readily be investigated by neutron scattering.

In a sample containing polarized nuclei, the polarized moments give rise to a pseudomagnetic field B^*, and the precession of polarized neutrons in this field provides a very accurate method for determining the nuclear spin incoherent scattering parameters. The pseudomagnetization in such a sample may be expressed by

$$M^* = N_0 \mu_N^* P_N, \tag{6.3.12}$$

where N_0 is the nuclear density, and P_N is the nuclear polarization. A polarized neutron passing through such a polarized target therefore experiences a pseudomagnetic field,

$$B^* = 4\pi M^* = 4\pi N_0 \mu_N^* P_N, \tag{6.3.13}$$

and this results in a shift of its Larmor frequency, in comparison with the value in vacuum, which is given by

$$\Delta\omega_L = -\gamma B^* = -4\pi\gamma N_0 \mu_N^* P_N, \tag{6.3.14}$$

where $\gamma = -1.833 \times 10^8$ rad s^{-1}T^{-1} is the neutron gyromagnetic ratio. Remembering that γ may be expressed as

$$\gamma = \boldsymbol{\mu}_n/(\hbar \hat{\mathbf{s}}') = (\gamma_n e)/(mc), \tag{6.3.15}$$

and then substituting equn (6.3.10) for μ_N^*, we obtain

$$\Delta\omega_L = \frac{4\pi\hbar}{m} N_0 \frac{i\mathscr{B}P_N}{2} = \frac{4\pi\hbar}{m} N_0 \frac{i}{(2i+1)}(b_+ - b_-)P_N. \tag{6.3.16}$$

This equation was first derived by Baryshevskii and Podgoetskii (1964) from the standpoint of the refractive index of the neutron, rather than by using the magnetic scattering analogy. The first experimental confirmation of the validity of this theory was obtained in an experiment reported by Abragam *et al.* (1972b), in which polarized neutrons passing through an LMN polarized proton target (see Section 3.1.2) underwent a resonant depolarization from the pseudomagnetic induction associated with the proton polarization.

6.3.2 The neutron spin precession method for determining the spin-dependent nuclear scattering parameter $\mathscr{B}$

A knowledge of the signs and magnitudes of the spin-independent (b) and spin-dependent (b_{NS}) nuclear scattering amplitudes is of fundamental importance to the neutron scattering experimentalist; the latter, for example, determines the spin incoherent scattering. We first recall from Section 2.2.1 how these amplitudes are related to the compound nuclear scattering amplitudes b_+ and b_-. The spin-independent amplitude is given by

$$b = [(i+1)b_+ + ib_-]/(2i+1), \tag{6.3.17}$$

and the spin-dependent amplitude is

$$b_{NS} = \mathscr{B}[i(i+1)]^{1/2}/2 \tag{6.3.18}$$

with

$$\mathscr{B} = (b_+ - b_-)/(i + \tfrac{1}{2}). \tag{6.3.19}$$

There are several excellent neutron diffraction and optical methods which can be used to give reliable estimates of both the signs and magnitudes of the spin-independent scattering amplitude b. These can be used together with total scattering cross-section measurements, i.e. $\sigma_s = 4\pi(b^2 + b_{NS}^2)$, to supply the magnitudes of b_{NS} (or $\mathscr{B}$), but they cannot give their signs. The signs and magnitudes of $\mathscr{B}$ can only be determined unequivocally by using polarized neutrons, and it is then possible to deduce the individual values of b_+ and b_- by combining equns (6.3.17) and (6.3.19).

In Section 4.5 we described how polarized neutron diffraction from single crystals containing polarized nuclei can be used to evaluate the signs and magnitudes of the $\mathscr{B}$ parameters for those nuclei. The neutron precession method in targets of polarized nuclei, first introduced by Abragam *et al.* (1973), provides an alternative method which is applicable without the need for single-crystal specimens. In the experiment, which is based on the two-coil beam method of Ramsey (1949), instead of measuring the change of the precession frequency $\Delta\omega_L$ as given in equn (6.3.16), the change in the integrated precession angle caused by the pseudomagnetic field of the polarized nuclei is determined.

The experimental set-up is shown in Fig. 6.28. Two r.f. coils, C′ and

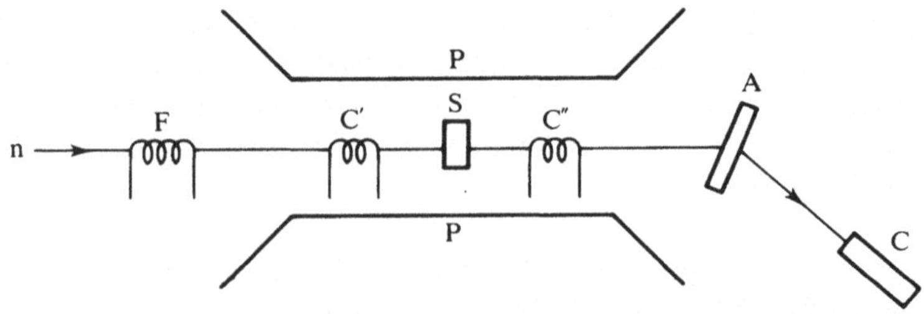

Fig. 6.28. Schematic representation of the two-coil neutron spin precession experiment: n, polarized monochromatic neutrons; C′ and C″, r.f. coils; S, sample; A, polarizing crystal spin analyser; P, magnet pole pieces; C, neutron detector, from Abragam *et al.* (1973).

C″, of length l and separated by a distance $L \gg l$, are placed in the gap of a magnet which supplies a steady field B_0. The coils produce two magnetic r.f. fields $2B_1' \cos(\omega t)$ and $2B_1' \cos(\omega t + \phi_L)$ perpendicular to B_0, and their frequencies can be adjusted to coincide with the neutron Larmor precession frequency, i.e. $\omega = \omega_L = -\gamma B_0$. The monochromatic incident beam with velocity v has a polarization P which can be flipped by the spin-flipper F, and its polarization (in the direction of B_0) after it leaves the second coil C″ is measured by the polarizing crystal analyser A. The precession angles in the two coils are given by

$$\phi_L' = -\gamma B_1' l/v \quad \text{and} \quad \phi_L'' = -\gamma B_1'' l/v, \tag{6.3.20}$$

and the extra precession angle in the sample S of thickness d, which contains the polarized nuclei that generate a pseudomagnetic field B^*, is

$$\alpha = -\gamma B^* d/v. \tag{6.3.21}$$

After accumulating these precession angles, the net polarization P' measured at the analyser becomes

$$P' = P[\cos \phi_L' \cos \phi_L'' - \sin \phi_L' \sin \phi_L'' \cos(\alpha + \phi_L)] \tag{6.3.22}$$

It is usually arranged that $\phi_L' = \phi_L'' = \pi/2$, and this relationship then simplifies to

$$(\alpha + \phi_L) = \cos^{-1}(P/P') \tag{6.3.23}$$

or

$$B^* = (-v/\gamma d)[\cos^{-1}(-P'/P) - \phi_L]. \tag{6.3.24}$$

These equations provide the basis for determining μ_N^* by using the two-coil precession method. For example, for statically polarized nuclei where the nuclear polarization P_N is low, we have $P_N \propto T^{-1}$, and, since $B^* \propto P_N$, a plot of $\cos^{-1}(-P'/P)$ against the inverse temperature T^{-1} gives a straight line whose slope provides the value of μ_N^*. Abragam *et al.* (1973) used statically polarized ^1H and ^{51}V targets to test the viability of the method. Their samples were a statically polarized 6 mm thick single crystal of LMN, doped with Pr^{3+} to shorten the nuclear relaxation time, and a 10 mm thick sample of metallic ^{51}V. The precession angles $\psi = \alpha + \phi_L$ determined for the LMN sample as a function of temperature are shown in Fig. 6.29. The μ_P^* value extracted from the slope of the straight line was in excellent agreement with the known value of this parameter assuming that the proton polarization was given by

$$P_H = [(i+1)/3i](\mu_P B_0)/(k_B T), \tag{6.3.25}$$

where μ_P is the proton magnetic moment and $i = \frac{1}{2}$. The equivalent ^{51}V experiment also gave excellent agreement with expectations.

It has already been remarked that this technique can be used equally well with polycrystalline or amorphous samples. Another advantage of

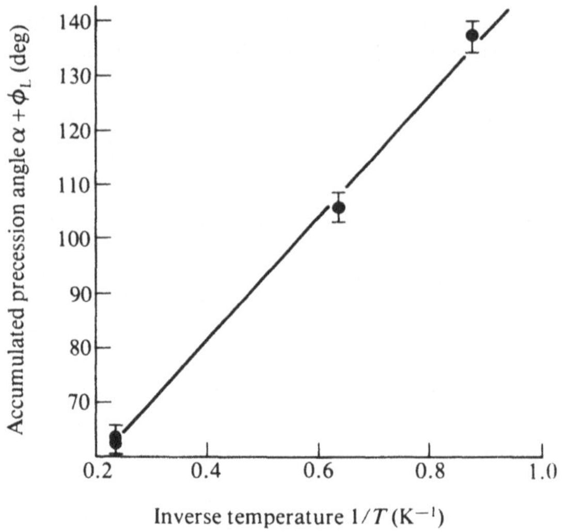

Fig. 6.29. Accumulated precession angle as a function of inverse temperature in a 6.34 mm thick statically polarized LMN single crystal. The straight line is a calculated curve using known polarization and neutron scattering parameters for protons (from Abragam *et al.* 1973).

the two-coil precession method is that, in an analogous way to neutron spin echo, it is very insensitive to the wavelength or velocity spread of the neutron beam. It is also of interest to compare its sensitivity relative to the polarized-beam Bragg diffraction method. Let us take as an example the flipping ratio measurement by Abragam *et al.* (1972a) of the (111) reflection from a polarized single-crystal specimen of LiH. Under the conditions of their experiment, the statically generated proton polarization was ~0.18%, and the change in the flipping ratio for this reflection was ~6%. In the spin precession method, under the same applied field conditions, a 6 mm thick sample would give a 90° precession angle change between 1 K and 4 K, and this gives a change in the analysed polarization P' of about 100%. The precession method is therefore far more sensitive than the Bragg diffraction method, and we now proceed to present some of the results obtained using this technique.

6.3.3 Results from spin precession experiments with polarized nuclei

The experimentally determined parameter is the accumulated precession angle $(\alpha + \phi_L)$, as given in equn (6.3.23), where only the α term is sample-dependent. We may now omit the ϕ_L term, since the quantity of interest is the difference in $(\alpha + \phi_L)$ due to the sample term alone. The extra precession angle due to the pseudomagnetic field in the sample is evaluated from the observed flipping ratio $R = I$(flipper 'off')/(flipper 'on') using

$$\cos\alpha = (1 - R)/\{D_s P P_A[1 - R(1 - 2f)]\}, \qquad (6.3.26)$$

where P in equn (6.3.23) is replaced by $D_s P P_A$. Here, D_s is a sample depolarization factor which has to be evaluated for each experiment. The other parameters are as previously defined in this book.

By combining equns (6.3.14) and (6.3.21) we see that α is related to the pseudomagnetic nuclear moment μ_N^* as follows:

$$\alpha = \Delta\omega_L(d/v) = -4\pi\gamma N\mu_N^* P_N(d/v). \qquad (6.3.27)$$

In all the experiments carried out so far, μ_N^*, and hence $\mathscr{B}$ or $(b_+ - b_-)$, has been determined by measuring α as a function of the nuclear polarization P_N. Sizeable nuclear polarizations are required, and these have been achieved both statically by cooling the samples to millikelvin temperatures, or dynamically using magnetic resonance techniques such as the 'solid-state effect'. In the former, the main experimental problem is to achieve thermal equilibrium through thermal contact between the sample and its coolant, and then between the lattice and the nuclear spins. For statically polarized nuclei in the high-temperature approximation ($|\mu_N B_0| \ll k_B T$), the nuclear polarization is given by equn (4.5.15) which, using $\mu_N = \gamma_N \hbar i$, we now express as

$$P_N = (i + 1)(\gamma_N \hbar B_0)/(3k_B T). \qquad (6.3.28)$$

Following the method described by Glättli *et al.* (1979), we may write

$$\alpha = sd(1/T), \qquad (6.3.29)$$

with

$$s = (-4\pi\gamma/v)(\hbar B_0/3k_B)\gamma_N(i + 1)N\mu_N^*. \qquad (6.3.30)$$

The experimental results for statically polarized nuclei are therefore analysed by plotting α versus $(1/T)$; the slope gives sd, which is directly proportional to μ_N^*. The measurements made by Roubeau *et al.* (1974) on ^{27}Al are shown in Fig. 6.30. At temperatures below 50 mK, for which the arrows indicate the time sequence of the measurements, there is clear evidence that the sample does not respond to the temperature changes in the dilution refrigerator mixing chamber. In the experimental set-up used by Glattli *et al.* (1979),

$$s = 2.108 \times 10^{-26}\gamma_N(i + 1)N(\mu_N^*/\beta)\ \mathrm{deg\ cm^{-1}\ K}, \qquad (6.3.31)$$

and for samples in which there are several polarizable isotopes j,

$$s = 2.108 \times 10^{-26}\sum_j \gamma_{N_j}(i_j + 1)N_j(\mu_{N_j}^*/\beta)\ \mathrm{deg\ cm^{-1}\ K}. \qquad (6.3.32)$$

This expression proved to be particularly useful when correcting data

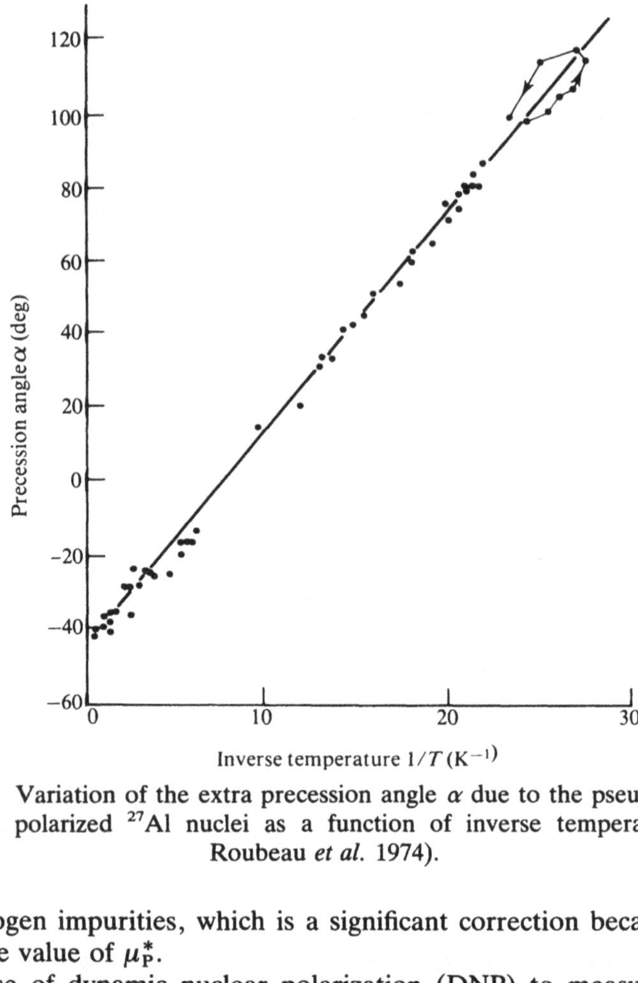

Fig. 6.30. Variation of the extra precession angle α due to the pseudomagnetic field from polarized ^{27}Al nuclei as a function of inverse temperature (from Roubeau *et al.* 1974).

for hydrogen impurities, which is a significant correction because of the very large value of μ_P^*.

The use of dynamic nuclear polarization (DNP) to measure μ_N^* has been developed by Glättli and Coustham (1983). In contrast to the 'brute force' static polarization method, DNP works well for insulators when they are doped with a small quantity ($N \sim 10^{19}$ at cm^{-3}) of a suitable paramagnetic impurity. The basic idea of the method is described for protons in Section 3.1.2, and illustrated in Fig. 3.8; the same principle can be applied to many other isotopes. Briefly, the nuclear spin reservoirs in the sample are cooled by being in good thermal contact with the electron spin system of the paramagnetic dope, the latter being highly polarized under the high-field and low-temperature conditions of the experiment. Microwave saturation at a frequency close to the paramagnetic resonance frequency of the dope causes the high electron spin polarization to be transferred to the nuclear spin system. The spin temperatures of the nuclei can therefore become much lower than the lattice temperature, and, depending on whether the microwave irradiation takes place above or below the paramagnetic resonance frequency, both positive and negative nuclear polarizations P_N can be produced. Much lower spin temperatures have been reached by DNP than by the 'brute force' polarization method.

Of particular relevance to the measurement of μ_N^* by the spin precession method is that it is possible to selectively depolarize the polarization of specific nuclei by radiofrequency irradiation at their respective nuclear magnetic resonance frequencies in the applied field B_0 of the experiment. In general, DNP polarizes many of the nuclei in the sample, though seldom to the same degree, and if the nuclear spin relaxation and mixing times are long compared with the time to measure the precession angles, then the precession angle α_j for nucleus j may be obtained from

$$\alpha_j = \alpha(P_{N_j}) - \alpha(P_{N_j} = 0), \qquad (6.3.33)$$

where $\alpha(P_{N_j})$ is the value following DNP, and $\alpha(P_{N_j} = 0)$ is its value following nuclear magnetic resonance (NMR) saturation. The procedure can be repeated for several nuclear species in the sample, provided that their NMR frequencies are sufficiently different. This ability to isolate the pseudomagnetic moments of different nuclei gives DNP, besides its higher polarization, two distinct advantages over the 'brute force' polarization method: it obviates the need for costly isotope separation, and also the method is insensitive to hydrogen impurities in the sample.

Following the determination of α_j, equn (6.3.27) then yields μ_j^* and therefore $\mathcal{B}_j$ and $(b_{j+} - b_{j-})$. The main difficulty with the DNP method lies in the evaluation of the nuclear polarization P_{N_j}, which must be

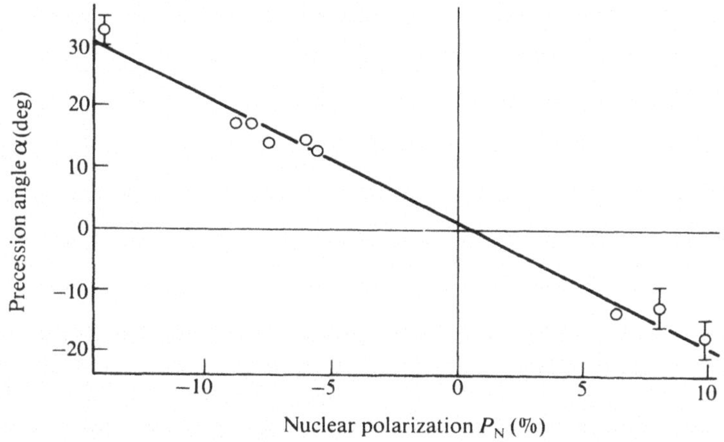

Fig. 6.31. Neutron spin phase shift as a function of ^{13}C nuclear polarization measured by the spin precession technique (from Glättli and Coustham 1983).

TABLE 6.3 Pseudomagnetic moments and $(b_+ - b_-)$ values for various isotopes measured by the neutron spin precession method. This compilation is a combination of the data tabulated by Glättli *et al.* (1979), supplemented by later results provided by Coustham, Glättli and Malinovski (1986, private communication)

Isotope	i	μ_N^*/β	$b_+ - b_-(10^{-12}$ cm)
^{1}H	1/2	+5.4	+5.82
^{6}Li	1	−0.49 ± 0.01	−0.40 ± 0.01
^{7}Li	3/2	−0.67 ± 0.01	−0.48 ± 0.01
^{9}Be	−3/2	+0.034 ± 0.010	+0.024 ± 0.007
^{13}C	1/2	−0.11 ± 0.02	−0.12 ± 0.02
^{19}F	1/2	−0.018 ± 0.002	−0.019 ± 0.002
^{23}Na	3/2	+1.02 ± 0.02	+0.735 ± 0.005
^{25}Mg	−5/2	+0.47 ± 0.03	+0.30 ± 0.02
^{27}Al	5/2	+0.085 ± 0.003	+0.055 ± 0.002
^{35}Cl	3/2	+1.74 ± 0.06	+1.25 ± 0.04
^{39}K	3/2	+0.39 ± 0.09	+0.28 ± 0.07
^{45}Sc	7/2	−2.00 ± 0.11	−1.23 ± 0.07
^{51}V	7/2	+2.074 ± 0.016	+1.278 ± 0.010
^{59}Co	7/2	−2.03 ± 0.06	−1.25 ± 0.04
^{63}Cu	3/2	+0.063 ± 0.006	+0.045 ± 0.005
^{65}Cu	3/2	+0.51 ± 0.02	+0.37 ± 0.02
^{67}Zn	5/2	−0.472 ± 0.024	−0.305 ± 0.015
^{69}Ga	3/2	−0.243 ± 0.016	−0.175 ± 0.011
^{71}Ga	3/2	−0.235 ± 0.027	−0.169 ± 0.019
^{75}As	3/2	−0.199 ± 0.017	−0.143 ± 0.012
^{79}Br	3/2	−0.30 ± 0.05	−0.22 ± 0.04
^{81}Br	3/2	+0.17 ± 0.04	+0.12 ± 0.03
^{89}Y	1/2	+0.23 ± 0.01	+0.25 ± 0.01
^{91}Zr	5/2	−0.34 ± 0.05	−0.22 ± 0.03
^{93}Nb	9/2	−0.046 ± 0.004	−0·028 ± 0.002
^{105}Pd	−5/2	−0.8 ± 0.5	−0.52 ± 0.32
^{107}Ag	1/2	+0.21 ± 0.03	+0.23 ± 0.03
^{109}Ag	1/2	−0.34 ± 0.03	−0.37 ± 0.03
^{133}Cs	7/2	+0.42 ± 0.04	+0.26 ± 0.03
^{139}La	7/2	+0.99 ± 0.07	+0.61 ± 0.04
^{165}Ho	7/2	−0.555 ± 0.026	−0.342 ± 0.016
^{169}Tm	−1/2	+0.22	+0.20 ± 0.06
^{181}Ta	7/2	−0.096 ± 0.01	−0.059 ± 0.006
^{195}Pt	1/2	−0.21 ± 0.04	−0.23 ± 0.04
^{197}Au	3/2	−0.53 ± 0.03	−0.38 ± 0.02
^{203}Tl	1/2	+0.227 ± 0.030	+0.245 ± 0.032
^{205}Tl	1/2	−0.052 ± 0.004	−0.056 ± 0.004
^{207}Pb	1/2	+0.031 ± 0.012	+0.033 ± 0.013
^{209}Bi	9/2	+0.087 ± 0.005	+0.052 ± 0.003

estimated independently in order to obtain μ_j^*. Several methods, which all rely on measuring NMR signals, have been used, and are described by Glättli and Coustham (1983). In practice, however, P_{N_j} is usually the worst-defined parameter in equn (6.3.27), and it is the limiting factor which determines the accuracy of the scattering length measurements.

The precession angles measured by Glättli and Coustham (1983) for

^{13}C in a dynamically polarized $Ba^{13}CO_3$ sample are shown in Fig. 6.31 as a function of both positive and negative nuclear polarization values. A large (200°) phase shift was observed in the sample which was attributed to hydrogen impurities, though selective saturation still allowed the phase shift due to ^{13}C to be measured separately. The slope of the straight line in the figure gave $\mu_N^*(^{13}C) = (-0.11 \pm 0.02)\beta$ or $(b_+ - b_-)(^{13}C) = (-0.12 \pm 0.02) \times 10^{-12}$ cm.

Table 6.3 shows a compilation of the μ_N^* and $(b_+ - b_-)$ values determined by the Saclay group using the neutron spin precession method, in which the nuclei were polarized both statically and dynamically.

References

Abragam, A., and Goldman, M. (1982). *Nuclear magnetism: Order and disorder.* Oxford University Press, Oxford.

Abragam, A., Bacchella, G. L., Long, C., Meriel, P., Piesvaux, J., and Pinot, M. (1972a). *Phys. Rev. Lett.*, **28,** 805.

Abragam, A., Bacchella, G. L., Glättli, H., Meriel, P., Piesvaux, J., and Pinot, M. (1972b). *C.r. hebd. Séanc. Acad. Sci. Paris,* **274 B,** 423.

Abragam, A., Bacchella, G. L., Glättli, H., Meriel, P., Pinot, M. and Piesvaux, J. (1973). *Phys. Rev. Lett.*, **31,** 776.

Abrahams, K., Ratynski, W., Stecher-Rasmussen, F. and Warming, E. (1966). *Nucl. Instrum. Methods,* **45,** 293.

Abrahams, K., Steinsvoll, O., Bongaarts, P. J. M., and de Lange, P. W. (1962). *Rev. Sci. Instrum,* **33,** 524.

Achieser, A., and Pomeranchuck, J. (1948). *J. Expt. Theor. Phys.*, **18,** 475.

Agamalian, M. M., Schweizer, J., Otchik, Y. M., and Khavronin, V. P. (1979). *Nucl. Instrum. Methods,* **158,** 395.

Ahmed, N. and Hicks, T. J. (1975). *J. Phys. F: Metal Phys.*, **5,** 2168.

Ahmed, N., Campbell, S. J., and Hicks, T. J. (1974). *J. Phys. E: Sci. Instrum.*, **7,** 199.

Aldred, A. T., Dunlap, B. D., Lam, D. J., Lander, G. H., Meuller, M. H., and Nowik, I. (1975). *Phys. Rev.*, **B11,** 530.

Alefeld, B., Badurek, G., and Rauch, H. (1981). *Z. Phys. B.*, **41** 231.

Allen, D. R., Mitchell, E. W. J., and Sinclair, R. N. (1980). *J. Phys. E: Sci. Instrum.* **13,** 639.

Alperin, H. A. (1962). *J. Phys. Soc. Japan,* **17,** BIII, 12.

Alperin, H. A., Brown, P. J., Nathans, R., and Pickart, S. J. (1961). *Phys. Rev. Lett.*, **8,** 237.

Alvarez, L. W. and Bloch, F. (1940). *Phys. Rev.*, **57,** 111.

Amoretti, G., Blaise, A., Bonnett, M., Boucherle, J. X., Delapalme, A., Fournier, J. M., and Vigneron, F. (1982). *J. Phys. (Paris),* **43,** C7-293.

Bacon, G. E. (1975). *Neutron diffraction.* Oxford University Press, Oxford.

Badurek, G. (1981). *Nucl. Instrum. Methods,* **189,** 543.

Badurek, G. and Westphal, G. P. (1975). *Nucl. Instrum. Methods,* **128,** 315.

Ballou, R., Deriagin, A. V., Givord, F., Lemaire, R., Levitin, R. Z., and Tasset, F. (1982). *J. Phys. (Paris),* **43,** C7-279.

Bampton, P. and Meardon, B. H. (1977). Rutherford Laboratory, Report RL-77-061/A.

Baryshevskii, V. G. and Podgoretskii, M. I. (1964). *Zh. éksp. teor. Fiz.*, **47,** 1050. [Trans *Soviet Phys. JETP,* **20,** 704.]

Bauer, G. S. (1979). In *Neutron scattering*', (ed. G. Kostorz). Academic Press, New York.
Beckurts, K. H. and Wirtz, K. (1964). *Neutron physics*. Springer-Verlag, Heidelberg.
Bednarski, S., Dobrzynski, L., and Steinsvoll, O. (1980). *Phys. Scripta*, **21,** 217.
Berndorfer, K. (1971). *Z. Physik*. **243,** 188.
Blaise, A., Boeuf, A., Delapalme, A., Forte, M., and Fournier, J. M. (1982). *J. Phys. (Paris)*, **43,** C7-287.
Blétry, J. and Sadoc, J. F. (1975). *J. Phys. F. Metal Phys.*, **5,** L110.
Bloch, F. (1936). *Phys. Rev.*, **50,** 259.
Blume, M., Freeman, A. J., and Watson, R. E. (1962). *J. chem. Phys.*, **37,** 1245.
Böni, P. and Shirane, G. (1985). *J. appl. Phys.*, **57,** 3012.
Böni, P., Shirane, G., Wicksted, J. P., and Stassis, C. (1985). *Phys. Rev.* **B31,** 4597.
Born, M. and Woolf, E. (1970). *Principles of optics*. Pergamon Press, Oxford.
Borowitz, S. and Hamermesh, M. (1948). *Phys. Rev.* **74,** 1285.
Boucher, B., Lienard, A., Rebouillat, J. P., and Schweizer, J. (1979). *J. Phys. F: Metal Phys.*, **9,** 1421.
Boucher, J. P., Mezei, F., Renault, L. P., and Renard, J. P. (1985). *Phys. Rev. Lett.*, **55,** 1778 [See also Erratum *Phys. Rev. Lett.* **55,** 2370.]
Boucherle, J. X., Gillon, B., Maruani, J., and Schweizer, J. (1982a). *J. Phys. (Paris)*, **43,** C7-227.
Boucherle, J. X., Givord, D., and Schweizer, J. (1982b). *J. Phys. (Paris)*, **43,** C7-199.
Boucherle, J. X., Fillion, G., Floquet, J., Givord, F., Lejay, P., and Schweizer, J. (1986). *J. Magn. magn. Mater.* **54–57,** 421.
Brockhouse, B. N. (1961). In '*Inelastic scattering of neutrons in solids and liquids*' p. 147. IAEA, Vienna.
Brown. P. J. (1970). *Thermal neutron diffraction*, (ed. B. T. M. Willis). Oxford University Press, Oxford.
Brown, P. J. (1979). In *Neutron scattering*, (ed. G. Kostorz). Academic Press, New York.
Brown, P. J. and Forsyth, J. B. (1964). *Br. J. Appl. Phys.*, **15,** 1529.
Brown, P. J. and Forsyth, J. B. (1967). *Proc. Phys. Soc.*, **92,** 125.
Brown, P. J. and Forsyth, J. B. (1977). *J. Phys. C: Solid St. Phys.*, **10,** 3157.
Brown, P. J., Forsyth, J. B., and Lander, G. H. (1968). *J. appl. Phys.* **39,** 1331.
Brown, P. J., Capiomont, A., Gillon, B., and Schweizer, J. (1979). *J. Magn. magn. Mater.*, **14,** 289.
Brown, P. J., Capellmann, H., Déportes, J., Givord, D., and Ziebeck, K. R. A. (1982a). *J. Magn. magn. Mater.*, **30,** 243.
Brown, P. J., Day, P., Fischer, P., Güdel, H. U., Herren, F., and Ludi, A. (1982b). *J. Phys (Paris)*, **43,** C7-235.
Brown, P. J., Capellmann, H., Déportes, J., Givord, D., and Ziebeck, K. R. A. (1983). *J. Magn. magn. Mater.*, **31–34,** 295.
Brown, P. J., Ziebeck, K. R. A., Déportes, J., and Givord, D. (1984). *J. appl. Phys.*, **55,** 1881.
Burgy, M., Hughes, D. J., Wallace, J. R., Heller, R. B., and Woolf, W. E. (1950). *Phys. Rev.* **80,** 953.

Burke, S. K., Stirling, W. G., and McEwen, K. A. (1981). *J. Phys. C: Solid St. Phys.*, **14,** L967.

Cable, J. W. and Wollan, E. O. (1973). *Phys. Rev.*, **B7,** 2005.

Cable, J. W., Werner, S. A., Felcher, G. P., and Wakabayashi, N. (1984). *Phys. Rev.*, **B29,** 1268.

Campbell, S. J., Ahmed, N., Hicks, T. J., Ebdon, F. R., and Wheeler, D. A. (1974). *J. Phys. E: Sci. Instrum.*, **7,** 195.

Carlile, C. J. and Ross, D. K. (1975). *J. appl. Crystallogr.*, **8,** 292.

Carlile, C. J., Taylor, A. D., and Williams, W. G. (1985). Rutherford Appleton Laboratory, Report RAL-85-052.

Chandler, G. S., Figgis, B. N., Phillips, R. A., Reynolds, P. A. and Mason, R. (1982). *J. Phys. (Paris)*, **43,** C7-323.

Childs, F., Hayter, J. B., Penfold, J and Williams, W. G. (1981). Rutherford Appleton Laboratory, Report RL-81-082.

Cocking, S. J., and Webb, F. J. (1965). In *Thermal neutron scattering,* (ed. P. A. Egelstaff). Academic Press, New York.

Collins, M. F. and Forsyth, J. B. (1963). *Phil. Mag.*, **8,** 401.

Collins, M. F., Haywood, B. C., and Stirling, G. C. (1969). *Disc. Faraday Soc.*, **48,** 163.

Collins, S. R. (1979). PhD. thesis. Massachusetts Institute of Technology.

Convert, P. and Forsyth, J. B. (ed.) (1983). *Position sensitive detection of thermal neutrons.* Academic Press, London.

Coupland, J. H. and Stovold, R. V. (1977). Rutherford Laboratory Report RL-77-096/B.

Cowley, J. M. (1950). *Phys. Rev.*, **77,** 669.

Currat, R. (1973). *Nucl. Instrum. Methods,* **107,** 21.

Cywinski, R. (1983). Rutherford Appleton Laboratory, Report NDR/P15/83.

Cywinski, R. and Hicks, T. J. (1980). *J. Phys. F: Metal Phys.*, **10,** 693.

Cywinski, R. and Williams, W. G. (1985). *Nucl. Instrum. Methods,* **228,** 406.

Cywinski, R., Mayers, J., and Williams, W. G. (1981). Rutherford Laboratory, Report RL-81-043.

Dabbs, J. W. T., Roberts, L. D., and Bernstein, S. (1955). Oak Ridge National Laboratory, Report ORNL-CF-55-5-126.

Dagleish, P. A., Hayter, J. B., and Mezei, F. (1980). *Lecture Notes in Physics* **128,** 66. Springer-Verlag, Berlin.

Daniels, J. M. (1965). *Oriented nuclei.* Academic Press, New York.

Davidson, P. D. and Wroe, H. (1983). In *Position-sensitive detection of thermal neutrons,* (ed. P. Convert and J. B. Forsyth), p. 166. Academic Press, London.

Davis, H. and Cooke, J. F. (1972). *AIP Conf. Proc.* No. 5, 1441.

Davis, J. R., Burke, S. K., and Rainford, B. D. (1980). *J. Magn. magn. Mater.*, **15–18,** 151.

Davis, J. R., Cywinski, R., Moze, O., and Hicks, T. J. (1982). *J. Phys. (Paris)*, **43,** C7-65.

de Gennes, P. G. (1976). *Macromolecules,* **9,** 594.

Delapalme, A. (1979). *Nukleonika,* **24,** 723.

Delapalme, A., Schweizer, J., Couderchon, G., and Perrier de la Bathie, R. (1971). *Nucl. Instrum. Methods,* **95,** 589.

Deraman, M., Dore, J. C., and Schweizer, J. (1985). *J. Magn. magn. Mater.*, **50,** 178.
Dore, J. C., Clarke, J. H., and Wenzel, J. T. (1976). *Nucl. Instrum. Methods,* **138,** 317.
Drabkin, G. M. (1963). *Sov. Phys. JETP,* **16,** 781.
Drabkin, G. M., Trunov, V. A., and Runov, V. V. (1968). *Sov. Phys. JETP.*, **27,** 194.
Drabkin, G. M., Zabidarov, E. I., Kasman, Ya. A., and Okorokov, A. I. (1969). *Sov. Phys. JETP,* **29,** 261.
Drabkin, G. M., Okorov, A. I., Schebetov, A. F., Borovikova, N. V., Gukasov, A. G., Yegorov, A. I., and Runov, V. V. (1975). Leningrad Institute of Nuclear Physics, Report FTI-183.
Drabkin, G. M., Okorokov, A. I., Schebetov, A. F. Borovikova, N. V., Gukasov, A. G., Kudriashov, V. A., Runov, V. V., and Korneev, D. A. (1976). *Nucl. Instrum. Methods,* **133,** 453.
Draghicescu, D., Luschchikov, V. I., Nikolenko, V. G., Taran, Yu. V., and Shapiro, F. L. (1964). *Phys. Lett.*, **12,** 334.
Dubois-Violette, E. and de Gennes, P. G. (1967). *Phys.* (*USA*), **3,** 181.
Edwards, D. M. (1983). *J. Magn. magn. Mater.*, **36,** 213.
Enderby, J. E. (1968). In *Physics of simple liquids,* (ed. H. N. V. Temperley, J. S. Rowlinson and G. S. Rushbrooke). North-Holland, Amsterdam.
Enderby, J. E., North, D. M., and Egelstaff, P. A. (1966). *Phil. Mag.*, **14,** 961.
Endoh, Y. and Ishikawa, Y. (1986). *Physica B,* **136,** 64.
Endoh, Y., Ikeda, S., Mitsuda, S., and Fujimoto, H. (1985). *Nucl. Instrum. Methods,* **A240,** 115.
Farnoux, B. (1985). *Neutron scattering in the 'nineties.* IAEA-CN-46/20, Vienna.
Felcher, G. P. (1981). *Phys. Rev.*, **B24,** 1595.
Felcher, G. P., Lander, G. H., Arai, T., Sinha, S. K. and Spedding, F. H. (1976). *Phys. Rev.*, **B13,** 3034.
Felcher, G. P., Kleb, R., Hinks, D. G. and Potter, W. H. (1984). *Phys. Rev.*, **B29,** 4843.
Felcher, G. P., Felici, R., Kampwirth, R. T., and Gray, K. E. (1985). *J. Appl. Phys,* **57,** 3789.
Fender, B. E. F., Hobbis, L. C. W., and Manning, G. (1980). *Phil. Trans. R. Soc.*, **B290,** 657.
Fleeman, J., Nicodemus, D. B., and Staub, H. H. (1949). *Phys. Rev.*, **76,** 1774.
Forsyth, J. B. (1979). *Atomic Energy Review* **17,** 345.
Freeman, A. J. and Desclaux, J. P. (1972). *Int. J. Magn.*, **3,** 311.
Freeman, A. J. and Watson, R. E. (1961). *Acta. Crystallogr.*, **14,** 231.
Freeman, F. F. and Williams, W. G. (1978). *J. Phys. E: Sci. Instrum.*, **11,** 459.
Freeman, F. F., Meardon, B. H., and Williams, W. G. (1977). Rutherford Laboratory, Report RL-77-080/A.
Freeman, F. F., Meardon, B. H., and Williams, W. G. (1978). *J. appl. Crystallogr.*, **11,** 80.
Freund, A. and Forsyth, J. B. (1979). In *Neutron Scattering,* (ed. G. Kostorz). Academic Press, New York.
Freund, A., Pynn, R., Stirling, W. G., and Zeyen, C. M. E. (1983). *Physica B,* **120,** 86.

Friesleben, H. and Rauch, H. (1972). *Nucl. Instrum. Methods*, **98**, 61.
Gerlach, P., Schärpf, O., Prandl, W., and Dorner, B. (1982). *J. Phys.* (*Paris*), **43**, C7-151.
Gilbert, E., Hanley, P., Hayter, J. B., and White, J. W. (1973). *J. Phys. E: Sci. Instrum.*, **6**, 714.
Givord, D., Givord, F., Gignoux, D., Koehler, W. C., and Moon, R. M. (1976). *J. Phys. Chem. Solids*, **37**, 567.
Givord, D., Laforest, J., Schweizer, J., and Tasset, F. (1979). *J. appl. Phys.*, **50**, 2008.
Givord, D., Liénard, A., Rebouillat, J. P., and Schweizer, J. (1985). *Physica*, **130B**, 351.
Glättli, H. and Coustham, J. (1983). *J. Phys.* (*Paris*), **44**, 957.
Glättli, H., Bacchella, G. L., Fourmond, M., Malinovski, A., Meriel, P., Pinot, M., Roubeau, P., and Abragam, A. (1979). *J. Phys.* (*Paris*), **40**, 629.
Goldberger, M. L. and Seitz, F. (1947). *Phys. Rev.*, **71**, 294.
Gukasov, A. G., Ruban, V. A., and Bedrizova, M. N. (1977). *Sov. Tech. Phys. Lett.*, **3**, 52.
Gukasov, A. G., Deriglazov, V. V., Kezerashvili, V. Y., Krutov, G. A., Kudryashov, V. A., Peskov, B. G., Syromyatnikov, V. G., Trunov, V. A., Kharchenkov, V. P. and Shchebetov, A. F. (1979). *Sov. Phys. JETP*, **50**, 862.
Guoan, W., Cowlam, N., Davies, H. A., Cowley, R. A., Paul, D. McK., and Stirling, W. G. (1982). *J. Phys.* (*Paris*), **43**, C7-71.
Gupta, R. P. and Freeman, A. J. (1976). *Phys. Rev. Lett.*, **36**, 613.
Hahn, E. L. (1950). *Phys. Rev.*, **80**, 850.
Halperin, B. I. and Hohenberg, P. C. (1969). *Phys. Rev.*, **177**, 952.
Halpern, O. (1949). *Phys. Rev.*, **75**, 343.
Halpern, O. and Holstein, T. (1941). *Phys. Rev.*, **59**, 960.
Halpern, O. and Johnson, M. H. (1937a). *Phys. Rev.*, **51**, 992.
Halpern, O. and Johnson, M. H. (1937b). *Phys. Rev.*, **52**, 52.
Halpern, O. and Johnson, M. H. (1939). *Phys. Rev.*, **55**, 898.
Halpern, O., Hamermesh, M., and Johnson, M. H. (1941). *Phys. Rev.*, **59**, 981.
Hamermesh, M. (1949). *Phys. Rev.*, **75**, 1766.
Hayter, J. B. (1978a). *Topics in current physics*, Vol. 6. Springer-Verlag, Heidelberg.
Hayter, J. B. (1978b). *Z. Physik. B.*, **31**, 117.
Hayter, J. B. (1980). *Lecture notes in physics* **28**, 53. Springer-Verlag, Berlin.
Hayter, J. B. and Penfold, J. (1979). *Z. Phys. B.*, **35**, 199.
Hayter, J. B. and Penfold, J. (1981a). *Mol. Phys.*, **42**, 109.
Hayter, J. B. and Penfold, J. (1981b). *J. Chem. Soc. Faraday Trans.* I, **77**, 1851.
Hayter, J. B., Jenkin, G. T., and White, J. W. (1974). *Phys. Rev. Lett.*, **33**, 696.
Hayter, J. B., Penfold, J., and Williams, W. G. (1978). *J. Phys. E: Sci. Instrum.*, **11**, 454.
Heeringa, W. (1985). *Hyperfine Interactions*, **22**, 263.
Hicks, T. J. (1979a). *Nukleonika*, **24**, 795.
Hicks, T. J. (1979b). In *Neutron scattering*, (ed. G. Kostorz). Academic Press, New York.
Höhne, P. (1961). *Ann. Phys. Lpz.*, **7**, 50.
Holden, T. M. and Stirling, W. G. (1977). *J. Phys. F: Metal Phys.*, **7**, 1901.

Holden, T. M., Comly, J. B., and Low, G. G. (1967). *Proc. Phys. Soc.*, **92,** 726.
Hood, M. (1983). *J. Phys. C: Solid St. Phys.* **16,** 4655.
Howe, R. A. (1978). In *Progress in liquid physics,* (ed. C. A. Croxton), Chap. 13. Wiley, Chichester.
Hubbard, J. and Marshall, W. C. (1965). *Proc. Phys. Soc.*, **86,** 561.
Hughes, D. J. (1953). *Pile neutron research.* Addison–Wesley, Reading, MA.
Hughes, D. J. and Burgy, M. T. (1951). *Phys. Rev.*, **81,** 498.
Hughes, D. J., Wallace, J. R., and Holtzman, R. H. (1948). *Phys. Rev.*, **73,** 1277.
Ishida, M., Ishikawa, Y., Ishimoto, S., Kohgi, M., Masaike, A., Masuda, Y., Morimoto, K., and Nakajima, T. (1985). 'Development of polarized epithermal neutron spectrometer PEN'. *Proc. Int. Collab. Adv. Neut. Sources ICANS VIII,* Oxford. Rutherford Appleton Laboratory, Report RAL-85-110, Vol. II, p. 612.
Ishikawa, Y., Noda, Y., Fincher, C., and Shirane, G. (1982). *Phys. Rev.*, **B25,** 254.
Ishikawa, Y., Noda, Y., Uemura, Y. J., Majkrzak, C. F., and Shirane, G. (1985). *Phys. Rev.*, **B31,** 5884.
Ito, Y. and Shull, C. G. (1969). *Phys. Rev.*, **185,** 961.
Izyumov, Yu. A. and Maleev, S. V. (1962). *Sov. Phys. JETP,* **14,** 1168.
Jacobson, A. J. and Fender, B. E. F. (1975). *J. Phys. C: Solid St. Phys.*, **8,** 844.
Jacrot, B. (1970). *Instrumentation for neutron inelastic scattering research,* p. 225. I.A.E.A., Vienna.
James, R. W. and Brindley, G. W. (1932). *Phil. Mag.*, **12,** 81.
Jeavons, A. P., Ford, N. L., Lindberg, B., and Sachot, R. (1978). *Nucl. Instrum. Methods,* **148,** 29.
Jones, T. J. L. and Williams, W. G. (1977). Rutherford Appleton Laboratory, Report RL-77-079/A.
Jones, T. J. L. and Williams, W. G. (1978). *Nucl. Instrum. Methods,* **152,** 463.
Jones, T. J. L. and Williams, W. G. (1980). *J. Phys. E: Sci. Instrum.*, **13,** 227.
Kajzar, F. and Parette, G. (1980). *Phys. Rev.*, **B22,** 5471.
Kakurai, K., Pynn, R., Dorner, B., and Steiner, M. (1984). *J. Phys. C: Solid St. Phys.*, **17,** L123.
Kendrick, H., King, J. S., Werner, S. A., and Arrott, A. (1970). *Nucl. Instrum. Methods,* **79,** 82.
Koehler, W. C. and Moon, R. M. (1972). *Phys. Rev. Lett.*, **29,** 1468.
Koehler, W. C. and Moon, R. M. (1976). *Phys. Rev. Lett.*, **36,** 616.
Koester, L. and Yelon, W. B. (1982). *Neutron cross section data.* Netherlands Eenergy Research Foundation, Petten.
Korenman, V. (1985). *J. appl. Phys.*, **57,** 3000.
Korneev, D. A. (1980). *Nucl. Instrum. Methods,* **169,** 65.
Kraan, W. H. and Rekveldt, M. Th. (1977). *J. Magn. magn. Mater.*, **5,** 247.
Landau, L. D. and Khalatnikov, I. M. (1948). *Sov. Phys. JETP,* **12,** 216.
Lander, G. H. (1976). *AIP Conf. Magn. magn. Mater.*, **29,** 311.
Lander, G. H., Brun, T. O., Desclaux, J. P., and Freeman, A. J. (1973). *Phys. Rev.*, **B9,** 3237.
Lander, G. H. Delapalme, A., Brown, P. J., Spirlet, J. C. Rebizant, J., and Vogt, O. (1984). *Phys. Rev. Lett.*, **53,** 2262.

Lander, G. H., Brown, P. J. Spirlet, M. R., Rebizant, J., Kanellakopulos, B., and Klenze, R. (1985). *J. chem. Phys.*, **83,** 5988.

Langley, K. H. and Jefferies, C. D. (1966). *Phys. Rev.*, **152,** 358.

Lebech, B., Rainford, B. D., Brown, P. J., and Wedgwood, F. A. (1979). *J. Magn. magn. Mater.*, **14,** 298.

Leslie, M., Jenkin, G. T., Hayter, J. B., White, J. W., Cox, S. and Warner, G. (1980). *Phil. Trans. R. Soc.*, **B290,** 497.

Lovesey, S. W. (1986). *Theory of neutron scattering from condensed matter.* Oxford University Press, Oxford.

Low, G. G. and Collins, M. F. (1963). *J. appl. Phys.*, **34,** 1195.

Lowde, R. D., Moon, R. M., Pagnosis, B., Perry, C. H., Sikoloff, J. B., Vaughan-Watkins, R. S., Wiltshire, M. C. K., and Crangle, J. (1983). *J. Phys. F: Metal Phys.*, **13,** 249.

Lushchikov, V. I., Taran, Yu. V., and Shapiro, F. L. (1970). *Sov. J. nucl. Phys.*, **10,** 669.

Lynn, J. E. (1980). *Contemp. Phys.*, **21,** 483.

Lynn, J. W. (1975). *Phys. Rev.*, **B11,** 2624.

Lynn, J. W. and Mook, H. A. (1981). *Phys. Rev.*, **B23,** 198.

Lynn, J. W. and Mook, H. A. (1986). *J. Magn. magn. Mater.*, **54–57,** 1169.

Lynn, J. W., Kjems, J. K., Passell, L., Saxena, A. M., and Schoenborn, B. P. (1976a). *J. appl. Crystallogr.*, **9,** 454.

Lynn, J. W., Shirane, G., and Blume, M. (1976b). *Phys. Rev. Lett.*, **37,** 154.

Maier, B. (ed.) (1983). *Neutron research facilities at the ILL high flux reactor.* Institute Laue–Langevin, Grenoble.

Maier-Leibnitz, H. (1972). *Proc. Symp. Neutron Inelastic Scattering, Grenoble,* p. 681. I.A.E.A. Vienna.

Majkrzak, C. F. and Passell, L. (1985). *Acta. Crystallogr.*, **A41,** 41.

Majkrzak, C. F. and Shirane, G. (1982). *J. Phys. (Paris),* **43,** C7-215.

Maleyev, S. V. (1982). *J. Phys. (Paris),* **43,** C7-23.

Mangin, P., Brown, P. J., and Tête, C. (1986). *J. Magn. magn. Mater.*, **62,** 127.

Marshall, W. (1968). *J. Phys. C: Solid St. Phys.*, **1,** 88.

Martinez, J. L., Böni, P. and Shirane, G. (1985). *Phys. Rev.* **B32,** 7037.

Mayers, J. (1982a). Rutherford Appleton Laboratory, Report RL-82-046.

Mayers, J. (1982b). Rutherford Appleton Laboratory, Report RL-82-104.

Mayers, J. and Cywinski. R. (1985). *Nucl. Instrum. Methods,* **A241,** 519.

Mayers, J. and Williams, W. G. (1981). Rutherford Appleton Laboratory, Report RL-81-047.

Mayers, J., Cywinski. R., Jones, T. J. L., and Williams, W. G. (1984). Rutherford Appleton Laboratory, Report RAL-84-118.

Mayers, J., Cywinski, R., Jones, T. J. L., and Williams, W. G. (1985). *J. Phys. D: Appl. Phys.*, **18,** 1843.

Mayers, J., Cywinski, R., Harris, I. R., Hixon, M., and Williams, W. G. (1986). *Proc. Ninth Int. Conf. on Advanced Neutron Sources, ICANS IX, SIN, Willigen, Switzerland, Sept. 1986.*

Medina, R. A. and Cable, J. W. (1977). *Phys. Rev.*, **B15,** 1539.

Menzinger, F. and Paoletti, A. (1966). *Phys. Rev.*, **143,** 365.

Menzinger, F. and Sacchetti, F. (1979). *Nukleonika,* **24,** 737.

Meyerhof, W. E. and Nicodemus, D. B. (1951). *Phys. Rev.*, **82,** 5.
Mezei, F. (1972). *Z. Phys.*, **255,** 146.
Mezei, F. (1976a). *Commun. Phys.*, **1,** 81.
Mezei, F. (1976b). *Proc. Conference on Neutron Scattering, Gatlinberg, Tennessee,* (ed. R. M. Moon), p. 670.
Mezei, F. (1977). *Proc. Int. Symp. on Neutron Scattering.* IAEA-SM-219/56, Vienna.
Mezei, F. (1979). *Nucl. Instrum. Methods,* **164,** 153.
Mezei, F. (1980a). *Lecture notes in physics* **128,** 3. Springer-Verlag, Berlin.
Mezei, F. (1980b). *Phys. Rev. Lett.*, **44,** 1601.
Mezei, F. (1982). *Phys. Rev. Lett.*, **49,** 1096.
Mezei, F. (1986). *Physica B,* **137,** 295.
Mezei, F. and Dagleish, P. A. (1977). *Commun. Phys.*, **2,** 41.
Mezei, F. and Murani, A. P. (1979). *J. Magn. magn. Mater.*, **14,** 211.
Mezei, F. and Richter D. (1984). *Neutron Scattering Instrumentation for SNQ, Proc. Workshop, Maria Laach, Sept. 1984.* KFA-Jülich Rep. Jül-1954.
Mirebeau, I., Parette, G., and Cable, J. W. (1987). *J. Phys. F: Metal Phys.* **17,** 191.
Mitsuda, S. and Endoh, Y. (1985). *J. Phys. Soc. Japan.*, **54,** 1570.
Mook, H. A. (1966). *Phys. Rev.*, **148,** 495.
Mook, H. A. and Lynn, J. W. (1984). *Phys. Rev.*, **B29,** 4056.
Mook, H. A. and Lynn, J. W. (1985). *J. Appl. Phys.*, **57,** 3006.
Mook, H. A. and Tsuei, C. C. (1977). *Phys. Rev.*, **B16,** 2184.
Mook, H. A., Lynn, J. W., and Nicklow, R. M. (1973). *Phys. Rev. Lett.*, **30,** 556.
Moon, R. M. (1964). *Phys. Rev.*, **A136,** 195.
Moon, R. M. (1970). *J. appl. Phys.*, **41,** 883.
Moon, R. M. (1971). *Int. J. Magn.*, **1,** 219.
Moon, R. M. (1979). *Nukleonika,* **24,** 787.
Moon, R. M. (1982). *J. Phys. (Paris),* **43,** C7-187.
Moon, R. M. and Koehler, W. C. (1975). *Phys. Rev.*, **B11,** 1609.
Moon, R. M., Riste,T., and Koehler, W. C. (1969a). *Phys. Rev.*, **181,** 920.
Moon, R. M., Riste, T., Koehler, W. C., and Abrahams, S. C. (1969b). *J. appl. Phys.*, **40,** 1445.
Moon, R. M., Koehler, W. C., Cable, J. W., and Child, H. R. (1972). *Phys. Rev.*, **B5,** 997.
Moon, R. M., Koehler, W. C., and Shull, C. G. (1975). *Nucl. Instrum. Methods,* **129,** 515.
Moon, R. M., Koehler, W. C., and Cable, J. W. (1976). *Proc. Conf. on Neutron Scattering, Gatlinberg, Tennessee,* (ed. R. M. Moon), p. 577.
Moon, R. M., Koehler, W. C., McWhan, D. B., and Holtzberg, F. (1978). *J. appl. Phys.*, **49,** 2107.
Moon, R. M., Koehler, W. C., Sinha, S. K., Stassis, C., and Kline, G. R. (1979). *Phys. Rev. Lett.*, **43,** 62.
Moriya, T. (1983). *J. Magn. magn. Mater.*, **31–34,** 10.
Moss, J. and Brown, P. J. (1972). *J. Phys. F: Metal Phys.*, **2,** 258.
Moze, O. and Hicks, T. J. (1982). *J. Phys. F: Metal Phys.*, **12,** 1.
Nathans, R. and Paoletti, A. (1959). *Phys. Rev. Lett.* **2,** 254.

Nathans, R., Shull, C. G., Shirane, G., and Andresen, A. (1959). *J. Phys. Chem. Solids,* **10,** 138.
Nathans, R., Alperin, H. A., Pickart, S. J., and Brown, P. J. (1963). *J. appl. Phys.,* **34,** 1182.
Newport, R. J., Penfold, J., and Williams, W. G. (1984). *Nucl. Instrum. Methods,* **224,** 120.
Okorokov, A. I., Runov, V. V., Volkov, V. I. and Gukasov, A. G. (1976). *Sov. Phys. JETP,* **42,** 300.
Parkin, S. S. P., Sigsbee, R., Felici, R., and Felcher, G. P. (1986). *Appl. Phys. Lett.,* **48,** 604.
Paul, D. McK., Cowley, R. A., Stirling, W. G., Cowlam, N., and Davies, H. A. (1982). *J. Phys. F: Metal Phys.,* **12,** 2687.
Penfold, J. and Williams, W. G. (1985). Rutherford Appleton Laboratory, Report RAL-85-045.
Penfold, J., Sutherland, J. C., Williams, W. G., and Hayter, J. B. (1975). Rutherford Laboratory, Report RL-75-112.
Placzek, G. (1952). *Phys. Rev.,* **86,** 377.
Postma, H., Marshak, H., Sailor, V. L., Shore, F. J., and Reynolds, C. A. (1962). *Phys. Rev.,* **126,** 979.
Pusey, P. N. (1975). *J. Phys. A: Math. Gen.,* **8,** 1433.
Pynn, R. (1978). *J. Phys. E: Sci. Instrum.,* **11,** 1133.
Pynn, R. (1984). *Rev. sci. Instrum.,* **55,** 843.
Rainford, B. D. (1982). *J. Phys. (Paris),* **43,** C7-33.
Rakhecha, V. R. and Satya Murthy, N. S. (1978). *J. Phys. C: Solid St. Phys.,* **11,** 4389.
Ramsey, N. F. (1949). *Phys. Rev.,* **76,** 996.
Reed, R. E., Bolling, E. D., and Harmon, H. E. (1973). Oak Ridge National Laboratory, Solid State Division Annual Report, 129.
Rekveldt, M. Th. (1971). *J. Phys. (Paris),* **C1,** 579.
Rekveldt, M. Th. (1973). *Z. Physik.,* **259,** 391.
Rekveldt, M. Th. (1977). *J. Phys. (Paris),* **C1,** 23.
Rekveldt, M. Th. and Kraan, W. H. (1983). *Physica,* **B120,** 81.
Rekveldt, M. Th. and van Schaik, F. J. (1979a). *J. Appl. Phys.,* **50,** 2122.
Rekveldt, M. Th. and van Schaik, F. J. (1979b). *J. Magn. magn. Mater.,* **14,** 325.
Rekveldt, M. Th., van Schaik, F. J., and Kraan, W. H. (1979). *Nukleonika,* **24,** 809.
Rekveldt, M. Th., Stüsser, N., and Spruijt, T. (1986). *J. Magn. magn. Mater.,* **54–57,** 855.
Richardson, R. M. (1982). Rutherford Appleton Laboratory, Report RL-82-035.
Richter, D., Hayter, J. B., Mezei, F., and Ewan, B. (1978). *Phys. Rev. Lett.,* **41,** 1484.
Rimmer, D. E. (1970). *Thermal neutron diffraction,* (ed. B. T. M. Willis). Oxford University Press, Oxford.
Riste, T., Moon, R. M., and Koehler, W. C. (1968). *Phys. Rev. Lett.,* **20,** 997.
Roberts, L. D., Bernstein, S., Dabbs, J. W. T., and Stanford, C. P. (1954). *Phys. Rev.,* **95,** 105.
Rose, M. E. (1949). *Phys. Rev.,* **75,** 213.
Roubeau, P., Abragam, A., Bacchella, G. L., Glättli, H., Malinovski, A., Meriel, P., Piesvaux, J., and Pinot, M. (1974). *Phys. Rev. Lett.,* **33,** 102.

Sáenz, A. W. (1960). *Phys. Rev.*, **125,** 1940.
Schärpf, O. (1982). *Neutron Scattering*-1981, (ed. J. Faber), p. 182. American Institute of Physics, New York.
Schärpf, O. (1985). *Neutron scattering in the 'nineties.* IAEA-CN-46/70, Vienna.
Schermer, R. I. and Blume, M. (1968). *Phys. Rev.*, **166,** 554.
Schoenborn, B. P., Caspar, D. L. D., and Kammerer, O. F. (1974). *J. appl. Crystallogr.*, **7,** 508.
Schweizer, J. (1982). *Nucl. Instrum. Methods,* **199,** 115.
Schwinger, J. S. (1937). *Phys. Rev.*, **51,** 544.
Shabalin, E. P. (1979). *Fast pulsed and burst reactors.* Pergamon Press, Oxford.
Sherwood, J. E., Stephenson, T. E., and Berstein, S. (1954). *Phys. Rev.*, **96,** 1546.
Shirane, G. (1984). *J. Magn. magn. Mater.*, **45,** 33.
Shirane, G., Nathans, R., Pickart, S. J., and Alperin, H. A. (1964). *Proc. Int. Conf. Magn, Nottingham,* p. 223.
Shirane, G., Axe, J. D., Majkrzak, C. F., and Mizoguchi, T. (1982). *Phys. Rev.*, **B26,** 2575.
Shirane, G., Uemura, Y. J., Wicksted, J. P., Endoh, Y., and Ishikawa, Y. (1985). *Phys. Rev.*, **B31,** 1227.
Shirane, G., Böni, P., and Wicksted, J. P. (1986). *Phys. Rev.*, **B33,** 1881.
Shull, C. G. (1963). *Phys. Rev. Lett.*, **10,** 297.
Shull, C. G. and Ferrier, R. P. (1963). *Phys. Rev. Lett.*, **10,** 295.
Shull, C. G. and Mook, H. A. (1966). *Phys. Rev. Lett.*, **16,** 184.
Shull, C. G. and Nathans, R. (1967). *Phys. Rev. Lett.*, **19,** 384.
Shull, C. G. and Wilkinson, M. K. (1955). *Phys. Rev.*, **97,** 304.
Shull, C. G. and Yamada, Y. (1962). *J. Phys. Soc. Japan,* **17,** (BIII), 1.
Shull, C. G., Strauser, W. A., and Wollan, E. A. (1951a). *Phys. Rev.*, **83,** 333.
Shull, C. G., Wollan, E. A., and Koehler, W. C. (1951b). *Phys. Rev.*, **84,** 912.
Stanford, C. P., Stephenson, T. E., Cochran, L. W., and Bernstein, S. (1954). *Phys. Rev.*, **94,** 374.
Stassis, C. (1970). *Phys. Rev. Lett.*, **24,** 1415.
Stassis, C. (1979). *Nukleonika,* **24,** 765.
Stassis, C., Kline, G. R., and Sinha, S. K. (1975). *Phys. Rev.*, **B11,** 2171.
Stassis, C., Kline, G. R., Freeman, A. J., and Desclaux, J. P. (1976). *Phys. Rev.*, **B13,** 3916.
Stassis, C., Kline, G. R., Loong, C.-K., and Harmon, B. R. (1977). *Solid St. Commun.*, **23,** 159.
Stassis, C., Loong, C.-K., Kline, G. R., McMaster, O. D., and Gschneider Jnr., K. A. (1978). *J. appl. Phys.*, **49,** 2113.
Stassis, C., Loong, C.-K., Harmon, B. N., and Liu, S. H. (1979). *J. appl. Phys.*, **50,** 7567.
Steigenberger, U. and Schärpf, O. (1986). *Physica,* **B136,** 87.
Steinberger, J. and Wick, G. C. (1949). *Phys. Rev.*, **76,** 994.
Steiner, M. (1985). *Neutron scattering in the 'nineties.* IAEA-CN-46/032, Vienna.
Steiner, M., Kakurai, K., Knop, W., Pynn, R., and Kjems, J. K. (1982). *Solid St. Commun.*, **41,** 329.
Steinsvoll, O., Moon, R. M., Koehler, W. C., and Windsor, C. G. (1981). *Phys. Rev.*, **B24,** 4032.

Stirling, W. G., Pynn, R., and Burke, S. K. (1982). *Proc. workshop on instrumentation*. Hahn-Meitner Institut, Berlin.

Stuhrmann, H. B., Schärpf, O., Krumpole, M., Niinikoski, T. O., Rieubland, J. M., and Rijllart, A. (1985). European Organization for Nuclear Research, Report CERN-EP/85-119.

Stüsser, N., Rekveldt, M. Th., and Spruijt, T. (1986). *J. Magn. magn. Mater.*, **54–57,** 723.

Taylor, A. D. (1984). Rutherford Appleton Laboratory, Report RAL-84-120.

Tofield, B. C. (1975). *Structure and Bonding,* Vol. 21. Springer-Verlag, Berlin.

Turchin, V. F. (1967). *Soviet Atomic Energy,* **22,** 124. Translated from *Atomnaya Énergia,* **22,** 119.

van Laar, B., Maniawski, F., Kaprzyk, S., and Dobrzynski, L. (1979). *J. Magn. magn. Mater.*, **14,** 94.

van Laar, B., Maniawski, F., Kaprzyk, S., and Dobrzysnki, L, (1980). *Nukleonika,* **25,** 835.

van Schaik, F. J. and Rekveldt, M. Th. (1978). *Phys. Rev. Lett.*, **41,** 767.

Von Jan, R. and Scherm, R. (1970). *Nucl. Instrum. Methods,* **80,** 69.

Watson, R. E. and Freeman, A. J. (1961). *Acta. Crystallogr.*, **14,** 27.

Wedgwood, F. A. (1972). *J. Phys. C: Solid St. Phys.*, **5,** 2427.

Wedgwood, F. A. (1976). *Proc. R. Soc.* (*Lond.*), **A349,** 447.

Weiss, R. J. and Freeman, A. J. (1959). *J. Phys. Chem. Solids.*, **10,** 147.

Werner, S. A., Arrott, A., and Atoji, M. (1969). *J. appl. Phys.*, **40,** 1447.

White, R. M. (1983). *Quantum theory of magnetism,* Springer Series in Solid-State Sciences No. 32. Springer Verlag, Berlin.

Wicksted, J. P., Böni, P., and Shirane, G. (1984). *Phys. Rev.*, **B30,** 3655.

Williams, W. G. (1973). Rutherford Laboratory, Report RL-73-034.

Williams, W. G. (1974). Rutherford Laboratory, Report RL-74-036.

Williams, W. G. (1975). Reactor Centrum Nederland, Report RCN-234, 103.

Williams, W. G. and Penfold, J. (1973). Rutherford Laboratory, Report NBRU/73-1.

Williams, W. G. and Rainford, B. D. (1974). Rutherford Laboratory, Report RL-74-130.

Windsor, C. G. (1981). *Pulsed neutron scattering.* Taylor and Francis, London.

Windsor, C. G. and Heenan, R. K. (1980). *Nucl. Instrum. Methods,* **171,** 115.

Zeyen, C. M. E. (1982). *Neutron scattering*-1981, (ed. J. Faber). American Institute of Physics, New York.

Zhuckenko, N. K., Krutov, G. A., Kurbakov, A. I., Suponev, N. P., Tarovik, E. G., Trunov, V. A., and Yagud, R. Z. (1984). *Sov. Tech. Phys. Lett.*, **10,** 483.

Ziebeck, K. R. A. and Brown, P. J. (1980). *J. Phys. F: Metal Phys.*, **10,** 2015.

Ziebeck, K. R. A., Webster, P. J., Brown, P. J., and Bland, J. A. C. (1981). *J. Magn. magn. Mater.*, **24,** 258.

Ziebeck, K. R. A., Capellmann, H., Brown, P. J., and Booth, J. G. (1982). *Z. Phys.*, **B48,** 241.

Index

Printed and bound by CPI Group (UK) Ltd, Croydon, CR0 4YY

The manufacturer's authorised representative in the EU for product safety is Oxford University Press España S.A. of El Parque Empresarial San Fernando de Henares, Avenida de Castilla, 2 - 28830 Madrid (www.oup.es/en or product.safety@oup.com). OUP España S.A. also acts as importer into Spain of products made by the manufacturer.
Printed and bound by CPI Group (UK) Ltd, Croydon, CR0 4YY
14/09/2026
02216307-0016